R. Furlong

New Uses for New Phylogenies

New Uses for New Phylogenies

Edited by

PAUL H. HARVEY
Department of Zoology, University of Oxford, UK

ANDREW J. LEIGH BROWN
Institute of Cell, Animal and Population Biology
University of Edinburgh, UK

JOHN MAYNARD SMITH
School of Biological Sciences, University of Sussex, UK

SEAN NEE
Department of Zoology, University of Oxford, UK

Oxford · New York · Tokyo

OXFORD UNIVERSITY PRESS

OXFORD
UNIVERSITY PRESS

Great Clarendon Street, Oxford OX2 6DP

Oxford University Press is a department of the University of Oxford and furthers the University's aim of excellence in research, scholarship, and education by publishing worldwide in

Oxford New York

Athens Auckland Bangkok Bogotá Buenos Aires Calcutta Cape Town Chennai Dar es Salaam Delhi Florence Hong Kong Istanbul Karachi Kuala Lumpur Madrid Melbourne Mexico City Mumbai Nairobi Paris São Paulo Singapore Taipei Tokyo Toronto Warsaw

and associated companies in Berlin Ibadan

Oxford is a registered trade mark of Oxford University Press

Published in the United States by Oxford University Press Inc., New York

First published 1996

Reprinted 1998

British Library Cataloging in Publication Data
(Data applied for)

Library of Congress Cataloging in Publication Data

ISBN 0–19–854985–7 (Hbk)
ISBN 0–19–854984–9 (Pbk)

1 3 5 7 9 10 8 6 4 2

Printed in Great Britain
on acid-free paper by Biddles Ltd,
Guildford & King's Lynn

Preface

Gene sequence data are accumulating at an ever-increasing pace, and phylogenetic trees are routinely being constructed from those data. Usually the phylogenies are not produced by scientists using a stamp collector's perspective, and nor are they produced as a convenient way of summarizing the data. Instead, they are being used to answer questions in diverse areas of biology. This book arises from the conviction of its editors that, even though the fields of enquiry may be disparate, the ways that phylogenies are used should be the same. There are just two basic procedures: (1) drawing inferences from the structure of a tree and (2) drawing inferences from the way character states map on to the tree. As a consequence, research workers using phylogenetic information for very different purposes should be united by a set of common methods. So long as the language is kept simple, communication should be straightforward.

Accordingly, three of us (PHH, AJLB, JMS) organized a Royal Society Discussion Meeting in December 1994 where we brought together a group of scientists who were making use of these new phylogenies. The contributors included people interested in tracing epidemics, understanding development, revealing population dynamic histories, making decisions in conservation biology, untangling the effects of natural selection and biogeographic history, explaining the structure of the genome, re-creating micro- and macro-evolutionary events, and assessing the extent of parasite–host cospeciation. We were surprised at the extent to which our expectations were fulfilled: the methods being used by the different investigators had many common elements, and the speakers from different disciplines were able to communicate freely about the interpretation of their results. After editing a short record of the proceedings of the Meeting, we decided to produce a more comprehensive book aimed at graduate-course level. To that end, most authors rewrote their articles to provide more contextual material for the presentation of their results. We also invited another six chapters on specific topics that introduce new material and make the coverage of this volume more complete.

We are grateful to our original authors as well as to those included here who did not present material at the original meeting. Everybody has pulled together, taking time out from busy schedules to ease our editorial tasks. We also thank the staff at Oxford University Press for everything they have done to help us along the way. Paul Harvey thanks the Wellcome Trust for supporting his research leave during the period in which this book was being prepared.

Paul H. Harvey
Andrew J. Leigh Brown
John Maynard Smith
Sean Nee

Contents

Combining phylogenetic evidence

Character evolution

Contributors

C. Gregory Anderson Department of Ecology and Evolutionary Biology, The University of Tennessee, Knoxville, TN 37996–0610, USA

N. H. Barton Institute of Cell, Animal and Population Biology, University of Edinburgh, King's Buildings, Edinburgh EH9 3JT, UK

M. J. Benton Department of Geology, University of Bristol, Bristol BS8 1RJ, UK

H. Black School of Biological Sciences, University of Wales, Bangor, Gwynedd LL57 2UW, UK

Paul L. Bollyky Wellcome Centre for the Epidemiology of Infectious Disease, Department of Zoology, University of Oxford, South Parks Road, Oxford OX1 3PS, UK

Keith A. Crandall Department of Zoology, Brigham Young University, Provo, UT 84602–5255, USA

Michael P. Cummings Department of Botany and Plant Sciences, University of California, Riverside, CA 92521–0124, USA

J. C. Daltry School of Biological Sciences, University of Wales, Bangor, Gwynedd LL57 2UW, UK

Walter M. Fitch Department of Ecology and Evolutionary Biology, University of California at Irvine, Irvine, CA 92717, USA

Geoff P. Garnett Department of Zoology, University of Oxford, South Parks Road, Oxford OX1 3PS, UK

John L. Gittleman Department of Ecology and Evolutionary Biology, The University of Tennessee, Knoxville, 37996–0610, USA

Mark S. Hafner Museum of Natural Science, Louisiana State University, Baton Rouge, LA 70803, USA

Beatrice H. Hahn Departments of Medicine and Microbiology, University of Alabama at Birmingham, Birmingham, AL 35294, USA

Thomas F. Hansen Department of Biology, University of Oregon, Eugene, OR 97403, USA

Rosalind M. Harding MRC Molecular Haematology Unit, Institute of Molecular Medicine, John Radcliffe Hospital, Headington, Oxford OX3 9DU, UK

Paul H. Harvey Department of Zoology, University of Oxford, South Parks Road, Oxford OX1 3PS, UK

Eddie C. Holmes Wellcome Centre for the Epidemiology of Infectious Disease, Department of Zoology, University of Oxford, South Parks Road, Oxford OX1 3PS, UK

R. R. Hudson Department of Ecology and Evolutionary Biology, University of California, Irvine, CA 92717, USA

N. L. Kaplan Statistics and Biomathematics, National Institute of Environmental Health Sciences, Research Triangle Park, NC 27709, USA

Mark Kot Department of Mathematics, 121 Ayres Hall, University of Tennessee, Knoxville, TN 37996–1300, USA

D. T. J. Littlewood Department of Palaeontology, The Natural History Museum, Cromwell Road, London SW7 5BD, UK

Jonathan B. Losos Department of Biology, Campus Box 1137, Washington University, St Louis, MO 63130–4899, USA

Hang-Kwang Luh Department of Mathematics, The University of Tennessee, Knoxville, TN 37996–1300, USA

A. Malhotra School of Biological Sciences, University of Wales, Bangor, Gwynedd LL57 2UW, UK

Emília P. Martins Department of Biology, University of Oregon, Eugene, OR 97403, USA

Axel Meyer Department of Ecology and Evolution and Program in Genetics, State University of New York, Stony Brook, NY 11794–5245, USA

Craig Moritz Department of Zoology and Centre for Conservation Biology, The University of Queensland, Qld 4072, Australia

Sean Nee Department of Zoology, University of Oxford, South Parks Road, Oxford OX1 3PS, UK

Sarah P. Otto Department of Zoology, University of British Columbia, 6270 University Boulevard, Vancouver, BC, Canada V6T 1Z4

Roderic D. M. Page Department of Environmental and Evolutionary Biology, University of Glasgow, Glasgow G12 8QQ, UK

Andy Purvis Department of Biology, Imperial College, Ascot, Berkshire SL5 7PY, UK

Andrew Rambaut Department of Zoology, University of Oxford, South Parks Road, Oxford OX1 3PS, UK

David L. Robertson Department of Genetics, University of Nottingham, Queens Medical Centre, Nottingham NG7 2UH, UK

Paul M. Sharp Department of Genetics, University of Nottingham, Queens Medical Centre, Nottingham NG7 2UH, UK

A. B. Smith Department of Palaeontology, The Natural History Museum, Cromwell Road, London SW7 5BD, UK

Alan R. Templeton Department of Biology, Washington University, St Louis, MO 63130–4899, USA

R. S. Thorpe School of Biological Sciences, University of Wales, Bangor, Gwynedd LL57 2UW, UK

John Wakeley Nelson Biological Laboratories, Busch Capmus, Rutgers University, Piscataway, NJ 08855–1059, USA.

I. Wilson Institute of Cell, Animal and Population Biology, University of Edinburgh, King's Buildings, West Mains Road, Edinburgh EH9 3JT, UK

G. A. Wray Department of Ecology and Evolution, State University of New York, Stony Brook, NY 11794–5245, USA

W. Wüster School of Biological Sciences, University of Wales, Bangor, Gwynedd LL57 2UW, UK

1

What this book is about

Paul H. Harvey and Sean Nee

From time to time scientists working in quite different areas find themselves united by common knowledge, methods, or theories. This volume shows how biologists working on a set of seemingly different problems are entering such a time. The revolution in gene sequencing technology is resulting in the production of more accurate phylogenies or gene genealogies which can be used to help understand biological processes occurring at many different levels of life's hierarchy. As different chapters demonstrate, genetics, development, behaviour, epidemiology, ecology, conservation biology, and evolution are examples of fields that can be illuminated by such information.

This volume is divided into four parts. The first consists primarily of chapters that introduce and exploit the power of the coalescence approach which has brought gene trees to the forefront of theoretical population genetics. The second part exploits results developed in the first part to show how inferences derived from the structure of genealogies can inform different areas of biological investigation including epidemiology, biogeography, genome structure, macroevolution, and conservation biology. The third part considers how molecular information can be combined with other sources for phylogenetic inference to date nodes more accurately, to reveal different modes of evolution for different characters, and to analyse patterns of cospeciation. The fourth and final part considers how character state evolution can be mapped on to tree structure to reveal correlates and precursors of character evolution, to test ideas about the evolutionary lability of different types of character, to suggest that similar evolutionary transitions are repeated when similar ecological conditions pertain, and to show how the evolution of developmental systems can be unravelled by phylogenetic analysis.

We must emphasize that this book is not about phylogeny reconstruction. The phylogenies used are really working hypotheses and, pragmatically, have been reconstructed using a variety of methods, usually maximum likelihood, neighbour joining, or maximum parsimony. There is a substantial literature summarized in several useful reviews on methods of tree reconstruction (see, for example, Swofford and Olson 1990), which is itself a rapidly developing field. While the aim of this volume is to show how inferences can be made from the trees we have, it is essential that uncertainties of tree structure are taken into account when doing so.

1.1 The coalescent

Coalescence theory has been developed as an approach for addressing questions in population genetics. Consider the following question about a sample of homologous molecular sequences from a population: what is the expected distribution of numbers of pairwise differences between the sequences, under the assumption that they have been evolving neutrally in a population that has always been of the same size? The coalescent approach to such a question is based on the following ideas. First, it recognizes that the molecules in the sample are actually related to each other, in a way that could, in principle, be depicted as a family tree, or gene genealogy. Second, it models a mutational process occurring along the lineages in this tree which will generate our observed sequence variation. As long as the mutation rate is not zero, more closely related sequences will tend to be more similar. So, to address the question posed above, we need to know (i) what kinds of trees we expect to exist under our assumptions that the population has been of constant size and that evolution is neutral (for example, what is the distribution of times between the nodes) and (ii) what sort of mutational process has been occurring. Knowledge from other branches of biology is required to address the second question, and we will not discuss it further. It is the first question that has given coalescence theory its name: it turns out that it is often much easier to address this question, either analytically or by simulation, if we adopt a backward temporal perspective. So, instead of thinking of the tree as growing, with bifurcations occurring at nodes, we model it as collapsing, with lineages coalescing at nodes.

For a population of constant size and a genome with constant mutation rates, analytical results can be used to provide the distribution of pairwise differences. Alternatively, we can simulate genealogies that would arise for sequences sampled from a constant-sized population, and we can simulate a mutational process occurring along the lineages in this genealogy. Each run of the simulation will generate hypothetical molecular sequences, and it is a straightforward matter to analyse the properties of their distributions of pairwise differences. This is the coalescent approach to the question posed in the previous paragraph.

Harding's essay (Chapter 2) provides a thoughtful overview of the issues involved with the coalescent approach, including the reasons for choosing it and its current limitations. The subsequent three chapters show coalescence theory in action. Each chapter extends the theory to deal with a different factor, resulting in the production of methods for evaluating how gene genealogies can be analysed to recognize histories of population dispersal, natural selection, and population dynamics.

Barton and Wilson (Chapter 3) explore the extent to which, in principle, the history of dispersal of a population in a landscape is retained in the gene genealogy and, hence, is visible to us. Although they point out that genealogies are more informative about population spatial structure than pairwise statistics such as Wright's F_{ST}, their analysis of coalescence times in a two-dimensional habitat indicates that the increase in spatial dispersion with lineage age is, unfortunately, a poor estimator of gene flow; most of the information in the tree concerns events in the distant past. Hudson and Kaplan (Chapter 4) describe and develop the theory for coalescence times under a model with background deleterious mutation. Loci that are more closely linked to those in which deleterious mutations occur will, on average, show less standing variability in a population. When the coalescent model incorporating background selection and distance from points of recombination is applied to predict the level of nucleotide diversity along chromosome 3 of *Drosophila melanogaster*, the fit to actual data is remarkably good. In the third extension of coalescent theory, Nee *et al.* (Chapter 5) ask how we might go about making inferences about past population dynamical history from gene trees. Here, theory is developed to reveal what tree structure would look like under two particular scenarios: when a population has been of constant size and when it has been growing exponentially. Each of these three chapters sets a framework for further research, highlighting the need for new theory that will allow yet more insights into population dispersal, modes of selection, and population dynamics.

A gene tree for human mitochondria could, in principle, be used to locate the geographic origins of the most recent common ancestor of modern human mitochondria. However, as the 'mitochondrial Eve' saga unfolded, it became apparent that the mitochondrial DNA sequences being analysed contained so few variable sites that many equally parsimonious historical scenarios could be constructed. Most such trees are inferred from comparison of sequences less than 340 base pairs in length from the control region, and the best from sequences 670 base pairs long. Even in these latter cases, we can expect that only about eight sites will have changed since the root of the tree, thus allowing very little discrimination among more closely related individuals. When mutational events have been uncommon, we expect to find different individuals yielding identical sequences of DNA. Each set of identical sequences is frequently referred to as a haplotype. If different haplotypes produce phenotypes with equal genetic fitness, then it is possible to infer relationships among haplotypes with reasonable accuracy. Templeton has been at the forefront of that exercise and his chapter (Chapter 6), co-authored with Crandall, provides a useful review of the methods, based on statistical parsimony rather than evolutionary models, which have been developed for reconstructing haplotype phylogenies. Crandall and Templeton also describe how haplotype trees are being used to further understanding in areas as diverse as epidemiology, population genetics, and conservation biology.

1.2 Inferences from trees

When drawing inferences from a tree, we should always remember that the tree itself is an inference which may be incorrect. We have already mentioned that the mitochondrial trees for samples from human populations are not well characterized because too few sites vary. However, we might not expect the same to be true for the mitochondrial tree inferred from comparisons among the full 16 000 base-pair genomes of more distantly related taxa, such as vertebrates from different orders and classes. There are two reasons that we might feel more confident in our tree. First, more time has occurred since their most recent common ancestor and that time has allowed informative mutations to accumulate. Second, we are comparing sequences which are more than an order of magnitude longer than those used in the comparisons involving humans, so more sites have the potential to vary. Otto *et al.* show that we should probably be correct in our inference (Chapter 7). But, in fact, their analysis also presents a much more cautionary tale.

Nowadays, most authors assess uncertainties of tree structure using the bootstrap (Felsenstein 1985). New sets of data are created from the original dataset by sampling with replacement, and several hundreds or thousands of trees are thereby created to examine the percentage of nodes that remain identical (in terms of specifying the same set of descendent species). Otto *et al.* use bootstrapping to ask how many data would be necessary before we could be reasonably sure that the phylogenetic tree is correct. If we are using molecular data, the lengths of sequences to be compared will depend on the rates at which base pairs change over evolutionary time. If rates are high, there is danger of saturation or homoplasy so that the phylogenetic information content per base is low. But if rates of change are very low, then taxa may not have diverged at all. Different areas of the genome evolve at different rates and it is, of course, sensible to choose an area for sequence comparison that has diverged by the optimum amount. Otto *et al.*'s chapter illustrates methods for determining areas of the genome with more or less variable sites; the nature of bootstrap support for clades; the power of different tree reconstruction methods; and the number of base pairs that must be compared, together with their distribution along the genome, before tree inference can be made with a specified degree of accuracy. Their analyses provide very sobering, straightforward, and useful insights into the problems associated with tree reconstruction from molecular data, not least of which is that, for the genome analysed and the species used 'Much longer sequences than are characteristic of single genes were necessary before the whole genome tree was reliably inferred.' Such analyses also reveal the strengths and weaknesses, which will differ according to circumstance, of the many assumptions which go into phylogeny reconstruction. As such, they can be used to help decide which are likely to be the most informative sequence data to use in tree reconstruction and which are the most reliable summary statistics for describing the informative components of tree structure. For example, some methods such as those discussed in Chapter 5, 10, and 11 of this volume utilize information on the rate of growth in the number of lineages through time in the tree. It may be that such methods are fairly robust to failures in describing the correct tree topology. Simulation and resampling analyses will answer that question.

Once we have our tree, we can use it make inferences about biological processes. While appreciating new uses for this phylogenetic information, it is important to realize that many phylogenetic relationships have been known for a long time, and have already provided important insights into a whole variety of biological processes. In a delightful romp through a disparate literature stretching back over several decades, Fitch (Chapter 8) shows how gene genealogies have been used to (i) reveal the origins of diversity in genome structure through such processes as gene duplication, recombination, and reassortment, (ii) determine rates of genetic change and date certain events, (iii) test theories about the reasons for evolutionary change, and (iv) reveal geographic and other biological correlates of tree structure which give insight into historical processes. Sharp *et al.* (Chapter 9) employ those same approaches to tackle a problem of present interest and practical urgency: where did the HIV viruses that cause AIDS in humans come from, and when? The HIV viruses are primate lentiviruses and so Sharp *et al.* explain what is known about those primate lentiviruses that are likely to be the closest relatives of HIV-1 and HIV-2. Using phylogenetic trees and parsimony arguments, they describe the most likely host–species transfers, and present clear evidence for recombination between viral lineages occurring when even highly divergent strains infect the same host individual. Their clear-minded analysis not only reveals unexpected routes of cross-species transmission, but it also shows where new data are needed that will allow us to distinguish between the various alternative explanations they present.

The theory and examples discussed so far do not require that a full phylogeny is available. For example, coalescent theory is generally applied to a very small random sample of extant lineages, say a sample of sequences from a population. Increasingly, however, for some comparisons complete phylogenies are becoming available. Purvis (Chapter 10) has produced from molecular, fossil, and modern morphological material a composite, dated phylogeny showing the relationships among extant primates. Given such a phylogeny, it has proved possible to estimate lineage splitting and lineage extinction rates. This is an exciting realization: it was far from obvious that lineage extinction rates could be estimated in a phylogeny that contained no explicit information on extinct lineages. In fact, as Purvis describes, complete phylogenies can be analysed to reveal evidence for mass extinctions, density-dependent cladogenesis, and to identify correlates of high speciation rates. Furthermore, Purvis's chapter has a non-obvious subtext to which we will here draw attention. The theoretical power of adopting a backward temporal perspective on trees, as in coalescent theory, does not mean that this is always the best way to proceed. The theory underlying the analyses of the Purvis chapter, which is theory appropriate for complete trees, was derived from a forward temporal perspective. Indeed, a coalescent approach to the theory of complete trees had previously been found intractable, except for the special cases of trees which have grown according to a pure birth process, or the trees of the extant lineages of a clade which has always had *exactly* the same number of lineages, so that each speciation event is accompanied by an extinction event (Hey 1992).

The coalescent approach, then, is probably most valuable for the inference of temporal trends from incomplete phylogenies. Excellent examples of such phylogenies are provided by the viruses responsible for a variety of infectious diseases including AIDS and hepatitis. Holmes *et al.* (Chapter 11) show how trees constructed from three such viruses (HIV-1, hepatitis B, and hepatitis C) can be used to trace their epidemiological histories. We make two points concerning this type of analysis. First, whenever possible it is desirable to perform similar but independent analyses on trees constructed from different genes sequenced from the same and from different populations of viruses. Results can then be compared and, if they are similar, we will be more confident that our patterns result from the inferred population processes rather than the effects of, say, different rates of evolution in different parts of the genome or selective sweeps of new variants spreading through populations. The second point is that reconstructing the history of epidemics of infectious disease is normally not possible in the absence of explicit historical information. Hepatitis C was not isolated until 1989, so it would probably not be possible to trace its spread prior to that time without the imprints of history retained in the structure of phylogenetic trees.

Chapters 5 and 10 show how the increased number of taxa incorporated into new molecular phylogenies have prompted the development and application of methods to test for temporal trends in cladogenesis. The same is also true for biogeographic analyses. It has long been common practice to map phylogenetic tree structure on to geographic maps, thereby suggesting lineage dispersal routes (for example Cavalli-Sforza *et al.* 1964). Thorpe *et al.* (Chapter 12) argue that the time has come to develop and apply statistical tests which will allow us to distinguish between alternative historical scenarios. For example, similarities between taxa may result from common inheritance, similar adaptations to equivalent selective pressures, or both. Thorpe *et al.* use partial Mantel tests to help tease apart these different factors. Some characters are highly correlated with environmental factors, some with phylogenetic relationships, and some with both. Of course it will often be possible that the relevant selective forces in the environment have not yet been identified, and are therefore not incorporated into the tests. Nevertheless, as Thorpe *et al.*'s chapter so clearly demonstrates, the combination of a phylogeny with nodes and geographic events that are accurately dated, together with measures of character and ecological variation in modern populations, frequently accords with only one of many historical scenarios.

The third section of this book returns to the issue of combining phylogenetic evidence from different sources, but before it does so Moritz (Chapter 13) develops the theme of using temporal and biogeographic analyses to help tackle particular applied problems. As the habitable area of the world becomes smaller, Moritz asks how phylogenetic information might be used to help inform decisions about priorities for the conservation of populations, species, and representatives of higher taxa. There is no doubt that several of the methods described in this volume and elsewhere have potential value for conservation biology but, as Moritz demonstrates with examples, the ecological and evolutionary time-scales may be so different that analysis of phylogenetic tree structure reveals a long-term population history that may have little relevance to present-day processes. This will, of course, not always be true as in the case of Holmes *et al.*'s analysis of epidemics of infectious disease, but it is a message that should be well taken. Another question, which has caused divisions between cladists and evolutionary biologists, is which particular combinations of species should be conserved if we are given a choice of how to allocate limited resources? Fortunately, that issue is now resolved, at least at an academic level: conserving those taxa that maximize the conserved branch length on a phylogeny will usually conserve most genetic diversity (see May 1994). Time will tell whether practical conservation decisions will actually be influenced by such concerns.

1.3 Combining phylogenetic evidence

Molecular phylogenies provide the working platform for most of this book but, as we have seen, they may be incorrect. Classically, fossil material has been used to reconstruct phylogenetic relationships and to help date nodes, and it would be wrong to let molecular data swamp the importance of fossil material. Benton (Chapter 14) provides a useful summary of the value of the fossil record, not only for establishing the temporal dimension of phylogenies, but also for correcting inaccuracies which will inevitably be present in molecular phylogenies. There can be no doubt that available fossil evidence is, at present, too rarely incorporated into molecular trees, probably because statistical methods (with few exceptions—for example, see Fisher in Maddison and Maddison (1992)) have been developed to deal with information from present-day taxa alone. As Benton points out, the increasing number of DNA sequences from fossil material make the development of appropriate analytical techniques an important task for the future.

In a radical new study, Smith *et al.* (Chapter 15) take on the challenge of combining fossil material with phylogenies derived from adult morphology, larval morphology, and ribosomal RNA derived from extant species. Their analysis of evolution in post-Palaeozoic echinoids faced various methodological problems, some concerning issues that are not yet fully resolved. Their pragmatic decision to use a parsimony-based consensus tree as a basis for testing hypotheses about concerted evolution, relative rates of homoplasy in different characters, and the correlation between genetic and morphological change is, as they point out, open to criticism from several quarters. For example, testing ideas about character change from a tree constructed in part from the characters under examination creates the possibility of circularity. And the reasons against using parsimony as a basis for phylogeny reconstruction are well rehearsed. Nevertheless, the data are now available and Smith *et al.*'s tentative conclusions are set out. We should be surprised if alternative methods of analysis produce substantially different results. That, of course, does not mean that improved methods cannot be developed and applied to the dataset, and we hope that they will be.

Both Benton and Smith *et al.* are concerned with combining different types of evidence to establish the one true tree which shows the relationships among the taxa they are interested in. They then show how that tree can be used to answer a variety of questions concerning, for example, evolutionary rates. Frequently, biologists deal with phylogenies which are not necessarily expected to have been congruent. For example, parasites often speciate with their hosts but sometimes they move on to new hosts. Comparing the structure of trees for parasites and their hosts may be expected to inform us about host–parasite fidelity and rates of evolution in independently evolving genomes. However, three stages of analysis are required before conclusions can be drawn. First, the two trees must be constructed. Second, the trees must be mapped on to each other. Third, the structures of the trees must be compared to produce biologically meaningful results. Each stage of the analysis is, as Page and Hafner (Chapter 16) write, 'fraught with theoretical and methodological challenges'. Their chapter shows how those challenges can be answered at each stage of analysis. Taking pocket gophers and chewing lice as their model host–parasite system, Page and Hafner lead us through each stage of the analysis pointing to the assumptions behind the different techniques available, why they chose the ones they did, and how the results can be interpreted. Their method of choice for comparing trees, component analysis (an alternative would have been Brooks' parsimony analysis), reveals that such tree comparisons have far wider applicability than just to parasite–host studies. Component analysis was, in fact, developed to study biogeographic relationships (mapping phylogenies on to, say, islands of an archipelago) and can also be used to compare gene trees with species trees. Any dissimilarities between trees used in the different contexts have different but analogous causes, so that familiarity with analysis of one type of system provides expertise for dealing with the others.

1.4 Character evolution

Phenotypes can and do evolve through time, so that more closely related taxa are, in general, phenotypically more similar than distantly related taxa. We also know that distantly related taxa can converge phenotypically, often as a consequence of living in similar environments. Given (i) a range of phenotypes in present-day taxa, (ii) a phylogeny relating those taxa to each other, and (iii) an evolutionary model for character evolution it is possible to trace character evolution and even reconstruct likely ancestral character states. Recent years have seen notable advances in the so-called 'comparative method' which tests the null hypothesis of independent evolution between characters, or between characters and environment. We have deliberately avoided going over the same ground again because several recent reviews cover the basic issues (for example

Harvey and Pagel 1991). Instead, Martins and Hansen (Chapter 17) take a much broader approach, providing a description of the general evolutionary modelling and statistical framework within which character evolution and character co-evolution can be examined. In particular, they highlight the interdependence of phenotypes, phylogeny, and evolutionary models, so that any two of these provide information about the third. Tree reconstruction itself uses phenotypes (or genotypes) and an evolutionary model to infer the most likely tree. The comparative method uses phenotypes and phylogenies to infer the most likely model of evolution (or co-evolution). And, of course, given a phylogeny and a model of evolution, it is possible to predict the likely distribution of phenotypes. Martins and Hansen focus much of their discussion on the phenotype variance–covariance matrix which describes the expected phenotypic relationships among taxa given a particular phylogeny and a model of evolution. Producing a model of character evolution which maximizes the fit between the actual and the expected phenotype variance–covariance matrix will, they believe, provide useful information about microevolutionary processes which have led to macroevolutionary patterns.

It is often argued that some types of phenotypic characters are likely to be more evolutionarily labile than others. Mapping character state change on to phylogenetic tree structure provides the potential to test that idea or, in Martins and Hansen's perspective, we can use the distribution of phenotypes and phylogenetic tree structure to determine the most likely model for character state evolution, thereby providing the potential to compare evolutionary rates. Gittleman *et al.* (Chapter 18) apply phylogenetic autocorrelation procedures for the first time to compare the evolutionary lability of life history, morphology, and behaviour among mammals. Phylogenetic autocorrelation and independent contrasts are the two methods now most commonly used for modelling the evolution and co-evolution of continuously varying traits, with independent contrasts methods probably being the more popular. Recent years have seen the development of many new comparative methods, followed by synthetic studies which show how they are related to each other and why one of them may be the preferred method for any particular study. Phylogenetic autocorrelation and independent contrasts are, indeed, logically connected (Martins 1995) but why or when one should be used in preference to the other remains a matter for debate.

The strength of comparative analysis rests on generalities. If something happens once during evolution then, given similar circumstances, a similar thing is likely to happen again. Unless such convergent or parallel evolution is reasonably common, the analysis of character change through evolutionary time provides a series of individual case studies, none of which informs any other. Most biologists, with the notable exception of Stephen Jay Gould, would probably accept the value of examining life's generalities rather than emphasizing life's contingencies. But what would really happen if we could play the tape of life through again? If the ancestral finch(es) were to reinvade the Galapagos Islands again under the same conditions would we get a similar radiation? The outcomes of such thought experiments are normally a matter of personal preference rather than scientific investigation. However, as Losos shows in Chapter 19, there are circumstances where the tape of life has, to all intents and purposes, actually been played twice. Anolis lizards in Jamaica and Puerto Rico comprise communities consisting of several species. Equivalent ecomorphs can be identified on the two islands and, when the phylogenetic relationships among the two independently evolved communities are compared, it seems that they have 'attained their similarity by evolving through a nearly identical sequence of ancestral communities'. Losos's ecological interpretation of the particular evolutionary sequences he obtained led to predictions about the outcome of transplant experiments and the types of ecomorphs which would be found on other islands. All in all, his studies, alone and with Tom Schoener, provide superb examples of the interplay between ecology and evolution. Each is a richer discipline when informed by the other.

One discipline that has had relatively little useful input from evolutionary biology over the past century is developmental biology. Time and again, models of development have been produced which seem to make little sense in an evolutionary context. While evolutionary biologists have been quick to criticize those models, the challenge from developmentalists has always been for evolutionary biologists to come up with better. The time is ripe. Meyer (Chapter 20) reviews the exciting recent developments of the discovery of homeobox genes, together with the subsequent analysis of their diversity and function. Suddenly, as Meyer shows, phylogenetic analysis can provide the key to answering vital questions about the evolution of development. The success of this enterprise will, as ever in phylogenetically based analyses, depend on getting the right phylogeny. In this case, as with other studies of character evolution, progress will also depend on the accurate description of characters in certain key taxa. Meyer describes some of the key phylogenetic relationships which must be resolved, and identifies key taxa in which the architecture of the HOM/hox clusters must be described. He also emphasizes, as did Martin and Hansen in the opening chapter of this section, the importance of using an appropriate model of evolution when analysing the relationship between character states and phylogenetic tree structure. To the reader of this volume, these emphases will come as no surprise because they will have recurred in the various chapters. It seems that experience with one area of phylogenetic analysis preadapts us to focus on the right questions in another area. This is because research scientists from very different areas of biology are using the same conceptual framework to answer seemingly disparate questions. That framework should be both recognized and built upon as ever newer uses are found for even newer phylogenies.

References

Cavalli-Sforza, L. L., Barrai, I., and Edwards, A. W. F. (1964). Analysis of human evolution under random genetic drift. *Cold Spring Harbor Symp. Quant. Biol.*, **29**, 9–20.

Felsenstein, J. (1985). Confidence limits on phylogenies: an approach using the bootstrap. *Evolution*, **39**, 783–91.

Fisher, D. C. (1992) Stratigraphic parsimony. In *MacClade: analysis of phylogeny and character evolution. Version 3*, (W. P. Maddison and R. D. Maddison), pp. 124–9, Sinauer, Sunderland, MA.

Harvey, P. H. and Pagel, M. D. (1991). *The comparative method in evolutionary biology*. Oxford University Press.

Hey, J. (1992). Using phylogenetic trees to study speciation and extinction. *Evolution*, **46**, 627–40.

Martins, E. P. (1995). Phylogenies and comparative data: an evolutionary perspective. *Proc. R. Soc.*, **B349**, 85–91.

May, R. M. (1994). Conceptual aspects of the quantification of the extent of biological diversity. *Phil. Trans. R. Soc.*, **B345**, 13–20.

Swofford, D. L. and Olsen, G. J. (1990). Phylogeny reconstruction. In *Molecular systematics*, (ed. D. M. Hillis and C. Moritz), pp. 411–501. Sinauer, Sunderland, MA.

The coalescent

2

New phylogenies: an introductory look at the coalescent

Rosalind M. Harding

It must be a reasonable assumption that anyone interested in this book will be well acquainted with phylogenies, though not necessarily with population genetics. A special kind of phylogeny that has recently come to prominence is a gene tree, which is a phylogeny constructed from DNA sequences, and of particular interest are gene trees not for sets of different taxa, but for samples of individuals from an interbreeding population. The coalescent is a population genetic model for the neutral evolution of gene trees (Griffiths 1980; Kingman 1982*a,b*; Tavaré 1984; Donnelly and Tavaré 1995), also known as gene genealogies (Tajima 1983; Hudson 1990) or allelic genealogies (Takahata 1993). This model is both powerful and efficient and has the following advantages. First, the coalescent is more powerful than diffusion theory formulations based on effective population size for modelling a range of demographic histories, such as those defining variable, expanding, and subdivided population size. Secondly, coalescent models can be used to simulate, and estimate from, gene trees, which provide information for stronger inferences on demographic history than summarizing statistics of DNA sequence data. Thirdly, the coalescent is more efficient than branching processes for simulating a wide variety of conditions in both mutation and demography.

2.1 Neutral models

Traditional phylogenies are studied and analysed with reference to Darwin's model for evolution by natural selection. To interpret the new phylogenies, however, a model for the neutral theory of DNA evolution (Kimura 1983) is more appropriate. Neutral models provide time-scaling. At the time of writing, the DNA sequences that have been used to construct gene trees have been sampled only from individuals still alive, or at least recently so, not from their fossilized, or otherwise preserved, ancestors. Without the equivalent of dated fossils for the ancestral sequences in a gene tree, application of a neutral model provides the only solution available to the problem of time-scaling. Time-scaling of all gene trees is based on the assumption that the turnover of variation at the DNA locus of interest ticks with the regularity of a molecular clock

(Kimura 1983; Jagers 1991). In other words, there should be strict functional neutrality among the DNA sequence variants, and the nucleotide mutation rate, which is the time-scaling factor, should be constant.

Compared with gene trees that relate different taxa, there is an added complication for interpreting gene trees of intraspecific DNA variation and this is due to their assembly from DNA polymorphisms rather than from fixed nucleotide differences. Whereas fixation is simply proportional to mutation rate, transient polymorphism results from a balance between the input of new variants by mutation and their loss by genetic drift (Kimura 1983). Genetic drift occurs as stochastic variation in reproductive success when each successive generation is sampled from the previous one. It is therefore a tracer of the history of population size. To understand and appreciate intraspecific gene trees, interaction between time and population size as demographic history must be considered. The payback from such investment in these new phylogenies is that details of the demographic history of a species can be inferred.

2.2 Genetic drift

At the end of the nineteenth century it was a profound observation for biologists to discover that different lineages in the present have common ancestors in the past, forming the 'coalescences' in a phylogeny. Since Darwin used his theory of evolution by natural selection to study history rather than predict the future, it has been natural for his successors to look backward in time rather than forward. Yet the related concept of coalescence in the ancestral lineages of a population of interbreeding individuals is not intuitive because tracing backward in a family tree from a single individual, the number of ancestors apparently doubles in every generation. However, were this so, within 34 generations an individual's expected number of ancestors would be greater than the current human population of the earth.

The number of ancestors in a family tree decreases because individuals alive now share (great)^n grandparents. In other words, while most of their contemporaries after many generations have no descendants, a few individuals in the past have, after many generations, large numbers of descendants.

Arguably, this may be for reasons of natural selection, but in models of neutral evolution the variance in reproductive success is attributed to random factors. The smaller the population size, the shorter the time interval between a successful founding ancestor and the generation when everyone in the population is a direct descendant. Going forward in time this process is recognized as genetic drift. The same process going backward in time is coalescence in ancestry. The coalescent is a stochastic model for the genealogical ancestry of a sample of individuals traced back to their most recent common ancestor. It was introduced by Kingman (1982*a*) as a continuous-time approximation, obtained in the limit of large population size, to this ancestral process. Kingman (1982*b*) also provided a very useful invariance principle showing that essentially all the exchangeable reproductive models can be approximated by this coalescent. In fact, coalescent models are appropriate for a wide range of demographic histories including subdivision (Herbots 1996), constancy, expansion, and fluctuation. This is one of the advantages of the coalescent over established population genetic models.

The coalescence model supercedes the traditional population genetics approach for modelling demographic history based on the concept of effective population size, N_e. N_e is the size of a theoretical population, not only constant but also panmictic and comprising individuals having equal probabilities of reproductive success over some fixed time parameter. It is substituted for a realistic population that is neither constant nor panmictic because the assumption of constant population size over evolutionary time is a necessary feature of Wright–Fisher demography. Consequently, a population that has expanded to a large census size may have the same N_e as a numerically smaller population that is subdivided. However, expanding and subdivided populations with the same N_e, have different demographic histories and, accordingly, different coalescence times.

Another advantage of coalescent models is that the lineage history of a sample of DNA sequences is simulated as a sufficient and efficient alternative to modelling the whole population forward in time. The realization that the reproductive contribution of individuals, other than those ancestral to a sample of interest from the current generation, can be ignored, has enabled the simulation of large populations for a range of demographic histories (Slatkin and Hudson 1991; Marjoram and Donnelly 1994), and as well, for a variety of mutation and recombination processes (Hudson and Kaplan 1988; Hudson 1990). The new willingness by mathematical geneticists to accept simulation as a viable alternative to analytical modelling demonstrates the power of the coalescent. When applied to estimation problems this approach is, however, even more computer intensive, and has yet to be embraced whole-heartedly by biologists.

To summarize so far, intraspecific gene trees relating DNA sequences contain information about demographic history, provided that it is reasonable to assume a neutral model of evolution. The most appropriate model for studying gene trees is the coalescent. Compared with established population genetics models the coalescent is both efficient and powerful. The relationships between coalescent trees and gene trees will be explained in greater detail below, and finally, a warning will be offered. The coalescent is a stochastic model. Simulating the coalescent a thousand times generates a thousand gene trees, astounding in their variety of topology and time-depth. There is no clearer demonstration of the magnitude of the evolutionary variance inherent in neutral models than the graphical output of a coalescent model.

2.3 Coalescent trees

The coalescent is a model of genetic drift run backward in time. A coalescent tree represents one run of evolution by genetic drift at a single DNA locus. In this model (without recombination) ancestral lines going backward in time coalesce when they share an ancestor, *i.e.* a parent, a grandparent, or a (great)^n grandparent (for graphical illustration of coalescent trees see Donnelly and Tavaré (1995)). Coalescences occur between pairs of individual lineages, and in each generation that a coalescence occurs the number of ancestral lines is reduced by one, generating a bifurcating coalescent tree. For a constant-sized diploid population having $2N$ lineages, the time taken until only two lineages are left is $2N$ generations, on average. After another $2N$ generations, again on average, the last two lineages coalesce as the single common ancestor. The most recent common ancestor (MRCA) defines the root. The expected total coalescence time is $4N$ generations. However, any single coalescent tree may be shorter or longer, and there is a large evolutionary variance in total coalescence time. Coalescent trees can also be constructed when population size has not remained constant in time. The rate that lineages coalesce in these trees is faster or slower than in a constant-sized population, depending on the demographic history.

To appreciate the flexibility of the coalescent for modelling a range of demographic histories, consider the following. Two populations have the same current population size, but while one has remained constant in size over time, the other has expanded to its current size from a smaller number of founders. Going back the same number of generations into the past in each population, there are fewer ancestors in the expanding population compared with the number of ancestors in the constant-sized population. If, looking backward, the number of ancestors decreases faster, pairs of lineages must coalesce more rapidly, reducing the total time-depth of the tree. Fluctuating population size also increases the rate of coalescence during the intervals of smaller size, likewise making the tree shorter. In contrast to population size variation, subdivision slows down the rate of coalescence compared with random mating, and stretches the tree further back into the past. This is because coalescence events between lineages from different subdivisions become dependent on migration, events that by definition bring mates together less often than expected with random mating. The range of demographic histories that can be modelled by the coalescent require no more than the choice of a rate of coalescence, in other words, an appropriate scaling factor.

Mutations occurring in lineages of the coalescent tree generate different types of DNA sequences. Because the coalescence of the last two ancestral lineages in a constant-sized population is expected to take half the total coalescence time, approximately half of the polymorphic mutations in the gene tree will segregate between these two primary lineages. The average number of different nucleotide sites between pairs of DNA sequences is a statistic that has been used to estimate the coalescence time to the MRCA (Tajima 1983). A problem is that this statistic is not consistent (Ewens 1979). Rather, the variance in pairwise differences decreases, not to zero but to an asymptotic value as the numbers of sequences sampled from a population increases (Tajima 1983). As a consequence, estimates of coalescence time and other parameters from pairwise differences appear not to be greatly improved by increasing numbers of sequences sampled greater than approximately 10 (Harding 1996).

2.4 Gene trees

Surveys of genetic variability have been conducted at the DNA level for more than a decade, and the use of techniques from molecular biology to score polymorphisms is routine. A more complex task is to score a set of polymorphisms linked in a sequence in order to provide data for constructing gene trees. Haploid genomes such as mitochondrial DNA (mtDNA) are considerably easier to survey than diploid nuclear DNA because without multiple heterozygous sites there are no ambiguous linkage relationships. This is one of several reasons why population genetic studies have favoured mtDNA analyses, although nuclear sequence data for large samples of individuals are becoming available.

A gene tree of DNA sequences sampled from a population in the current generation can be considered a subgraph of the coalescent tree (for graphical illustration see the gene tree presented by Griffiths and Tavaré (1994*a*)). The gene tree reconstructs the history of mutation events, describing some of the genealogical structure that occurred in the coalescent tree, in particular providing details of the order in which pairs of lineages coalesced. More information is available for estimating parameters of demographic history from a gene tree if mutations are consistent with an infinite-sites assumption than if they are recurrent (Griffiths and Tavaré 1994*a*). The infinite-sites assumption requires that each mutation occurs at a different nucleotide site, so that the mutations in the gene tree fully represent the mutation history of the sample since the MRCA. Genetic data which indicate recombination or recent parallel mutations due to either recurrent mutation or gene conversion are not infinite-sites compatible. Computational algorithms have been developed for finite-sites mutation (Griffiths and Tavaré 1994*b*) but they have to do a lot more work to find coalescent trees consistent with a given dataset, and are much slower.

Whereas the coalescent is a population genetics model that relates the process of neutral evolution to DNA sequence variation represented as a gene tree, earlier population genetics models relate neutral evolution to allelic variation represented by heterozygosity or numbers of alleles in a sample. The great advantage of coalescent models for simulating and analysing genetic data is that gene trees are more informative than heterozygosity or numbers of alleles in a sample. Interestingly, many population geneticists who use coalescence theory continue to reduce gene trees to heterozygosity statistics, or their modern derivative for DNA sequences, pairwise difference distributions. Heterozygosity statistics were originally favoured as an accommodation of slightly deleterious mutations in models assuming strict neutrality. In mtDNA studies, pairwise difference distributions appear to accommodate parallel mutations in models assuming infinite-sites mutation. Considering that population genetic models using allelic variation have not been very successful in providing discrimination between selection and neutrality, it is a tall order to ask for discrimination between different neutral models. To use the coalescent at its most powerful, DNA sequence variation should be represented as a gene tree.

In a population genetics context, a gene tree is not just a connectivity graph. The frequencies of each of the DNA sequences in a sample must be observed as well. By this definition, a single unrooted gene tree is a complete representation of the information in a sample of DNA sequences, assuming infinite-sites mutation. This tree is a maximum parsimony tree. If the infinite-sites assumption is not met, the data contain homoplasies and should be represented as a network rather than a tree. The opportunity to use all of the information contained in a gene tree is a major advantage of coalescent models. Furthermore, estimates of coalescence times are consistent and can be improved by increasing sample numbers of sequences. However, if a genetic study is limited to a trade-off between increasing the number of sequences or increasing the length of DNA sequenced, the important factor to consider for improving the informativeness of a tree is the number of mutations scored. The larger the number of mutations scored, the easier it is to deduce where, and when, in the original tree the coalescence events occurred that unite the various lineages represented in the sample of DNA sequences.

Time and population size scaling, present in the coalescent tree, are lost from the gene tree, but can be estimated, assuming infinite-sites mutation, using the information in the tree, if the mutation rate is known. Ideally, the gene tree provides sufficient information to fully resolve the ordering of coalescence events, but it may not if either the length of DNA sequenced is too short, or if coalescence events occurred more rapidly than mutation events. Unresolved ordering is typical of exponential population expansions from a very small source population, resulting in star-shaped gene trees; in these trees all the lineages coalesce at the root. The coalescent provides the scaling between the gene tree and the coalescent tree, with different scalings for different demographies. When scaling a star-shaped gene tree to fit a coalescent tree there will be many ways to resolve the ordering of coalescence events and the variance around estimates of the coalescence time will be large, particularly for an assumption of constant population size and the implication of insufficient data. Scaling such a gene tree to fit a coalescent tree for population expansion will give a shorter time back to the MRCA and also a smaller variance.

2.5 The coalescent is a stochastic model

Simulation studies show that the variability produced with each realization, or run, of a coalescent model is extensive almost beyond credibility. This is the evolutionary variance. For constant-sized and structured population models in particular, data presented as either pairwise difference distributions or gene trees rarely look like their expected values. Expected values can be computed, for instance, as averages of these distributions, but these distributions look nothing like normal distributions, and neither averages nor modes represent a substantial proportion of the individual realizations of the model. These large evolutionary variances imply that little confidence can be placed in inferences about demographic history that are made on a gene tree for a single locus, such as mtDNA, until confirmed by additional studies of further loci from the genome. In fact, if the anticipated evolutionary variance is not demonstrated by widely varying gene trees for different loci in the genome, the appropriateness of a neutral stochastic evolutionary model will become increasingly suspect. As analyses of gene trees for a species accumulates, much more evidence will be available for assessing the extent of selective constraint in the genome imposed by hitch-hiking (Hudson 1994) and recombinational DNA repair processes (Stephan and Langley 1992).

2.6 Conclusions

In the use and interpretation of phylogenies in the Darwinian tradition, population genetics has not made a great contribution over the last few decades. Nor in the field of population genetics has much use been made of phylogenetic data. But for the new phylogenies a merger is happening. The high information content preserved in a gene tree is being appreciated by population geneticists, and the value of stochastic population genetic models is providing new insights for evolutionary biologists. For both evolutionary biologists and population geneticists, the main cause for scepticism about coalescent models is their basis on the assumption of selective neutrality. It is unlikely that the coalescent as formulated by Kingman (1982*a*,*b*) can be used to model selection of any form. However, there are branching processes appropriate for studying selection as a stochastic model (Sawyer 1976). What biologists need, at least for simulation efficiency, is a genealogical process like the coalescent that runs backward.

Acknowledgements

The origins of this chapter trace to a discussion between mathematical geneticists and molecular population geneticists at a workshop held by the Institute for Mathematics and its Applications, 23–29 January 1994. I particularly wish to acknowledge Robert Adler, Peter Donnelly, Robert Griffiths, Norman Kaplan, Paul Marjoram, Neil O'Connell, Stanley Sawyer, Simon Tavaré, and Ryk Ward, whose ideas expressed at that meeting have greatly influenced and educated my understanding of coalescence modelling. I hope I have not misrepresented their ideas in any way. The proceedings of that meeting are published in Donnelly P. and Tavaré S. (ed.) (1996). *Progress in population genetics and human evolution*. Springer, Berlin.

References

Donnelly, P. and Tavaré, S. (1995). Coalescents and genealogical structure under neutrality. *Ann. Rev. Genet.*, **29**, 410–21.

Ewens, W. J. (1979). *Mathematical population genetics.* Springer, Berlin.

Griffiths, R. C. (1980). Lines of descent in the diffusion approximation of neutral Wright–Fisher models. *Theor. Pop. Biol.*, **17**, 37–50.

Griffiths, R. C. and Tavaré, S. (1994*a*). Ancestral inference in population genetics. *Stat. Sci.*, **9**, 307–19.

Griffiths, R. C. and Tavaré, S. (1994*b*). Simulating probability distributions in the coalescent. *Theor. Pop. Biol.*, **46**, 131–59.

Harding, R. M. (1996). Lines of descent from mitochondrial Eve—an evolutionary look at coalescence. In *Progress in population genetics and human evolution*, (ed. P. Donnelly and S. Tavaré). Springer, Berlin. (In press.)

Herbots, H. (1996). The structured coalescent. In *Progress in population genetics and human evolution*, (ed. P. Donnelly and S. Tavaré), Springer, Berlin. (In press.)

Hudson, R. R. (1990). Gene genealogies and the coalescent process. *Oxford surveys in evolutionary biology*, **7**, 1–44.

Hudson, R. R. (1994). How can the low levels of DNA sequence variation in regions of the Drosophila genome with low recombination rates be explained? *Proc. Natl. Acad. Sci. USA*, **91**, 6815–18.

Hudson, R. R. and Kaplan, N. L. (1988). The coalescent process in models with selection and recombination. *Genetics*, **120**, 831–40.

Jagers, P. (1991). The growth and stabilization of populations. *Stat. Sci.*, **6**, 269–74.

Kimura, M. (1983). *The neutral theory of molecular evolution*, Cambridge University Press.

Kingman, J. F. C. (1982*a*). On the genealogy of large populations. In *Essays in statistical science*, (ed. J. Gani and E. J. Hannan), pp. 27–43. Sheffield, Applied Probability Trust.

Kingman, J. F. C. (1982*b*) Exchangeability and the evolution of large populations. In *Exchangeability in probability and statistics*, (ed. G. Koch and F. Spizzichino), pp. 97–112. North-Holland, Amsterdam.

Marjoram, P. and Donnelly, P. (1994). Pairwise comparisons of mitochondrial DNA sequences in subdivided populations and implications for early human evolution. *Genetics*, **136,** 673–83.

Sawyer, S. (1976). Branching diffusion processes in population genetics. *Adv. Appl. Prob.*, **8,** 659–89.

Slatkin, M. and Hudson, R. R. (1991). Pairwise comparisons of mitochondrial DNA sequences in stable and exponentially growing populations. *Genetics*, **129**, 555–62.

Stephan, W. and Langley, C. H. (1992). Evolutionary consequences of DNA mismatch inhibited repair opportunity. *Genetics*, **132**, 567–74.

Tajima, F. (1983). Evolutionary relationship of DNA sequences in finite populations. *Genetics*, **105**, 437–60.

Takahata, N. (1993). Allelic genealogy and human evolution. *Mol. Biol. Evol.*, **10,** 2–22.

Tavaré, S. (1984). Line-of-descent and genealogical processes, and their applications in population genetic models. *Theor. Pop. Biol.*, **26**, 119–64.

3

Genealogies and geography

N. H. Barton and I. Wilson

3.1 Introduction

Any set of homologous genes can be traced back to a single common ancestor, and so their evolutionary relationship can be described by a genealogy, such as that shown in Fig. 3.1a. With asexual reproduction, every gene has the same ancestry. For example, since human mitochondrial DNA is inherited maternally, there is no opportunity for recombination, and so all mitochondria must trace back to one maternal ancestor. By contrast, sexual reproduction allows recombination, so that different genes, or even different segments of the same gene, have different genealogies (Fig. 3.1b). The ancestral lineages of each gene might coincide, so that even in a sexual population, one individual might by chance contribute the entire future ancestry (top of Fig. 3.1b). In the more distant past, all segments of the genome presumably trace back to one ancestral gene, *via* successive duplications. Here, however, we will be concerned with the immediate genealogy, and what this tells us about current evolutionary processes. The discordant genealogies generated by recombination allow natural selection to select efficiently on individual loci, and also give us very much more information about the whole population. However, it makes it difficult (if not impossible) to estimate the whole set of genealogies that give the relationships between genomes, and greatly complicates the inferences that can be made from that set. In this chapter we analyse the structure of genealogies where genes diffuse through a two-dimensional habitat, and discuss how genealogical information might be used to estimate the rate and pattern of gene flow.

Until recently, population genetics was based on allele frequencies, or occasionally on genotype frequencies. In particular, the relative rates of gene flow and genetic drift have traditionally been estimated using the standardized variance of allele frequencies across subpopulations (F_{st}—Wright *et al.* (1942)). DNA sequencing now presents us with data which are more naturally represented by genealogies. Every sequence may be unique, in which case allele frequencies tell us nothing; all the information is contained in the genealogical relationship between sequences. If recombination is rare enough relative to mutation, the genealogy can be seen more or less directly (as for example in bacteria where recombination occurs via occasional transformation (Maynard Smith 1990) or in viruses, where mutations are frequent (Sharp *et al.* 1996)).

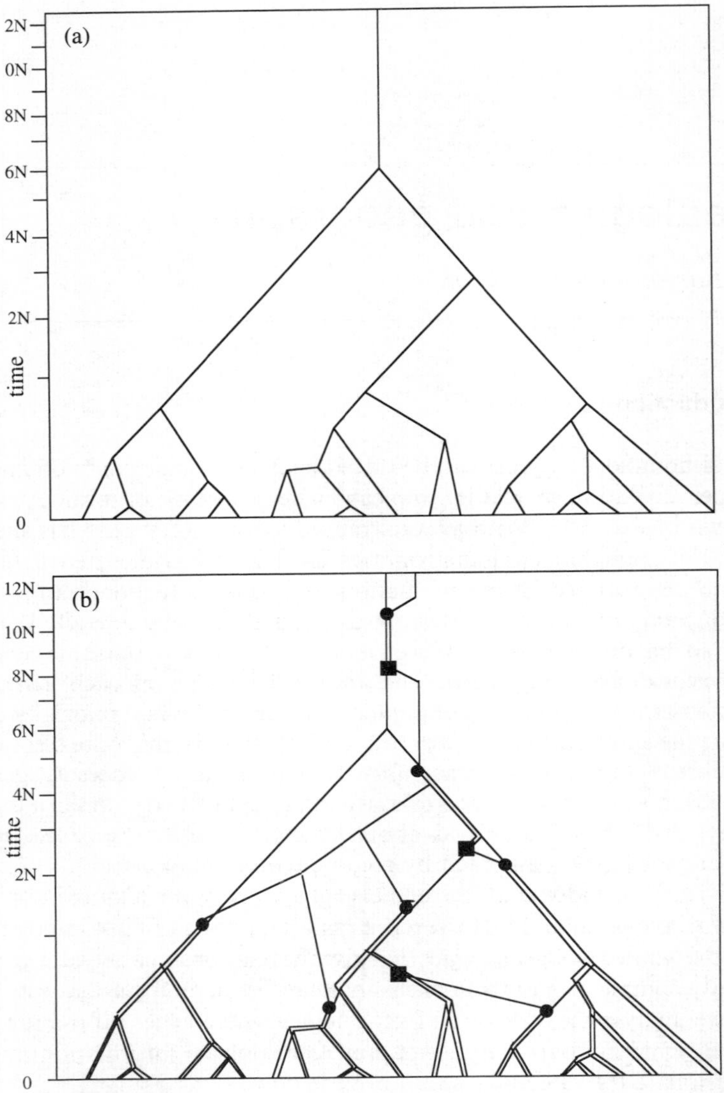

Fig. 3.1 (a) A typical genealogy for one locus in one panmictic population. Initially there are many pairs of lineages, and so coalescence events are very frequent; as lineages coalesce, further coalescence becomes rarer. For this example, it takes $2.77N$ generations for 20 lineages to coalesce into two, and then a further $3.14N$ generations for these two to meet in the common ancestor. Note that time is plotted on a square-root scale: on a linear scale, early coalescences are indistinguishable. (b) Genealogy for two loci in one panmictic population, illustrating the discordance between genealogies caused by recombination at a rate r. In each generation, there is a chance r that genes carried together will have been inherited from different parents (full circles). Lineages coalesce at random, and so may come back into association (full squares). All genes at the first locus become identical by descent at time $5.91N$, whilst those at the second become identical at $4.10N$. Further back in time, lineages at the two loci will sporadically come together on one chromosome for expected times of $1/r$ generations, separated by intervals of $2N$ generations (top of graph).

Even if we cannot be sure of the relationships, the theory may still best be described in terms of the underlying genealogies. For example, if we can calculate the likelihood of parameters such as the population size or the rate of gene flow for a particular genealogy, the overall likelihood can be found as a weighted sum over the set of plausible genealogies (Felsenstein 1992). Even if our statistical inferences are based on allele frequencies, it may still give us more insight to derive the theoretical predictions *via* a genealogical approach.

Here we will assume that we have the full genealogy of each segment of the genome, including the times at which different lines of descent coalesce. Of course, estimating the genealogy itself is hard. However, by considering the ideal data we can at least set an upper limit on what can be inferred about the processes of evolution. How much more valuable is it to use information from the full set of genealogies, rather than from allele frequencies, or pairwise relationships between genes? What is the best sampling scheme—is it better to have small genealogies across many loci, or genealogies for a few loci, but each containing many individuals? We concentrate on the specific question of how to analyse genealogies with geography, and in particular, how to estimate rates of gene flow and genetic drift. We first show the close relationship between the classical theory of identity by descent and the genealogical structure, and then develop a simple diffusion approximation which describes how lines of descent coalesce. This approximation applies to a variety of local population structures, and so could be used to make robust inferences about effective population densities and rates of gene flow. However, we show that simple estimators based on the rate of dispersion of lineages over time can be misleading, and that in a two-dimensional population, the genealogical structure depends primarily on long-term history. Finally, we discuss possible ways of making inferences about this history.[1]

This chapter is to a large extent a synthesis of existing analyses: the classic work of Wright (1943*b*) and Malecot (1948) on identity by descent; Slatkin's (1991) work on the structure of genealogies in stepping-stone models, and its application to measuring gene flow (Slatkin and Maddison 1990; Hudson *et al.* 1992); Felsenstein's (1992) likelihood methods; and Neigel, Ball, and Avise's (Ball *et al.* 1990; Neigel *et al.* 1991; Neigel and Avise 1993) use of genealogies to infer population histories. The methods here aim primarily at estimating rates of gene flow and genetic drift, rather than distinguishing qualitatively different population histories. Thus, they complement Crandall and Templeton's methods, which make qualitative distinctions between alternative hypotheses (Crandall and Templeton 1993; Templeton 1992; see Chapter 6).

3.2 The coalescent process

First, consider a single panmictic population containing $2N$ genes. If a small fraction of these genes are sampled, then their relationship can be approximated

[1] A condensed version of this chapter is to be found in Barton and Wilson (1995).

very simply: there is a probability $1/2N$ that any two lines of descent will coalesce in a common ancestor in each generation (Kingman 1982). Thus, if there are k genes, there are $k(k-1)/2$ pairs which might coalesce, and the time back to the first coalescence is exponentially distributed with expectation $4N/k(k-1)$ generations. There are then $(k-1)$ lines of descent remaining, and so the expected time back to the previous coalescence is $4N/(k-1)(k-2)$ generations. The expected age of the whole genealogy (that is, the expected time back to the common ancestor) is thus $4N\,[1/k(k-1)+1/(k-1)(k-2)\ldots+1/2]=4N(1-1/k)$ generations, which tends to $4N$ generations for a large sample (Felsenstein 1992).

This calculation shows that the structure of any genealogy is very variable. Looking back, the average time taken for a large number of lineages to coalesce to two ancestors is the same as the time these two remaining lineages take to merge (Fig. 3.2a). Since the latter time is exponentially distributed, and since any two randomly chosen genes have a chance of 1/3 of being related via the last common ancestor, the average divergence time between randomly chosen pairs of genes has a high variance, regardless of how large a sample is taken. Felsenstein (1992) uses this argument to show that pairwise statistics are much less efficient than estimators which include the genealogical structure. Essentially the same consideration applies with geographical structure, and so we summarize the argument here. The population size could be estimated from the average pairwise divergence time $\bar{t}$, using the relationship $\hat{N}=E(\bar{t})/2$. (In practice, $\bar{t}$ could itself be estimated from the average pairwise sequence divergence.) However, the variance of $\hat{N}$ tends to $2N^2/9$ for large samples, rather than to zero (Felsenstein 1992, eqn 18). By contrast, the maximum likelihood estimator (m.l.e.) is

$$\hat{N}=\sum_{j=2}^{k} j(j-1)\,t_j/4(k-1),$$

where t_j is the time for which there are j lines of descent. The variance of the m.l.e. is $N^2/(k-1)$, which does tend to zero as sample size increases (Felsenstein 1992, eqn 7). This basic argument suggests that genealogies will also give much more information about spatial structure than will pairwise measures such as Wright's F_{st}, or the statistics proposed by Slatkin and co-workers (Slatkin 1989; Slatkin and Hudson 1991; Hudson *et al.* 1992). However, it also suggests that information from any one genealogy may be misleading, so that reliable estimates may require data from many loci.

This theory of the coalescent process is equivalent to the classical theory of identity by descent. Two genes are said to be *identical by descent*, relative to an ancestral population t generations in the past, if they derive from the same gene in that population (Malecot 1948). Clearly, the probability that two genes coalesce at time t, f_t, is just the difference between the probability of identity by descent relative to ancestral populations at times t and $t-1$ ($f_t=\tilde{F}_t-\tilde{F}_{t-1}$, where $\tilde{F}_t$ denotes the probability of identity by descent *via* a population t

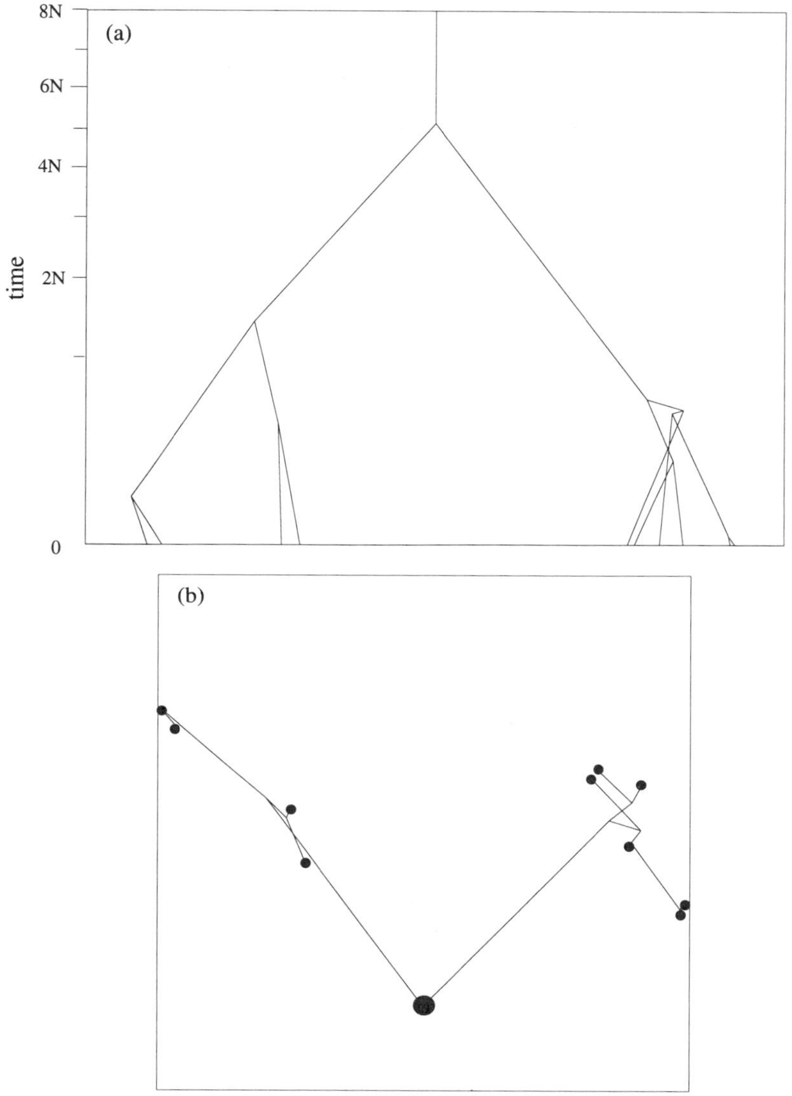

Fig. 3.2 (a) A genealogy connecting 10 genes, plotting time on one axis, and one spatial axis on the other. These trees are drawn assuming random reproduction, with no local density regulation. Time is drawn on a square-root scale. (b) The same tree in two dimensions, plotting locations in space, and drawing lines to indicate the relationships. Small circles indicate genes, and the large circle the common ancestor. The species' range is assumed to be infinite, corresponding to zero density and zero neighbourhood size. With finite range, and hence non-zero density and neighbourhood size, these plots would be wrapped around, confusing the relation between geography and genealogy for large times.

generations back). It is important to realize that the probability of identity by descent is purely a description of the genealogy, and does not depend on the allelic state of the genes. The latter is described by the probability of *identity in state*, which is the chance that two randomly chosen genes share the same allele. This depends on how alleles are identified, and on the mutational process, but is not defined relative to any base population. The probabilities of identity by descent and identity in state are closely related, and indeed are often confused with each other. If there are infinitely many alleles, and a rate μ per generation of mutation to a novel allele, then the probability of identity in state is

$$F=\sum_{t=1}^{\infty}(1-\mu)^{2t}(\tilde{F}_t-\tilde{F}_{t-1})=\sum_{t=1}^{\infty}(1-\mu)^{2t}f_t. \tag{3.1}$$

(Throughout, we assume discrete generations.) Equation (3.1) applies to any population structure, and shows the close relation between identity in state ($F(\mu)$), identity by descent ($\tilde{F}_t$), and the distribution of coalescence times (f_t). Moreover, eqn (3.1) shows that if F is considered as a function of $z=(1-\mu)^2$, then $F(z)$ is the generating function for the distribution of coalescence times. Then, $\partial^t F/\partial z^t|_{z=0}=t!f_t$, and $\partial^k F/\partial z^k|_{z=1}=k!E(t^k)$. The wealth of existing results on identity in state in spatially structured populations therefore leads directly to the distribution of coalescence times.

3.3 Gene flow and the coalescent process

The pattern of neutral genetic variation can be found by superimposing random changes in allelic state on the genealogy. This is possible because, by definition, neutral mutations do not affect reproduction, and so are independent of the genealogy. It is tempting to treat the movement of genes in an analogous way: just as genes change their allelic state through random mutation, so their location changes as the organisms which carry them disperse. This idea was introduced by Cavalli-Sforza and Edwards (1964) as the 'Brownian/Yule process'; Edwards (1970) discusses the joint inference of ancestral locations and times. (The process was originally conceived as a model of the whole population, run forward in time, in contrast to the present application.) Here we use this model to derive a simple expression for the joint distribution of geographic location and genealogical structure. However, we will see that this expression cannot describe natural populations because it implicitly ignores local regulation of population density (Felsenstein 1975). Nevertheless, it leads us to a diffusion approximation which does apply to a variety of actual population structures.

Consider a large population of $2N$ genes, which are distributed over a two-dimensional habitat with area A. A sample of genes is taken at random from the population, without regard to their location. If we make the apparently innocent assumption that reproduction is independent of the position of the genes in the sample (or their ancestors), then their relationship should have the same distribution as in a single panmictic population, the intervals between coalescence times t_j being exponentially distributed with expectation $4N/j(j-1)$. If we now assume that genes move in a Gaussian random walk, with variance σ^2 per generation along each of the two axes, we can write the joint distribution of coalescence times and locations. For example, the probability that two randomly chosen genes are related through a common ancestor t generations ago, and are at positions $\underline{x}$, $\underline{x}'$ is:

$$\psi\,(\underline{x},\,\underline{x}',t)\;\mathrm{d}\underline{x}\mathrm{d}\underline{x}'=\frac{1}{2t\;Nb}\,\frac{\mathrm{d}\underline{x}\mathrm{d}\underline{x}'}{A^2}\,(1-1/2N)^t\exp(-|\underline{x}-\underline{x}'|^2/4\sigma t) \tag{3.2}$$

where $Nb=4\pi\rho\sigma^2$ is Wright's (1943*b*) neighbourhood size, and $\rho=N/A$ is the population density. Neighbourhood size plays a crucial role in determining the relative rates of gene flow and genetic drift in a two-dimensional population; roughly speaking, it is the number of individuals within one generation's dispersal range. The expression for a set of k genes would involve a similar product of the exponential distribution of coalescence times, with the Gaussian distribution of locations, given those times.

In fact, genes are sampled from particular locations. We therefore require the distribution of coalescence times conditional on location: $\psi(t|\;\underline{x},\,\underline{x}')=\psi(\underline{x},\,\underline{x}',\,t)/\psi(\underline{x},\,\underline{x}')$, where $\psi(\underline{x},\,\underline{x}')$ is the chance that genes will be found at $\underline{x}$, $\underline{x}'$. Summing over time:

$$\psi(\underline{x},\,\underline{x})\;\mathrm{d}\underline{x}\mathrm{d}\underline{x}'=\frac{\log(\sqrt{2N})}{Nb}\,\frac{\mathrm{d}\underline{x}\mathrm{d}\underline{x}'}{A^2}\qquad\text{for } |\underline{x}-\underline{x}'|=0,\; N\gg 1 \tag{3.3a}$$

$$\psi(\underline{x},\,\underline{x}')\;\mathrm{d}\underline{x}\mathrm{d}\underline{x}'=\frac{K_0(|\underline{x}-\underline{x}'|/\sqrt{2N\sigma^2})}{Nb}\,\frac{\mathrm{d}\underline{x}\mathrm{d}\underline{x}'}{A^2}\qquad\text{for } |\underline{x}-\underline{x}'|\gg\sigma \tag{3.3b}$$

$$\approx\frac{\log(\sigma\sqrt{2N}/|\underline{x}-\underline{x}'|)}{Nb}\,\frac{\mathrm{d}\underline{x}\mathrm{d}\underline{x}'}{A^2}\qquad\text{for } \sigma\sqrt{2N}\gg|\underline{x}-\underline{x}'|\gg\sigma \tag{3.3c}$$

where K_0 is the modified Bessel function.

This expression raises two difficulties. First, it implies extreme clumping: eqn (3.3a) shows that the density near to a randomly chosen individual is increased by a factor $\approx\log(\sqrt{2N})$ above the average. This tendency for randomly dispersing and reproducing populations to become clumped was first emphasized by Felsenstein (1975) as a criticism of classical models of identity by descent. It can be seen as a consequence of the heterogeneous structure of random genealogies discussed above. On average, any large genealogy takes $2N$ generations to coalesce into two clades, and then a further $2N$ generations for the remaining two lineages to meet. Hence, the population tends to evolve into distantly related and widely scattered clusters.

The second difficulty is that we have not properly accounted for the finite range of the species. At equilibrium, a random set of genes will have spread over an area[2] of $\approx 8N\sigma^2 = (2Nb/\pi)A$. Thus, unless neighbourhood size is very small, most sets of genes will be related by lineages that have crossed the species' range several times. This is true however large the range, because for given neighbourhood size, divergence time increases in proportion to area. Hence, to be consistent, we must allow for either the finite spatial range, or the finite age, of the species.

It is traditional to allow for finite range by assuming that the habitat is wrapped over the surface of a torus of size $D \times D = A$, so that a movement x is equivalent to a movement of $(\underline{x} + \underline{j}\mathrm{D})$, where i represents a pair of integers (j_1, j_2). Now, eqn (3.2) must be replaced by a sum over all equivalent locations:

$$\psi\,(\underline{x},\ \underline{x}',t)\ \mathrm{d}\underline{x}\mathrm{d}\underline{x}' = \frac{1}{2t\ Nb}\ \frac{\mathrm{d}\underline{x}\mathrm{d}\underline{x}'}{A^2}\ (1 - 1/2N)^t \sum_{\underline{j}} \exp(-|\underline{x} - \underline{x}' + \underline{j}D|^2/4\sigma^2 t). \quad (3.4)$$

Assuming a toroidal habitat should be seen as a convenient approximation to more realistic models, with complicated boundaries, and where individuals near the edge might be reflected back into the range. A square habitat with reflecting boundaries would be described in essentially the same way, by folding over the square rather than by wrapping around the torus.

The spatial distribution is found by summing over time, and applying the same approximations that led to eqns (3.3):

$$\psi\,(\underline{x},\ \underline{x})\ \mathrm{d}\underline{x}\mathrm{d}\underline{x}' = \frac{\mathrm{d}\underline{x}\mathrm{d}\underline{x}'}{Nb\ A^2}\left[\ \log(\sqrt{2N}\,) + \sum_{\underline{j}\neq 0} K_0(\sqrt{|\underline{j}|^2 2\pi Nb})\right]$$

$$\text{for } |\underline{x} - \underline{x}'| = 0,\ 2N \gg 1 \qquad (3.5a)$$

$$\psi\,(\underline{x},\ \underline{x}')\ \mathrm{d}\underline{x}\mathrm{d}\underline{x}' \approx \frac{\mathrm{d}\underline{x}\mathrm{d}\underline{x}'}{Nb\ A^2}\left[\ \log(\sigma\sqrt{2N}/|\underline{x} - \underline{x}'|) + \sum_{\underline{j}\neq 0} K_0(\sqrt{|\underline{j}|^2 2\pi Nb})\right]$$

$$\text{for } \sigma\sqrt{2N} \gg |\underline{x} - \underline{x}'| \gg \sigma. \quad (3.5b)$$

For large neighbourhood size, the sum over Bessel functions can be approximated by an integral, and tends to 1. Figure 3.3 shows examples of the clumping

[2] Recall that σ^2 is the variance of distance moved along each of the two axes.

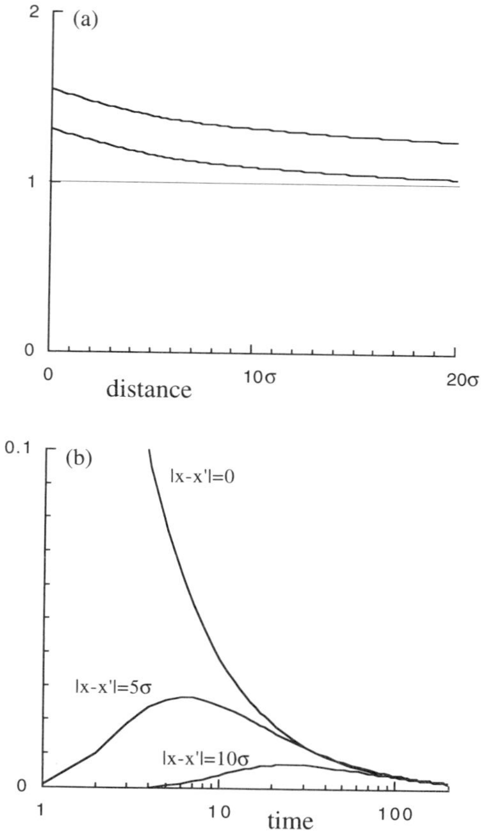

Fig. 3.3 (a) The correlation in density as a function of distance, with random reproduction ($A^2\psi(\underline{x},\underline{x}')$ from eqn (3.5). This gives the density at a point, conditional on the presence of a gene a distance $|\underline{x}-\underline{x}'|$ away. Neighbourhood size is $Nb=10$. The upper curve is for a species' range of $A=10^6\sigma^2$, whilst the lower curve is for $A=10^4\sigma^2$. (b) Conditional distribution of coalescence times for genes separated by $|\underline{x}-\underline{x}'|=0$, 5σ, 10σ (from the ratio of eqns (3.4) and (3.5)).

produced by random reproduction (Fig. 3.3a), and the distribution of coalescence times, conditional on location (Fig. 3.3b). Since the degree of clumping is small for large neighbourhood size, the conditional distribution does not differ much from the raw distribution.

Sedentary organisms may not have had time to diffuse over the species' range over the time since they occupied it. For example, the flightless alpine grasshopper *Podisma pedestris* has a dispersal range of $\sigma \approx 20$ m/generation$^{1/2}$ (Barton and Hewitt 1982) and an annual life cycle. It has occupied its present range since the glaciers retreated ($T<10^4$ generations). In that time, genes would diffuse only ≈ 2 km. In such cases ($\sigma^2 T \ll A$) clustering is limited by finite age rather than finite range. In general, we should consider a sporadic series of extinctions and recolonizations over the whole history of the organism. Instead, we make the simpler assumption that the range was occupied by N diploid individuals T generations ago, distributed uniformly without regard to their relationship. The joint distribution of locations and genealogies is then given by eqn (3.2) (or its multigene extension) for $t \ll T$, and by a uniform spatial distribution for more ancient relationships ($t > T$). For two genes

$$\psi\,(\underline{x},\,\underline{x}',t)\,\mathrm{d}\underline{x}\mathrm{d}\underline{x}' = \frac{1}{2t\;Nb}\,\frac{\mathrm{d}\underline{x}\mathrm{d}\underline{x}'}{A^2}\,(1-1/2N)^t \exp(-|\underline{x}-\underline{x}'|^2/4\sigma^2 t) \quad t<T \qquad (3.6\mathrm{a})$$

$$\psi\,(\underline{x},\,\underline{x}',t)\,\mathrm{d}\underline{x}\mathrm{d}\underline{x}' = g(t)\frac{\mathrm{d}\underline{x}\mathrm{d}\underline{x}'}{A^2} \quad t \geqslant T$$

$$(1-1/2N)^t \sum_{t=T}^{\infty} g(t) = (1-1/2N)^N \qquad (3.6\mathrm{b})$$

where $g(t)$ is the arbitrary distribution of relationships among pairs of colonists. Summing over time to obtain the distribution of spatial locations

$$\psi\,(\underline{x},\,\underline{x}',t)\,\mathrm{d}\underline{x}\mathrm{d}\underline{x}' = \frac{\mathrm{d}\underline{x}\mathrm{d}\underline{x}'}{A^2}\left[(1-1/2N)^T + \frac{1}{2Nb}\sum_{t=1}^{T}\frac{(1-1/2N)^t \exp(-|\underline{x}-\underline{x}'|^2/4\sigma^2 t)}{t}\right] \qquad (3.7\mathrm{a})$$

$$\approx \frac{\mathrm{d}\underline{x}\mathrm{d}\underline{x}'}{A^2}\left[1+\frac{\log T+\gamma}{2Nb}\right] \qquad \underline{x}=\underline{x}',\; T \ll 2N \qquad (3.7\mathrm{b})$$

$$\approx \frac{\mathrm{d}\underline{x}\mathrm{d}\underline{x}'}{A^2}\left[1+\frac{\log(4\sigma^2 T/|\underline{x}-\underline{x}'|^2)+\gamma}{2Nb}\right]$$
$$\sigma\sqrt{T} \gg |\underline{x}-\underline{x}'| \gg \sigma,\; T \ll 2N \qquad (3.7\mathrm{c})$$

where $\gamma = 0.5772$ is Euler's constant. Hence, the conditional distribution of coalescence times is

$$\psi(t|\underline{x},\,\underline{x}) \approx \frac{1}{t(2Nb+\log T+\gamma)} \qquad T \ll 2N \qquad (3.8\mathrm{a})$$

$$\psi(t|\underline{x},\,\underline{x}') \approx \frac{\exp(-|\underline{x}-\underline{x}'|^2/4\sigma^2 t)}{t[2Nb+\log(4\sigma^2 T/|\underline{x}-\underline{x}'|^2)+\gamma]} \qquad \sigma \ll |\underline{x}-\underline{x}'| \ll \sigma\sqrt{T}. \qquad (3.8\mathrm{b})$$

These models of random reproduction give a consistent conclusion for times which are short compared with the age of the population or the time taken for genes to diffuse across the species' range ($t \ll T$, A/σ^2). The distribution of coalescence times is then given by $\psi(t|\underline{x}, \underline{x}') \approx \exp(-|\underline{x}-\underline{x}'|^2/4\sigma^2 t)/(2t\ \widetilde{Nb})$, where $\widetilde{Nb}(|\underline{x}-\underline{x}'|)$ is the effective neighbourhood size, which is increased somewhat above $4\pi\rho\sigma^2$ by the clustering that occurs in the absence of local density regulation. While we have given explicit results only for pairs of genes, it is easy to generate the joint distribution of genealogies and locations for arbitrary numbers of genes, or to draw random realizations of this distribution (Fig. 3.2).

3.4 A diffusion approximation to the coalescent process in two dimensions

The model developed in the previous section is simple, but unrealistic: in nature, fitness must decrease with local density. Densities may vary from place to place, but in response to local carrying capacity, rather than to the unchecked accumulation of demographic fluctuations. The model of a locally unregulated population might describe the random evolution of morphology, and would give an explanation for the clustering of asexual phenotypes into species (Higgs and Derrida 1991). However, it is implausible as a description of population dynamics in two spatial dimensions.

There have been extensive treatments of the extreme case where individuals are grouped into demes whose density is absolutely regulated. Analyses of such stepping-stone models have dealt primarily with identity in state (Wright 1943*b*; Malecot 1948; Kimura and Weiss 1964; Maruyama 1972; Felsenstein 1976; Nagylaki 1974, 1986). However, since the identity in state is the generating function for the distribution of coalescence times eqn (3.1), it leads immediately to the distribution of coalescence times (see below). Slatkin (1991) and Nei and Takahata (1993) have derived the mean coalescence time. However, the results would be cumbersome to extend to higher moments. Here we develop a simple diffusion approximation which applies to all but local scales, and which extends to whole genealogies.

The approximation is based on Wright's (1943*b*) argument that ancestors can be considered as being drawn from a neighbourhood whose size increases with time into the past. Thus, the probability that two nearby genes are identical by descent in the previous generation is (by definition) $1/2Nb$; in two dimensions, the pool of ancestors is spread over an area whose area increases linearly with time, and so the probability of identity by descent t generations back is $1/2tNb$.

Let $f(t,\underline{z}|\underline{x},\underline{x}')$ be the probability that genes at $\underline{x}$ and $\underline{x}'$ are identical by descent *via* an ancestor t generations back, who lived at $\underline{z}$. (With two genes, we will not need to keep track of the positions of ancestors; however, this is necessary to extend the argument to more genes; see eqn (3.14).) Let $g_1(\underline{y},\underline{x})$ be the chance that a gene at $\underline{x}$ derived from an ancestor at $\underline{y}$ in the previous generation. We assume that g_1 depends only on the distance between parent and offspring, has zero mean, and has variance σ^2 along each of the two axes. Let

$f(1,\underline{y}|\underline{x},\underline{x}')=h(\underline{y},\underline{x},\underline{x}')/2Nb$ be the chance that two genes at $\underline{x}$, $\underline{x}'$ were identical by descent through an ancestor at $\underline{y}$ in the previous generation. There is a chance $g_1(\underline{y},\underline{x})g_1(\underline{y},\underline{x}')\delta\underline{y}^2$ that two lines of descent come from some small area $\delta\underline{y}$ in the previous generation; if they do, there is a chance $(1/2\rho\delta\underline{y})$ that they will be identical, and a chance $(1-1/2\rho\delta\underline{y})$ that they are not, but instead are identical *via* some more distant ancestor. Hence

$$f(1,\underline{y}|\underline{x},\underline{x}')\equiv\frac{1}{2Nb}h_1\,(\underline{y},\underline{x},\underline{x}')=\frac{g_1(\underline{y},\underline{x})g_1(\underline{y},\underline{x}')}{2\rho} \tag{3.9a}$$

$$f(t,\underline{z}|\underline{x},\underline{x}')=\int f(t-1,\underline{z}|\underline{y},\underline{y}')g_1(\underline{y},\underline{x})g_1(\underline{y}',\underline{x}')\mathrm{d}\underline{y}\mathrm{d}\underline{y}'-\frac{1}{2Nb}\int f(t-1,\underline{z}|\underline{y},\underline{y})h_1(\underline{y},\underline{x},\underline{x}')\mathrm{d}\underline{y}. \tag{3.9b}$$

This argument requires several approximations. It is assumed that the lines descend independently, so that the joint probability of movement of two genes is $g_1(\underline{y},\underline{x})g_1(\underline{y}',\underline{x}')$. It is assumed that one can choose an area $\delta\underline{y}$ large enough that $1/2\rho\delta\underline{y}<1$, but small enough that f is approximately constant within it. These assumptions hold for a demic structure with strict density regulation, in which case equations (3.9) give the coalescence times exactly. They are approximations to models of truly continuous populations; in the cases we consider, the approximation is remarkably good (see Fig. 3.4).

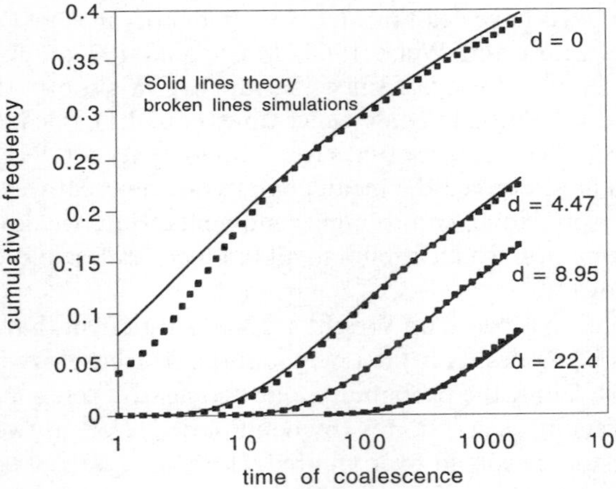

Fig. 3.4 Comparison of simulated distributions of coalescence times from a two-dimensional stepping-stone with 10 diploid individuals per deme and a migration rate of 0.05 in each direction (dotted curves), to a Gaussian approximation for the distribution of times (eqns (3.10); full curves). Simulations were based on a grid of 100 by 100 demes over 2000 generations. The simulated cumulative frequency is based on 8000 replicates for each distance. The distances are zero demes ($d=|\underline{x}-\underline{x}'|=0$), one deme in both directions ($d=|x-\underline{x}'|=4.47\sigma$), two demes in both directions ($d=|x-\underline{x}'|=8.95\sigma$), and five demes in both directions ($d=|\underline{x}-\underline{x}'|=22.2\sigma$). Neighbourhood size $=$ 6.28.

Equations (3.9) give a recursion across one generation which relates the probability of coalescence at time t to that at time $(t-1)$. They can be rewritten as a recursion across many generations, which leads naturally to the diffusion approximation. Let $g_t(\underline{y},\underline{x})$ be the chance that a gene at $\underline{x}$ descended from an ancestor at $\underline{y}$, t generations back. Extend the definition of h to

$$\frac{1}{2Nb} h_t(\underline{y},\underline{x},\underline{x}') = \frac{g_t(\underline{y},\underline{x}) g_t(\underline{y},\underline{x}')}{2\rho t}. \tag{3.10a}$$

This is the chance that two genes both descend from an ancestor at $\underline{y}$, t generations back, ignoring the possibility of more recent coancestry. Applying eqn (3.9b) recursively:

$$f(t,\underline{z}|\underline{x},\underline{x}') = \frac{1}{2Nb}\left[h_t(\underline{y},\underline{x},\underline{x}') - \sum_{i=1}^{t-1} \int f(t-i,\underline{z}|\underline{y},\underline{y}) \frac{h_i(\underline{y},\underline{x},\underline{x}')}{i} \,\mathrm{d}\underline{y}\right]. \tag{3.10b}$$

This is the chance that the two genes descend from a common ancestor at $\underline{z}$, subtracting the probabilities that they were identical by descent in any of the intervening generations.

By the central limit theorem, $g_t(\underline{y},\underline{x})$ tends to a Gaussian with variance $\sigma^2 t$ for large t; h_t tends to a Gaussian with variance $\sigma^2 t/2$, being the distribution of locations of the common ancestor. If we average over the location of the ancestors $(f(t|\underline{x},\underline{x}'))$, and use the fact that $f(t|\underline{x},\underline{x}')$ is independent of $\underline{x}$, eqn (3.10b) simplifies to

$$f(t|\underline{x},\underline{x}) = \frac{1}{2Nb}\left[\frac{1}{t} - \sum_{i=1}^{t-1} \frac{f(t-i|\underline{x},\underline{x})}{i}\right] \tag{3.11a}$$

$$f(1|\underline{x},\underline{x}) = \frac{1}{2Nb} \tag{3.11b}$$

$$f(2|\underline{x},\underline{x}) = \frac{1}{2Nb}\left(\frac{1}{2} - \frac{1}{2Nb}\right) \tag{3.11c}$$

$$f(3|\underline{x},\underline{x}) = \frac{1}{2Nb}\left[\frac{1}{3} - \frac{1}{2Nb\,2} - \frac{1}{2Nb}\left(\frac{1}{2} - \frac{1}{2Nb}\right)\right] = \frac{1}{2Nb}\left(\frac{1}{3} - \frac{1}{2Nb} + \frac{1}{(2Nb)^2}\right). \tag{3.11d}$$

The distributions of coalescence times predicted by this Gaussian approximation are shown in Fig. 3.4, together with simulation results. Agreement is close over all but short times and nearby genes. In most natural populations, F_{st} is low (< 0.10, say; Slatkin, (1987)), implying that neighbourhood size is large. In this limit, the distribution of coalescence times in this model of absolute density regulation converges to that developed in the previous section for the case of no local regulation.

These recursions for the distribution of coalescence times lead to parallel recursions for the identity in state. Applying eqn (3.1) to eqn (3.9b):

$$F(\underline{x},\underline{x}') = (1-\mu)^2 \int g_1(\underline{y},\underline{x}) g_1(\underline{y}',\underline{x}') \left[F(\underline{y},\underline{y}') + \frac{1}{2\rho}(1-f(0))\delta(\underline{y}-\underline{y}') \right] \mathrm{d}\underline{y}\mathrm{d}\underline{y}' \quad (3.12)$$

where $\delta(\underline{y})$ is the Dirac delta function, and the probability that nearby genes are identical has been rewritten as $F(\underline{x},\underline{x}) = F(0)$ to emphasize that it is independent of location. Equation (3.12) is identical to Malecot's (1948) model if (as Malecot assumed) the dispersal distribution is Gaussian. It applies exactly to stepping-stone models if the integrals are replaced by sums, and is an approximation to continuous models which neglect the interactions between nearby genes caused by local density dependence. The Appendix shows that for arbitrary dispersal distributions eqn (3.12) is approximated by

$$F(0) = \left[1 + \frac{Nb}{\log(\sigma/K\sqrt{1-(1-\mu^2)})} \right]^{-1} \quad (3.13a)$$

$$F(\underline{x},\underline{x}') = \frac{1-F(0))}{Nb} K_0 \left[\frac{|\underline{x}-\underline{x}'|}{\sigma} \sqrt{1-(1-\mu^2)} \right]$$

for $|\underline{x}-\underline{x}'| \gg \sigma$, K, and $\mu \ll 1$. (3.13b)

Equations (3.13) give the probability of identity in state, assuming infinitely many alleles; the same expressions give F_{st} measured from the variance in allele frequencies at loci with a finite number of alleles. K is a characteristic scale which depends on the local structure of the population. For a stepping-stone model with nearest-neighbour migration on a square grid, K = (deme spacing)/$\sqrt{32}$. For Malecot's model of Gaussian dispersal, $K = \sigma$. If mutation rates are low ($\mu \ll 1$), $1-(1-\mu)^2 \approx 2\mu$, and eqns (3.13) depend on the scale $l = \sigma/\sqrt{2\mu}$. There will be significant fluctuations over local scales ($\approx \sigma$, K), but there will be correlations between allele frequencies over the much longer scale l. Equation (3.13b) is a close approximation to continuously distributed populations, and breaks down only over local scales (Fig. 3.5). Agreement is similarly close for stepping-stone models, even for neighbouring demes. The distribution of coalescence times can be calculated by differentiating eqn (3.13b) with respect to $z=(1-\mu)^2$. The breakdown of eqn (3.13b) for small $|\underline{x}-\underline{x}'|/\sigma$ corresponds to the breakdown of the Gaussian approximation to eqn (3.9b) for small coalescence times (Fig. 3.4).

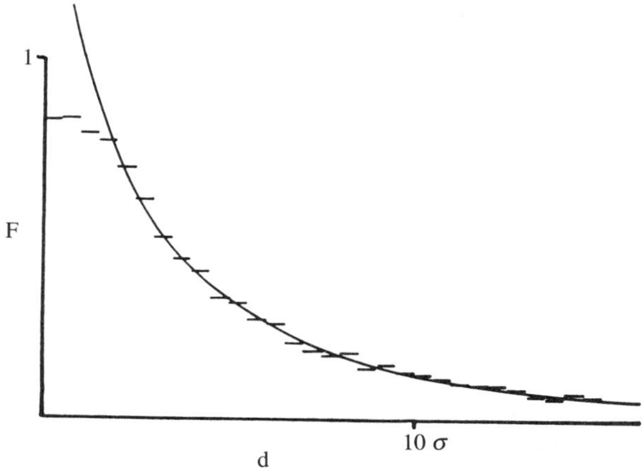

Fig. 3.5 Identity by descent in a continuously distributed population. The full curve shows the Bessel function approximation of eqns (3.13), whilst the bars show simulation results. The mutation rate is $\mu=0.01$. Haploid individuals disperse in a Gaussian distribution with variance σ^2. They reproduce in discrete generations, with a number of offspring drawn from a Poisson distribution with mean $4^{(1-\rho/\rho_k)}$, where $\rho_k=0.16$. This ensures density-dependent regulation towards ρ_k. The actual density averaged 0.106, and the effective neighbourhood size (based on identity by descent in the previous generation) was 0.66. ρ is the local density, measured by weighting neighbours according to a Gaussian curve with variance σ^2. Simulations were on a torus, $(100\sigma)^2$, for two replicates of 150 generations.

The approximation that led to eqns (3.10) and (3.11) extends to many genes. It becomes more complicated, however, because the recursions involve the locations of all the ancestors. Usually, we are concerned only with the present locations of the genes, and so must average over the locations of the ancestors. This involves an integral over a Gaussian, which can be carried through using a symbolic algebra package such as *Mathematica* (Wolfram 1991). By the same arguments that led to eqns (3.10):

$$f_{ij}(t,\underline{\underline{\hat{z}}}^{ij}|\underline{\underline{x}})=\frac{1}{2Nb}\left[h_{ij}(t,\underline{\underline{\hat{z}}}^{ij}|\underline{\underline{x}})-\sum_{kl}\sum_{\tau=1}^{t-1}\int f_{ij}(t-\tau,\underline{\underline{\hat{z}}}^{ij}|\underline{\underline{\hat{y}}}^{kl})\frac{h_{kl}(\tau,\underline{\underline{\hat{y}}}^{kl}|\underline{\underline{x}})}{\tau}\right]. \quad (3.14)$$

Here, $\underline{\underline{x}}$ represents a set of n locations $\{\underline{x}_1,\underline{x}_2,\ldots\underline{x}_n\}$. $\underline{\underline{\hat{z}}}^{ij}$ *represents a set of* n locations, with $z_i=z_j\equiv z_{ij}$. $f_{ij}(t,\underline{\underline{\hat{z}}}^{ij}|\underline{\underline{x}})$ *is the chance that out of a set of* n genes at locations $\underline{\underline{x}}$, the first coalescence is in generation t, between lineages i and j, at locations $\hat{\underline{\underline{z}}}^{ij}$. $h_{ij}(t,\hat{\underline{\underline{z}}}^{ij}|\underline{x})/2Nb$ is defined in the same way as in eqn (3.9a), as the chance that two lineages, starting at $\underline{x}_i$ and $\underline{x}_j$, coalesce t generations back, at $\underline{z}_i=\underline{z}_j$, and that the other lineages at $\underline{x}_k$ descend from $\underline{z}_k$. For large t, it tends to a Gaussian:

$$h_{ij}(t,\hat{\underline{\underline{z}}}^{ij}|\underline{x})=\frac{\exp[-(\underline{z}_{ij}-\underline{x}_{ij})^2/\sigma^2 t)]}{\pi\sigma^2 t}\prod_{k\neq i,j}\frac{\exp[-(\underline{z}_k-\underline{x}_k)^2/2\sigma^2 t)]}{2\pi\sigma^2 t} \tag{3.15}$$

where $\underline{x}_{ij}=(\underline{x}_i+\underline{x}_j)/2$. Though cumbersome, this recursion could be carried through symbolically.

3.5 Estimating the rate of gene flow

The recursions developed in the preceding sections allow calculation of the distribution of coalescence times amongst genes sampled from a two-dimensional population. However, it would not be easy to use this distribution to make statistical estimates. Here we outline possible methods for estimating the rate of gene flow (σ), assuming that genes diffuse through a stable and homogeneous population. The traditional method for inferring population structure from genetic data was introduced by Wright (Wright *et al.* 1942; Wright 1943*a*). Geographical variation in allele frequency is generated by genetic drift, and reduced by gene flow. In the island model, the balance between these processes is determined by Nm, the number of migrants exchanged between demes; the standardized variance of allele frequencies across demes is $F_{st}=1/(1+4Nm)$ for large deme size and low migration rates. In two dimensions (though not in one), the relationship is similar, with F_{st} decreasing with neighbourhood size ($Nb=4\pi\rho\sigma^2$; eqns (3.13)).

Almost all analyses of population structure infer Nm or Nb from F_{st}, or from some equivalent measure of variation in allele frequency, such as Slatkin's private allele method (Slatkin 1985; Barton and Slatkin 1986). However, the spatial pattern of allele frequencies also contains information. If (as is usually the case) F_{st} is small, allele frequencies will fluctuate rather little, and the whole distribution can be approximated by a multivariate Gaussian, which is defined by its mean and covariance. This covariance is given by eqns (3.13), and depends on two scales: the scale which describes local population structure, and the scale $l=\sigma/\sqrt{2\mu}$, which describes the balance between mutation and gene flow. Thus, if the mutation rate is known, the rate of gene flow (σ) can be estimated. This approach was first used by Sokal and Wartenberg (1983), and has been explored more recently by Epperson (1989, 1993). It is important to realize that it gives estimates of both the number of migrants or neighbourhood size (Nm or $4\pi\rho\sigma^2$), and the proportion of migrants or rate of diffusion (m or σ): confusingly, both are referred to as the 'rate of gene flow'.

To illustrate this approach, and to provide a benchmark for comparison with genealogical methods, consider a set of k samples, each of $2n$ genes, spread over a two-dimensional area. The frequency of alleles at j independent loci is measured, giving a set of jk allele frequencies. (Since we will only consider the limit of large sample size, the variance of the estimates scales inversely with the number of loci.) Assume that samples are large enough, and F_{st} small enough, that the joint distribution of allele frequencies is approximately Gaussian. Samples are taken from points separated by much more than one dispersal range, so that the approximation of eqns (3.13) is accurate, regardless of the local population structure. First, consider estimation of F_{st}, disregarding spatial information for the moment. The maximum likelihood estimate is $\hat{F}_{st} = (\mathrm{var}(p)/pq - 1/2n)$. Since the coefficient of variation (c.v.) of an estimate of a variance tends to $\sqrt{2/jk}$ for large samples, the c.v. of $\hat{F}_{st}$ tends to $\sqrt{2/jk}(1 + 1/2nF_{st})$ in large samples. For a given total number of genes ($2njk$), the best sampling scheme is to set $1/2n = F_{st}$; then, the c.v. is $\approx \sqrt{8/jk}$. The heavy line in Figs. 3.6a and 3.6b shows the coefficient of variation of $\hat{F}_{st}$, plotted against the number of sample sites; for these parameters, the greatest accuracy is when there are 100 samples of $1/F_{st} \approx 10$ genes. (There has been considerable discussion as to the best estimators of F_{st} and Nm (Weir and Cockerham 1984; Slatkin and Barton 1990; Cockerham and Weir, 1993); here, we consider only large samples, in which case the maximum likelihood estimate has minimum variance, and bias which tends to zero.)

An estimate based solely on the variance among samples neglects the correlations among samples, and cannot distinguish the three parameters Nb, K and $l = \sigma/\sqrt{2\mu}$. Figure 3.6 shows the c.v. of estimates based on fitting the full covariance matrix, calculated using eqns (3.13), and plotted against number of samples. Figure 3.6a shows results when samples are taken over an area large compared with l; in this case, all three parameters can be distinguished, although large samples are needed to give accurate estimates of the two scales K and l. Estimates of neighbourhood size are more accurate; their coefficient of variation is only slightly lower than $\sqrt{8/jk}$, because correlations between samples are weak. These calculations are encouraging, in that they show that data on allele frequencies can (with large enough samples, and with a time-scale calibrated by knowledge of μ) give separate estimates of neighbourhood size and dispersal rate. However, the method assumes that the population has been at equilibrium for at least $\approx 1/\mu$ generations, which is unlikely for most organisms. One should therefore sample over scales smaller than l and $\sigma\sqrt{T}$. Then, the distribution of allele frequencies becomes independent of mutation rate and long-term history. However, only two parameters can now be estimated, neighbourhood size and the local scale (Fig. 3.6b). Because the relation between K and σ depends in a complicated way on local population structure, there is a fundamental limitation to the use of allele frequencies to estimate gene flow.

Neigel, Ball, and Avise propose a straightforward method for estimating the long-term rate of gene flow from genealogies (Neigel *et al.* 1991; Neigel and Avise 1993). They suggest plotting the squared distance between genes against the time since they diverged, for all pairs in the sample, and using the slope of

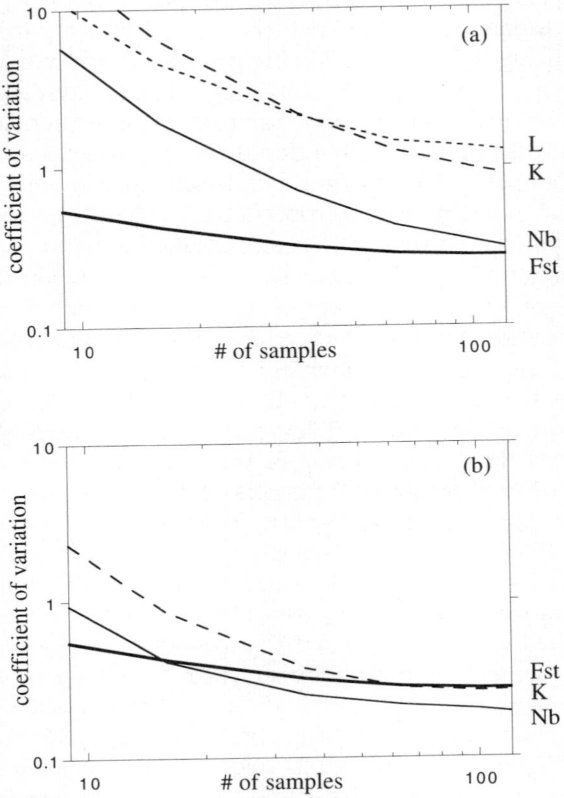

Fig. 3.6 The accuracy of estimates made from allele frequencies. 1000 diploid individuals were taken from a two-dimensional square grid of sampling locations. Ten loci were scored, each with two neutral alleles that mutate at $\mu = 5 \times 10^{-5}$; hence, the scale is $L = \sigma/\sqrt{2\mu} = 100\sigma$. Neighbourhood size is $Nb = 4\pi\rho\sigma^2 = 80$, and we assume Malecot's (1948) model, corresponding to $K = \sigma$. Hence $F_{st} = 0.054$. The matrix of covariances amongst allele frequencies sampled from these locations was calculated using eqn (3.13b). This was used to derive the expected information matrix, and hence the asymptotic variance of the maximum likelihood estimates. The heavy line shows the coefficient of variation (i.e. the standard deviation/mean) of $F_{st} = \log(L/K)/[Nb + \log(L/K)]$, estimated without taking into account any spatial autocorrelations. This is plotted as a function of the number of sampling locations. (a) The coefficients of variation of the maximum likelihood estimates of the three parameters Nb (thin line), L (dotted line), and K (dashed line), assuming that samples are taken from a square of size $(160\sigma \times 160\sigma)$. (b) Here, samples were taken from a square, $(20\sigma \times 20\sigma)$, much smaller than the scale L. Now, only Nb (thin line) and K (dashed line) can be estimated. As in (a) the heavy line shows the coefficient of variation of estimates of F_{st}, made without using any spatial information.

this relationship to estimate $4\sigma^2$ (Fig. 3.7). (In the notation used in that work, the squared distance in two dimensions increases as $2\sigma^2$ per unit time; two genes that shared a common ancestor t generations back are separated by $2t$ generations). Neigel *et al.* (1991) simulate a set of genes forward in time, assuming random dispersal over a square of size $(10\,000\sigma)^2$. Overall population size is regulated at 1000 individuals, so that the model corresponds to that described by eqn (3.2). Statistics are based on samples taken from this set, and support a linear relation between the variance of dispersal distance and lineage age, at least for closely related genes. Neigel and Avise (1993) run a wider range of simulations, some of which include local population regulation.

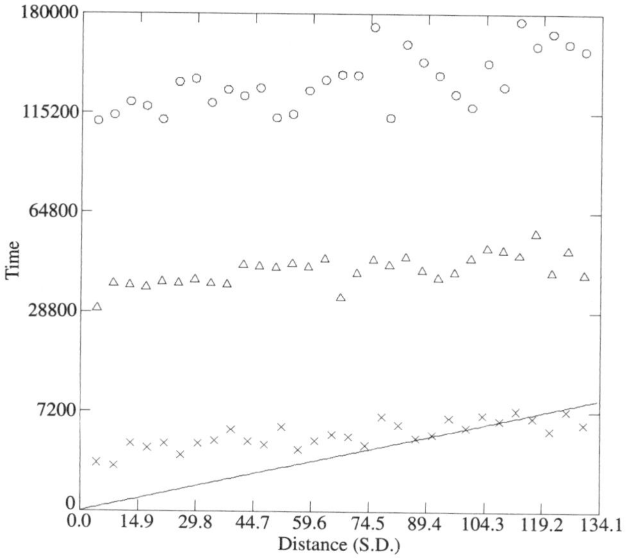

Fig. 3.7 Simulation results from a stepping-stone model compared with Neigel and Avise's (1993) results (straight line). The simulated results are based on all coalescence times (circles), times less than 100 000 generations (triangles), and times less than 10 000 generations (crosses). The distances were taken from a linear array of demes in a two-dimensional stepping-stone model of 100 by 40 demes with 20 diploid individuals per deme and a migration rate of 0.05 in both directions. Distances are in standard deviations, σ.

This approach is supported by the analytic results given here, which are also based on the idea that pairs of genes diffuse apart at a rate proportional to σ^2. However, there are several problems, which suggest some possible improvements. First, because the simulations are run forward in time, Neigel *et al.* must necessarily follow a small number of individuals, of which a relatively large proportion are sampled. Moreover, the neighbourhood size is much smaller than is typical of natural populations; for example, Fig. 5 of Neigel and Avise (1993) shows results from a population of $N = 10\,000$ genes, dispersing over an area of $A = 10^6\sigma^2$. This corresponds to a neighbourhood size of $Nb = 4\pi\rho\sigma^2 = 4\pi N\sigma^2/A = 0.13$. Hence most lineages coalesce before they have had time to spread over the whole species range, and nearby genes are likely to be close relatives. Thus, it is not clear how far their results would extend to small samples taken from a large population with moderate to large neighbourhood size. Simulations of the coalescent process for $Nb = 6.28$ show that although $|\underline{x} - \underline{x}'|^2$ does increase approximately linearly with time, at least initially, the slope of the relation is much smaller than $4\sigma^2$, and even nearby genes are separated by a long coalescence time (Fig. 3.7).

Secondly, Neigel *et al.* suggest regressing $|\underline{x} - \underline{x}'|^2$ against time. Now, the distribution of $|\underline{x} - \underline{x}'|^2$ is determined by the sampling geometry, and so $|\underline{x} - \underline{x}'|^2$ should be treated as the independent variable; the coefficient of regression of $|\underline{x} - \underline{x}'|^2$ against time will depend on the distribution of distances sampled, whereas the converse regression of time against $|\underline{x} - \underline{x}'|^2$ would not. Neigel *et al.* (1991, p. 425) avoid this difficulty by sampling uniformly over the species' range; however, this would not be easy in practice. Thirdly, because the distribution of coalescence times decreases with $1/t$ in two-dimensional populations, the expected coalescence time at equilibrium is infinite, for any $|\underline{x} - \underline{x}'|$. (This can be shown by treating F in eqns (3.13) as the generating function for $f(t)$; its differential with respect to $(1-\mu)^2$ at $\mu = 0$ is infinite.) In any actual sample the mean must be finite; however, if the distribution has infinite moments, the results will be very variable. This paradoxical behaviour of the distribution of coalescence times only applies to an infinitely large and indefinitely old population. If the population in fact colonized its present range T generations back, it would be reasonable to use only pairs that share a common ancestor before that time. However, the mean coalescence time would depend strongly on T (cf. Fig. 3.7), and would be very variable.

One might try to improve accuracy using the m.l.e., derived from the actual distribution of coalescence times, rather than the mean. If neighbourhood size is large, the distribution can be approximated by $f(t) = \exp(-|\underline{x} - \underline{x}'|^2/4\sigma^2 t)/(2Nb\,t)$. Then, the maximum likelihood estimate of σ^2 (counting only pairs up to time T), is given by the solution to

$$\frac{|\underline{x} - \underline{x}'|^2}{t} = |\underline{x} - \underline{x}'|^2 \frac{\sum_{\tau=1}^{T} \exp(-|\underline{x} - \underline{x}'|^2/4\sigma^2\tau)/\tau^2}{\sum_{\tau=1}^{T} \exp(-|\underline{x} - \underline{x}'|^2/4\sigma^2\tau)/\tau}. \tag{3.16}$$

This equation gives the m.l.e. for σ^2, given a single pair of genes, separated by $|\underline{x} - \underline{x}'|$ and related at time t. An estimate for many pairs of genes could be found by summing both sides of eqn (3.16). However, this will not be an m.l.e. because the pairs are not independent. We have run simulations of this estimator, and find that it performs poorly. The reason seems to be that the approximation $f(t) = \exp(-|\underline{x} - \underline{x}'|^2/4\sigma^2 t)/(2Nb\,t)$ breaks down for times beyond $t \approx Nb$. Using the full distribution of eqns (3.9) would not be feasible.

In a single panmictic population, pairwise estimates are inefficient because they include a disproportionately large contribution from a single event—the coalescence of two clades into the one common ancestor (Felsenstein 1992). This problem is less severe in samples from two dimensions, because the present rate of gene flow can be estimated only from closely related pairs of genes; the locations of more distantly related genes depend on the ancient history of the species. Unless neighbourhood size is unusually low, only a small proportion ($\approx 1/Nb$) of pairs of genes will be closely related. Hence, most information will come from independent pairs of genes rather than related clusters of genes. This raises a difficulty, however, in that most of the genealogy does not tell us about the rate of diffusion, but rather about the more distant history of the species. Genealogical data are therefore a fundamentally inefficient source of information about gene flow. The ideal solution to the problem would be to make an m.l.e. based on the structure of the whole tree, rather than on pairs of genes. Since only the recent part of the tree, which evolved after the species occupied its present range, can be used to infer σ^2, this method would only need to be applied to small clusters of genes. Nevertheless, it presents a daunting computational problem. In the next section we discuss ways of analysing the bulk of the tree, which tells us about the long-term history of the species.

3.3 Extinction and recolonization

Over long time-scales, populations cannot be adequately described by the uniform diffusion of genes from place to place. We know that (at least in temperate regions) most species have suffered drastic range changes in the last few thousand generations as a result of climatic change. If genes diffused at the current dispersal rate, they would not have had time to spread to fill the present range of the species. If lineages coalesced at the slow rate implied by current population sizes, then genealogies would be much deeper than is seen, and neutral heterozygosity would be much higher. The above analysis shows that unless neighbourhood size is very small, most of the information in a tree comes from distant relationships: only a small fraction $\approx 1/Nb$ of gene pairs are close relatives. The prime need in genealogical analysis is thus to find statistical methods for using this information to infer the distant history of the population, and the processes responsible for that history. In particular, we must explain how genes spread faster, and become more closely related, than is possible by diffusion through a large and stable population.

If the whole range were rapidly colonized from a randomly mating source at time T, there would be no association between genealogy and geography before that time. At the other extreme, spatial relations might be preserved despite expansions and contractions of the range. This is plausible if populations are adapted to a climatic gradient, and shift with that gradient (Coope 1979; Atkinson *et al.* 1987). Between these extreme possibilities, there might be expansion from a number of refugia, so that before time T genealogies would only reflect the source of the ancestral population and not more detailed spatial relations. We can imagine fitting data to a variety of such particular historical scenarios, and indeed, this is the usual approach to 'phylogeography' (Avise 1991). However, unless these scenarios can be constrained or corroborated by independent evidence, there is a danger of being able to explain too much. It is therefore attractive to seek ways of representing drastic changes as a statistical process—for example, by supposing that there is a low rate of expansions, in which the population in some area A_1 is replaced by individuals drawn from a smaller area A_2. The way allele frequencies are affected by random extinctions and recolonizations has received considerable attention, though mainly for the simplest case of the island model (for example Slatkin 1977; Wade and McCauley 1988; Whitlock and McCauley 1990; McCauley 1991). However, this theory has not led to ways of distinguishing random drift from random extinction (Slatkin 1987); the question is whether genealogical data may be more informative.

Changes in the species' distribution will cause older lineages to spread over larger areas than expected with diffusion alone, and cause lineages to coalesce faster then expected from the current population density. One could make an *ad hoc* estimate of some effective diffusion rate, σ_e^2, and effective neighbourhood size, Nb_e, as a function of time, by adapting the methods discussed above. Naively, an increase in σ_e^2 would lead to an increase in the scale over which allele frequency fluctuations are correlated, or equivalently, an increase in the rate of dispersion of lineages with age. Genetic distances do indeed often increase over scales much larger than can be explained by simple isolation by distance, suggesting sporadic range expansions. For example, in the alpine grasshopper *Podisma pedestris*, allele frequencies are correlated over all scales from 50 m to 3 km—a much flatter relationship than is consistent with isolation by distance with a σ rate of 20 m year$^{-1/2}$ and $T \approx 8000$ years (Fig. 3.8).

However, diffusion rates do not actually increase back into the past: what is needed is a model of the extinction/recolonization process itself. The crucial feature of this process is that it involves concerted movements, such that all the genes within an area tend to move together. This correlation across genes in turn generates correlations between the relationships of genes at different loci. How does this affect genetic variation? First, consider allele frequencies. Isolation by distance alone (i.e. diffusive gene flow and sampling drift), causes fluctuations which are independent across loci; by contrast, range expansions tend to produce correlations between the patterns at different loci. This idea has been applied with particular success to the reconstruction of the history of human populations in Europe. Consistent patterns across loci have shown that linguistic boundaries coincide with genetic boundaries (Sokal *et al.* 1990), and that genes from the originators of agriculture spread into the native European population with a degree of intermingling ('demic diffusion'; Ammermann and Cavalli-Sforza 1981; Sokal *et al.* 1991).

What degree of concordance is to be expected between genealogies at different loci? Even if dispersal is solely by the independent diffusion of genes, and even if

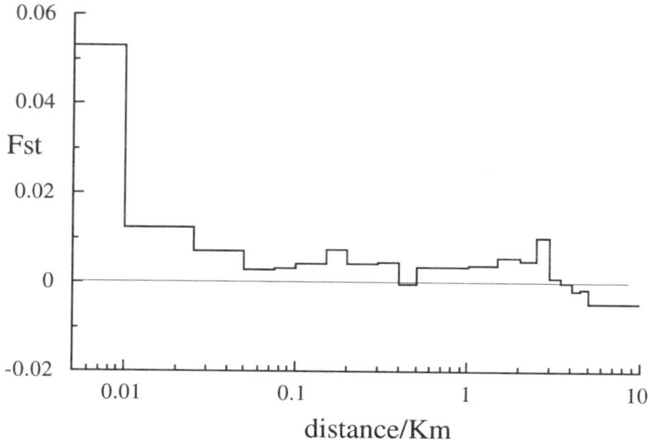

Fig. 3.8 Correlations in allele frequency versus geographical distance in *Podisma pedestris*, estimated from 1815 individuals from 273 sites, scored for 5 enzyme loci with 18 alleles (Est-1, Est-2, Amy-1, Amy-2, ME—data from Halliday *et al.* (1983, 1984)). F_{st} was calculated from the covariance of square-root transformed allele frequencies between pairs of sites in different distance classes, corrected for sampling error. Significant correlations are found, over all scales up to ≈ 3 km.

linkage is loose, genealogies at different loci are expected to be to some degree parallel, because they will mirror the geographical location of the samples. In contrast, allele frequencies should be independent across loci if linkage disequilibria are negligible. In order to interpret genealogical data from multiple loci, we therefore need to understand the outcomes expected under the null hypothesis of isolation by distance. Slatkin (1989) and Slatkin and Maddison (1990) use the proportion of concordant genealogies to set a bound on the number of migrants between two demes; here we illustrate the same idea for a population in two dimensions. It is simplest to find the chance that a genealogy will be concordant with the geographic location of the samples, since this determines the concordance of genealogies derived from independent genes with each other. Figure 3.9a shows the three types of relationship between two pairs of genes, drawn from two locations. The genealogy may perfectly match the geography (Type 1); one pair may match (Type 2); or no pair may match (Type 3). If the sample locations are very close, or if the neighbourhood size is very high, there will be no relation between geography and genealogy, and the three types of tree will be in the proportions 1:2:6. Figure 3.9b shows how the degree of concordance increases as the genes move further apart, whilst Fig 3.9c shows how concordance falls with increasing gene flow. When gene flow is low, and the samples are far apart, the concordance between geography and genealogy reaches a plateau, which depends on neighbourhood size. This is because in two dimensions only a proportion $\approx 1/Nb$ of lineage pairs coalesce early enough to preserve spatial information. Thus, under isolation by distance, only a fraction of the more closely related genealogies will be concordant with geography. Spatial patterns, and concordance across loci, among more distantly related genes therefore indicate large-scale changes in population structure.

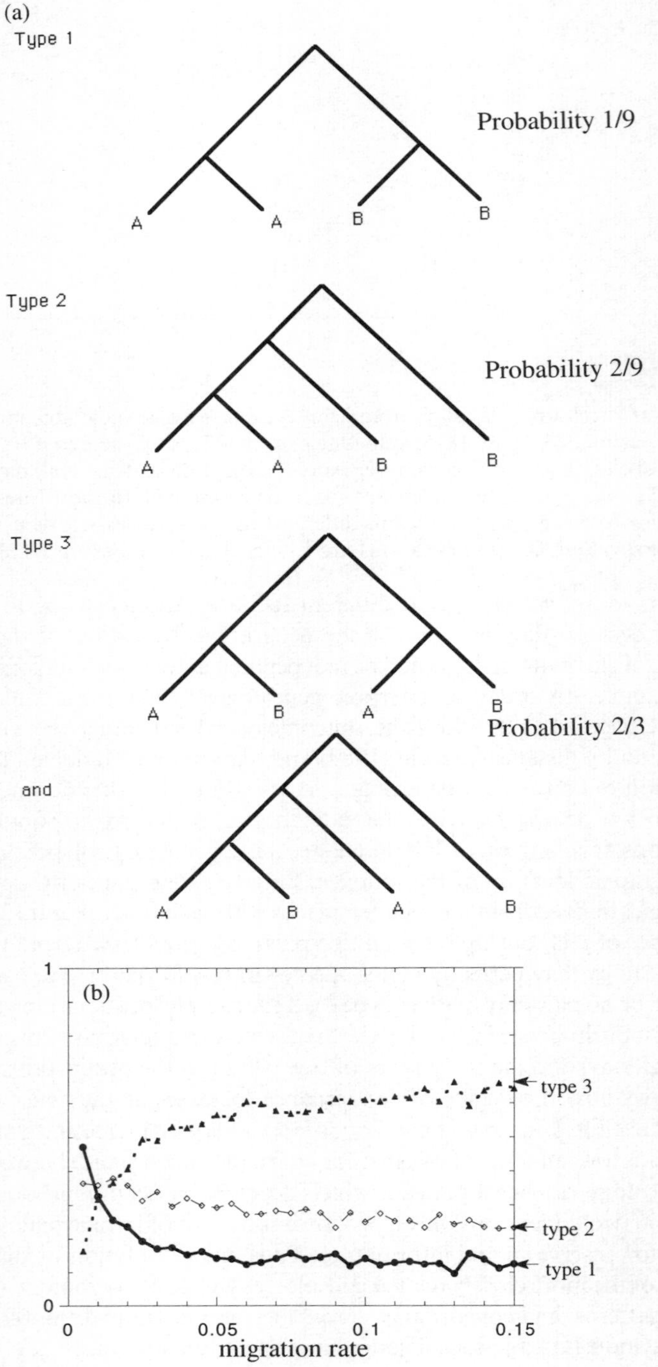
(a)
Type 1
Probability 1/9
A
A
B
B
Type 2
Probability 2/9
A
A
B
B
Type 3
A
B
A
B
Probability 2/3
and
A
B
A
B
(b)
type 3
type 2
type 1
1
0
0
0.05
0.1
0.15
migration rate

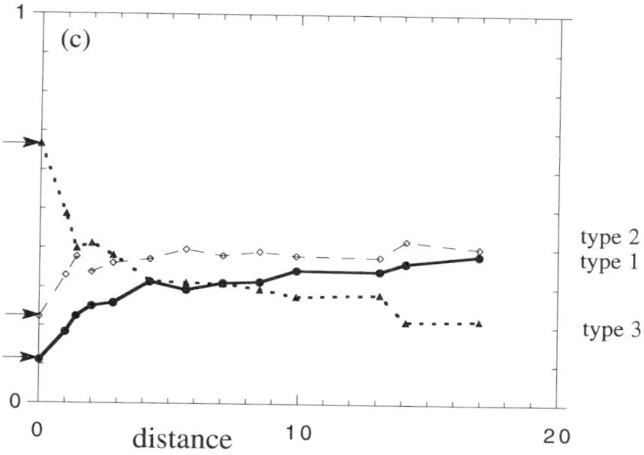

Fig. 3.9 (a) The three possible concordances for two pairs of genes; with no population subdivision, the null probabilities of the trees are (1/9, 2/9, 2/3). (b) The frequencies of the three different types of concordance possible for two sets of two genes. Simulations are from a stepping-stone model with 20 diploid individuals per deme. Pairs of genes are sampled from demes a distance two demes apart in both directions. Migration rates given are in both directions. (c) Simulation results for pairs of genes sampled from a stepping-stone model with 10 diploid individuals per site, and a migration rate of 0.06 per generation in both x and y directions. The distance is given in standard deviations of distance moved, σ. For distance 0 the results are exact.

The power of analyses of allele frequencies comes from having data from many samples and many loci. The same may be true for genealogies. As noted above, strong concordance with geography over large scales, and concordance between loci, indicate the degree of large-scale population movement as opposed to independent diffusion. However, discordance may arise for a variety of reasons, reducing the power of this approach. Contraction of the range into refugia will only leave a genetic trace if the populations are small enough, for long enough, for there to be appreciable coalescence. Otherwise, the only effect will be a randomization of ancestral locations. There are inevitably errors in estimating the tree: it is disturbing that even when using mitochondrial DNA to estimate the relationships among ten major groups of vertebrates, at least 8000 contiguous bases of sequence are needed to give a 95 per cent chance of inferring the correct tree (Cummings *et al.* 1995; see Chapter 7). Discordance may also arise through selection on particular loci. This may be a particular problem in using mitochondrial DNA for within-species analyses, since the genealogy can be distorted by selection on any of the genes it carries (Malhotra *et al.* 1996), or on other elements that are inherited maternally, such as *Wolbachia* (Turelli *et al.* 1992). *Drosophila* mitochondrial sequences show significant deviations from neutral expectations (Ballard and Kreitman 1994; Rand *et al.* 1994), and the frequent introgression of mitochondrial genomes across boundaries demarcated by nuclear alleles may also be a sign of selection on the mitochondrial genome (Harrison 1989). It remains to be seen whether the more detailed information that is contained in genealogies will compensate for the much smaller sample sizes, and whether the concordance between the genealogies for a few loci could allow similar inferences to those based on allele frequencies at many more loci.

3.7 Hybrid zones and barriers to gene flow

Species are often divided into a mosaic of distinct geographic races, which are separated by narrow zones of hybridization. Such parapatric distributions may be first identified through particular character—for example, different warning colours in *Heliconius* butterflies (Mallet 1993), different chromosome arrangements, as in the grasshopper *Podisma pedestris* (Barton and Hewitt 1981), or different plumage, such as define many avian subspecies (Remington 1968; Hall and Moreau 1970). However, these races are usually found to differ in many other respects; the average Nei's genetic distance between 34 examples of hybrid zones is $D=0.26$ (Barton and Hewitt 1989). In most cases, the present distribution is due to secondary contact between distinct populations, and the extensive genetic divergence reflects the long time over which the source populations have been diverging (Barton and Hewitt 1985; Hewitt 1993*a,b*). Thus, parapatric distributions are the most striking examples of concordant geographical patterns across many loci. Until recently, they have been studied primarily through analysis of Mendelian markers and quantitative traits. Here, we discuss whether genealogical information could, in principle, tell us more about the origins of hybrid zones, and about their effect on gene flow.

The boundary between hybridizing populations is often much narrower than would be expected if genes had simply been diffusing freely since the time of contact. For example, the two chromosome races in *Podisma pedestris* are thought to have met after the last glaciation, $T \approx 8000$ years ago. Random movement at the present annual rate of $\sigma \approx 20$ m/generation$^{-1/2}$ would give a cline $\sqrt{2/\pi\sigma^2 T} \approx 4.5$ km wide; yet, in most places the chromosomal cline is only ≈ 800 m wide (Barton and Hewitt 1981). Natural selection must act to balance diffusion and maintain the sharp boundary between the hybridizing populations. This selection acts to produce a barrier to gene exchange that impedes the flow of genes that are not themselves under selection: in order to pass from one population to the other, they must recombine onto the new genetic background before being eliminated by selection against the genes with which they are associated. Such barriers to gene flow have the same effect as physical obstacles to dispersal; both can be described by a parameter B, which has the dimensions of a distance. A barrier produces a sharp step in allele frequency, proportional to the gradient in allele frequency ($\Delta p = B(\mathrm{d}p/\mathrm{d}x)$; Fig. 3.10a), and reduces the correlation between allele frequencies on either side (Fig. 3.10b). Both effects can be used to give an estimate of barrier strength. For example, the two chromosomal races of *Podisma pedestris* meet abruptly at Lac Autier, and are separated by a stream a few metres wide. The pattern of chromosome frequencies, and of alleles at six enzyme loci, combine to give an estimate of

$B \approx 1.5$ km (Fig. 3.10). This is partly due to reduced dispersal across the stream and partly to a genetic barrier caused by selection against hybrids (Jackson 1992; Barton and Gale 1993).

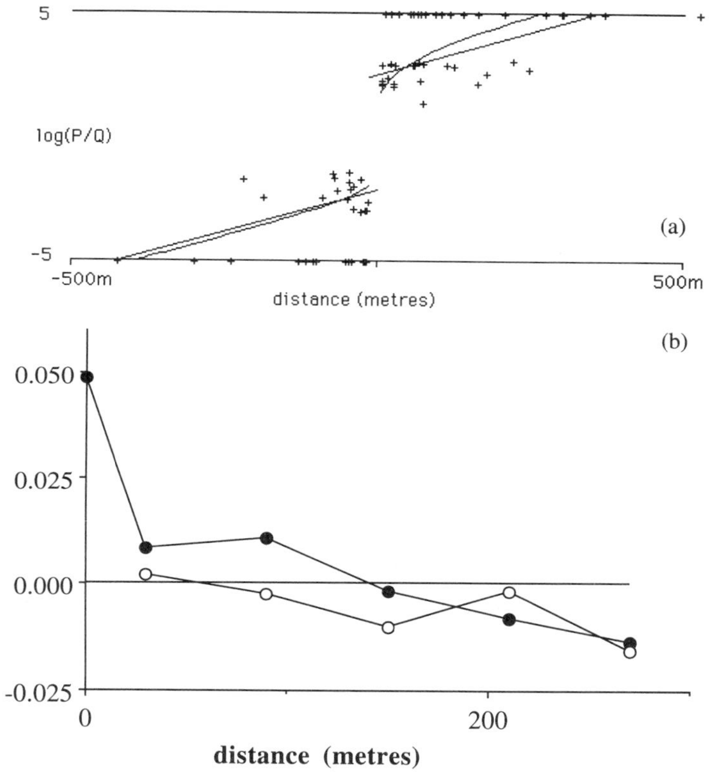

Fig. 3.10 (a) The frequency of the chromosomal fusion which distinguishes two races of the grasshopper *Podisma pedestris*, across a transect at Lac Autier in the Alpes Maritimes, plotted on a logistic scale. (Data courtesy of D. Currie; see Barton and Gale (1993)). In most places the cline is smooth, and would give a single straight line on this scale. Here, however, the frequency changes sharply across a stream. This could simply be due to a physical barrier to dispersal, disrupting a cline maintained by selection against chromosomal heterozygotes (straight lines). However, this requires a physical barrier an order of magnitude stronger than is measured directly from grasshopper movements ($B = 1390$ m for flow to the left, 2050 m to the right, *versus* 150 m measured directly). A model in which the physical barrier is augmented by selection against introgressing alleles at 100 loci predicts a further change just near the stream (curved lines; $S/R = 0.26$ to the left, 0.61 to the right, giving net barriers of 1670 m on the left, 3870 m on the right). (b) The covariance between fluctuations in allele frequency at six polymorphic enzyme loci, plotted against distance. The full circles are for sites on the same side of the stream, and the open circles are for sites on opposite sides. The net barrier estimated from the difference between the two lines is 1.5 km (support limits 500 m – ∞), consistent with the estimate from the cline in the chromosomal fusion (Fig. 3.10a).

The effect of a local barrier on genealogical relationships can be derived by treating the probability of identity by descent as a generating function (eqn 13.1). Nagylaki (1988) and Nagylaki and Barchilon (1988) have analysed the effect of a barrier on identity by descent in one dimension. Extending these results to two dimensions shows that fluctuations in allele frequency are more localized, and the effect of a local barrier is consequently greater (Fig. 3.11a). Successive differentiation of the expression for identity by descent gives the distribution of coalescence times. Figure 3.11b shows the difference in the distributions for genes immediately adjacent to the barrier, either on the same side or on different sides, for the limit of large neighbourhood size. Although there can be a substantial effect on coalescence times, this is only apparent for the first few generations. Since in practice genealogies can only be resolved over long time-scales, they are unlikely to be much help in detecting even strong barriers. By contrast, fluctuations in allele frequencies can reveal the immediate effects of quite weak barriers (Fig. 3.10b).

Over the long time-spans which are reflected in most genealogies, the barriers to gene flow generated by selection in hybrid zones are unlikely to have much evolutionary consequence. For example, the pattern of allozyme genotypes gives an estimate of a barrier to gene flow from the toad *Bombina bombina* into *B. variegata* of ≈ 50 km, relative to a dispersal rate of ≈ 1 km/generation$^{-1/2}$ (Szymura and Barton 1991). This would restrict the diffusion of neutral alleles for $(B/\sigma)^2 \approx 2500$ years since the toads met in their current distribution. However, the broad distribution of gene frequencies across Europe is dominated by successive expansion and contraction of populations over the $\approx 3 \times 10^6$ years for which these two taxa have been diverging. Thus, genealogical information would tell us about this long-term history rather than about current patterns of gene flow.

3.8 Conclusions

The main purpose of this chapter has been to emphasize the close relation between genealogical descriptions of spatially structured populations and the classical theory of identity by descent. A naive model of an unregulated population leads to an explicit formula for the joint distribution of locations and relationships, but also leads to unreasonable clumping. For populations subject to strict density regulation, we develop a diffusion approximation for the relation between genes. Both approaches give approximately the same distribution of coalescence times, $f_t = \exp(-|\underline{x} - \underline{x}'|^2/4\sigma^2 t)/(2Nb\,t)$, though only for short times ($t \approx Nb$; see eqns (3.11)). This mathematical complexity makes it hard to develop sound statistical estimators which make full use of genealogical information. For example, Neigel, Ball, and Avise's suggestion that the rate of dispersion of lineages with time gives the rate of gene flow fails for populations with large neighbourhood size.

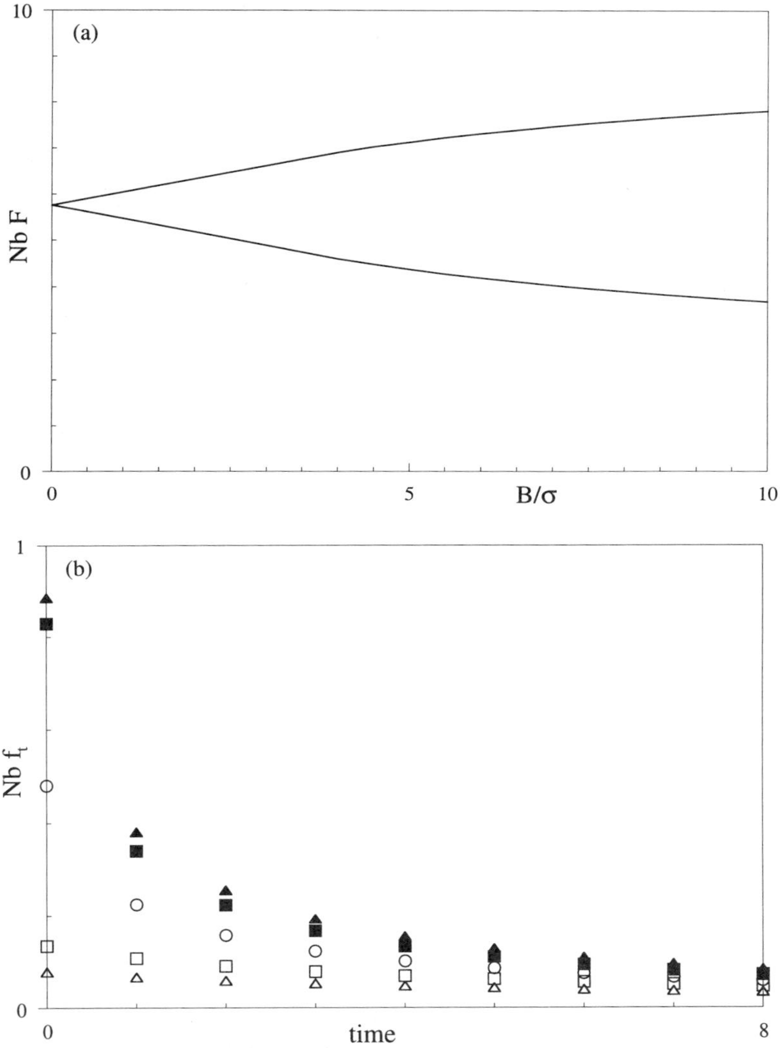

Fig. 3.11 (a) The variance of allele frequency fluctuations (F) immediately to the left of a barrier of strength B (upper curve), and the covariance of allele frequencies between genes immediately on either side (lower curve). Results are for the limit of large neighbourhood size, and so are given as the product ($Nb\ F$). The mutation rate is $\mu = 5 \times 10^{-5}$. (b) The distribution of coalescence times up to $t = 8$ generations, calculated by differentiating F with respect to $z = (1-\mu)^2$ at $\mu = 0$. Full symbols are for points on the same side of the barrier, and open symbols are points separated by the barrier. Circles: no barrier; squares: $B = 5\sigma$; triangles: $B = 10\sigma$.

These difficulties arise in part from the mathematical and computational complexities. However, there are fundamental problems in making inferences about evolutionary processes from genealogies and geography—even if the ideal data were available. First, in two dimensions only a small fraction of gene pairs are likely to be closely related, and so most of the information in the tree is about sporadic events in the distant past. This contrasts with the simpler case, where the population is divided across a few islands, all of which can be sampled (cf. Slatkin and Maddison 1990). Inferences may then be tested against geological history (for example Thorpe *et al.* 1996). Secondly, sporadic events are hard to fit to any quantitative model, and so we are left with the difficult task of judging the relative merits of a multitude of possible histories, rather than estimating any well-defined parameters. Thirdly, genealogies derived from one or a few loci can only inform us of the history of the whole population if they are all affected by population structure in the same way. If species really consist of competing geographic races, which hardly recombine, then genealogies may well be largely concordant. However, whether this is so is at present obscure. Here, we have sketched some possible solutions to the simplest case of isolation by distance. There is an urgent need for a better theoretical and empirical understanding of the distribution of genealogies across multiple loci, and of the effects of large-scale population restructuring.

Acknowledgements

This work was supported by BBSRC grant GR/H/09928, and by a MAFF studentship. We would like to thank A. W. F. Edwards and S. Otto for their helpful comments.

Appendix: Convergence to the Wright/Malecot model of identity in state

We define the Fourier transform of $f(\underline{x})$ as $\tilde{f}(\underline{\tilde{x}}) = (1/2\pi)\int f(\underline{x})\exp(\mathrm{i}\underline{x}\cdot\underline{\tilde{x}})\mathrm{d}\underline{x}$, where $\underline{x}$ is a two-dimensional vector. Then, if the dispersal is the same everywhere $(g_1(\underline{y},\underline{x}) = g_1(\underline{y}-\underline{x}))$ the Fourier transform of eqn. (3.12) is:

$$\tilde{F}(\underline{\tilde{x}},\underline{\tilde{x}}') = (1-\mu)^2(2\pi\tilde{g}_1(\underline{\tilde{x}}))^2\left[\tilde{F}(\underline{\tilde{x}},\underline{\tilde{x}}') + \frac{\delta(\underline{\tilde{x}}+\underline{\tilde{x}}')(1-F(0))}{2\rho}\right]. \quad (3.\mathrm{A}1)$$

Hence

$$\tilde{F}(\underline{\tilde{x}},\underline{\tilde{x}}') = \frac{(1-\mu)^2\delta(\underline{\tilde{x}}+\underline{\tilde{x}}')(1-F(0))}{2\rho\{[1/(2\pi\tilde{g}_1(\underline{\tilde{x}}))^2]-(1-\mu)^2\}}. \quad (3.\mathrm{A}2)$$

The Dirac delta function $\delta(\underline{\tilde{x}}+\underline{\tilde{x}}')$ arises because $F(\underline{x},\underline{x}')$ depends only on the displacement $(\underline{x}-\underline{x}')$. Since g_1 is a probability density, and is assumed to have mean zero, $2\pi\tilde{g}_1(\underline{\tilde{x}})$ tends to $(1-\sigma^2|\underline{\tilde{x}}|^2/2)$ as $\sigma|\underline{x}|$ tends to zero. For small μ

$$\tilde{F}(\underline{\tilde{x}},\underline{\tilde{x}}') = \frac{2\pi\delta(\underline{\tilde{x}}+\underline{\tilde{x}}')(1-F(0))}{Nb(|\underline{\tilde{x}}|^2+2\mu/\sigma^2)}\ . \tag{3.A3}$$

High spatial frequencies $(\sigma|\underline{\tilde{x}}|\approx 1)$ make a negligible contribution to the inverse Fourier transform for $|\underline{x}-\underline{x}'|\gg\sigma$. Hence

$$F(\underline{x},\underline{x}') \approx \frac{1-F(0)}{Nb}\ K_0\left(\frac{\sqrt{2\mu}|\underline{x}-\underline{x}'|}{\sigma}\right). \tag{3.A4}$$

To find the probability of identity in state between nearby genes $(F(0))$, we must integrate the full expression (eqn. (3.A2)):

$$F(0) = \frac{1}{1+[Nb/\log(\sigma/K\sqrt{2\mu})]}$$

where (3.A5)

$$\log(K/\sigma) = \frac{\sigma^2}{2\pi}\int\left(\frac{1}{\exp(|\underline{\tilde{x}}|^2\sigma^2)-1} - \frac{1}{[1/(2\pi\tilde{g}_1(\underline{\tilde{x}}))^2]-1}\right)\mathrm{d}\underline{\tilde{x}}.$$

K depends only on the local dispersal distribution; if it is Gaussian $K=\sigma$.

References

Ammermann, L. and Cavalli-Sforza, L. L. (1981). *The Neolithic transition and the genetics of populations in Europe*. Princeton University Press, Princeton, NJ.

Atkinson, T. C., Briffa, K. R., and Coope, G. R. (1987). Seasonal temperatures in Britain during the past 22,000 years, reconstructed using beetle remains. *Nature*, **325**, 587–93.

Avise, J. C. (1991). Ten unorthodox perspectives on evolution prompted by comparative population genetic findings on mitochondrial DNA. *Ann. Rev. Genet.*, **25**, 45–69.

Ball, R. M., Neigel, J. E., and Avise, J. C. (1990). Gene genealogies within the organismal pedigrees of random-mating populations. *Evolution*, **44**, 360–70.

Ballard, W. O. and Kreitman, M. (1994). Unravelling selection in the mitochondrial genome of *Drosophila*. *Genetics*, **138**, 757–72.

Barton, N. H. and Gale, K. S. (1993). Genetic analysis of hybrid zones. In *Hybrid zones and the evolutionary process*, (ed. R. G. Harrison), pp. 13–45. Oxford University Press.

Barton, N. H. and Hewitt, G. M. (1981). A chromosomal cline in the grasshopper *Podisma pedestris*. *Evolution*, **35**, 1008–18.

Barton, N. H. and Hewitt, G. M. (1982). A measurement of dispersal in the grasshopper *Podisma pedestris* (Orthoptera: Acrididae). *Heredity*, **48**, 237–49.

Barton, N. H. and Hewitt, G. M. (1985). Analysis of hybrid zones. *Ann. Rev. Ecol. Syst.*, **16**, 113–48.

Barton, N. H. and Hewitt, G. M. (1989). Adaptation, speciation and hybrid zones. *Nature*, **341**, 497–503.

Barton, N. H. and Slatkin, M. (1986). A quasi-equilibrium theory of the distribution of rare alleles in a subdivided population. *Heredity*, **56**, 409–16.

Barton, N. H. and Wilson, I. (1995). Genealogies and geography. *Phil. Trans. R. Soc.* **B**, **349**, 49–59.

Cavalli-Sforza, L. L. and Edwards, A. W. F. (1964). Analysis of human evolution. *Proc. Int. Congr. Genetics*, **3**, 923–33.

Cockerham, C. C. and Weir, B. S. (1993). Estimation of gene flow from *F*-statistics. *Evolution*, **47**, 855–63.

Coope, G. R. (1979). Late Cenozoic fossil Coleoptera: evolution, biogeography, and ecology. *Ann. Rev. Ecol. Syst.*, **10**, 247–67.

Crandall, K. A. and Templeton, A. R. (1993). Empirical tests of some predictions from coalescent theory with applications to intraspecific phylogeny reconstruction. *Genetics*, **134**, 959–69.

Cummings, M. P., Otto, S. P., and Wakeley, J. (1995). Sampling properties of DNA sequence data in phylogenetic analysis. *Mol. Biol. Evol.*, **12**, 814–22.

Edwards, A. W. F. (1970). Estimation of the branch points of a branching diffusion process. *J. R. Stat. Soc.*, **B32**, 155–74.

Epperson, B. K. (1993). Spatial and space–time correlations in systems of subpopulations with genetic drift and migration. *Genetics*, **133**, 711–27.

Epperson, B. K. and Allard, R. W. (1989). Spatial autocorrelation analysis of the distribution of genotypes within populations of lodgepole pine. *Genetics*, **121**, 369–77.

Felsenstein, J. (1975). A pain in the torus: some difficulties with the model of isolation by distance. *Amer. Nat.*, **109**, 359–68.

Felsenstein, J. (1976). The theoretical population genetics of variable selection and migration. *Ann. Rev. Genet.*, **10**, 253–80.

Felsenstein, J. (1992). Estimating effective population size from samples of sequences: inefficiency of pairwise and segregating sites as compared to phylogenetic estimates. *Genet. Res.* (*Camb.*), **59**, 139–47.

Hall, B. P. and Moreau, R. E. (1970). *Atlas of speciation in African Passerine birds*. British Museum (Natural History), London.

Halliday, R. B., Barton, N. H. and Hewitt, G. M. (1983). Electrophoretic analysis of a chromosomal hybrid zone in the grasshopper *Podisma pedestris*. *Biol. J. Linn. Soc.*, **19**, 51–62.

Halliday, R. B., Webb, S. F., and Hewitt, G. M. (1984). Genetic and chromosomal polymorphism in hybridizing populations of the grasshopper *Podisma pedestris*. *Biol. J. Linn. Soc.*, **24**, 299–305.

Harrison, R. G. (1989). Animal mitochondrial DNA as a genetic marker in population and evolutionary biology. *Trends Ecol. Evol.*, **4**, 6–12.

Hewitt, G. M. (1993*a*). After the ice: *parallelus* meets *erythropus* in the Pyrenees. In *Hybrid zones and the evolutionary process*, pp. 40–164. Oxford University Press.

Hewitt, G. M. (1993*b*). Postglacial distribution and species substructure: lessons from pollen, insects and hybrid zones. In *Evolutionary patterns and processes*, pp. 98–123. Linnean Society, London.

Higgs, P. G. and Derrida, B. (1991). Stochastic models for species formation in evolving populations. *J. Phys. A: Math. Gen.*, **24**, 985–92.

Hudson, R. R., Slatkin, M., and Maddison, W. P. (1992). Estimation of levels of gene flow from DNA sequence data. *Genetics*, **132**, 583–9.

Jackson, K. S. (1992). The population dynamics of a hybrid zone in the alpine grasshopper *Podisma pedestris*: an ecological and genetical investigation. Ph.D. thesis, University College London.

Kimura, M. and Weiss, G. H. (1964). The stepping stone model of population structure and the decrease of genetic correlation with distance. *Genetics*, **49**, 561–76.

Kingman, J. F. C. (1982). The coalescent. *Stoch. Proc. Appl.*, **13**, 235–48.

McCauley, D. E. (1991). Genetic consequences of extinction and recolonisation. *Trends Ecol. Evol.*, **6**, 5–8.

Malecot, G. (1948). *Les Mathematiques de l'Heredite*. Masson et Cie, Paris.

Malhotra, A., Thorpe, R. S., Black, H., Daltry, J. C., and Wüster, W. (1996). In *New uses for new phylogenies* (ed. P. H. Harvey, A. L. Brown, J. Maynard Smith, and S. Nee) pp.187–203. Oxford.

Mallet, J. (1993) Speciation, raciation and color pattern evolution in *Heliconius* butterflies: evidence from hybrid zones. In *Hybrid zones and the evolutionary process*, (ed. R. G. Harrison), pp. 226–60. Oxford University Press.

Maruyama, T. (1972). Rate of decrease of genetic variability in a two-dimensional continuous population of finite size. *Genetics*, **70**, 639–51.

Maynard Smith, J. (1990). The evolution of prokaryotes—does sex matter? *Ann. Rev. Ecol. Syst.*, **21**, 1–12.

Nagylaki, T. (1974). Continuous selective models with mutation and migration. *Theor. Pop. Biol.*, **5**, 284–95.

Nagylaki, T. (1986). Neutral models of geographic variation. In *Stochastic spatial processes*, (ed. P. Tautu), pp. 216–37. Springer, Berlin.

Nagylaki, T. (1988). The influence of spatial inhomogeneities on neutral models of geographical variation I Formulation. *Theor. Pop. Biol.*, **33**, 291–310.

Nagylaki, T. and Barchilon, V. (1988). The influence of spatial inhomogeneities on neutral models of geographic variation II The semi-infinite linear habitat. *Theor. Pop. Biol.*, **33**, 311–43.

Nei, M. and Takahata, N. (1993). Effective population size, genetic diversity and coalescence time in subdivided populations. *J. Mol. Evol.*, **37**, 240–4.

Neigel, J. C. and Avise, J. C. (1993). Application of a random walk model to geographic distributions of animal mtDNA variation. *Genetics*, **135**, 1209–20.

Neigel, J. E., Ball, R. M., and Avise, J. C. (1991). Estimation of single-generation migration distances from geographic variation in animal mitochondrial DNA. *Evolution*, **45**, 423–32.

Otto, S. P., Cummings, M. P., and Wakeley, J. (1996). Inferring phylogencies from DNA sequence data: the effects of sampling. In *New uses for new phylogenies*, (ed. P. H. Harvey, A. L. Brown, J. Maynard Smith, and S. Nee), pp. 103–15. Oxford University Press.

Rand, D. M., Dorfsman, M., and Kann, L. M. (1994). Neutral and non-neutral evolution of Drosophila mitochondrial DNA. *Genetics*, **138**, 741–56.

Remington, C. L. (1968). Suture-zones of hybrid interaction between recently joined biotas. *Evol. Biol.*, **2**, 321–428.

Sharp, P. M., Robertson, D. L., and Hahn, B. H. (1996). Cross-species transmission and recombination of 'AIDS' viruses. In *New uses for new phylogenies*, (ed. P. H. Harvey, A. L. Brown, J. Maynard Smith, and S. Nee), pp. 134–52. Oxford University Press.

Slatkin, M. (1977). Gene flow and genetic drift in a species subject to frequent local extinctions. *Theor. Pop. Biol.*, **12**, 253–62.

Slatkin, M. (1985). Rare alleles as indicators of gene flow. *Evolution*, **39**, 53–65.

Slatkin, M. (1987). Gene flow and the geographic structure of natural populations. *Science*, **236**, 787–92.

Slatkin, M. (1989). Detecting small amounts of gene flow from phylogenies of alleles. *Genetics,* **121**, 609–12.

Slatkin, M. (1991). Inbreeding coefficients and coalescence times. *Genet. Res. (Camb.),* **58**, 167–75.

Slatkin, M. and Barton, N. H. (1990). A comparison of three methods for estimating average levels of gene flow. *Evolution,* **43**, 1349–68.

Slatkin, M. and Hudson, R. R. (1991). Pairwise comparison of mitochondrial DNA sequences in stable and exponentially growing populations. *Genetics,* **129**, 555–62.

Slatkin, M. and Maddison, W. P. (1990). Detecting isolation by distance using phylogenies of genes. *Genetics,* **126**, 249–60.

Sokal, R. R., Oden, N. L., Legendre, P., Foster, M.-J., Kim, J., Thomson, B. A., *et al.* (1990). Genetics and language in European populations. *Am. Nat.,* **135**, 157–75.

Sokal, R. R., Oden, N. L., and Wilson, C. (1991). Genetic evidence for the spread of agriculture in Europe by demic diffusion. *Nature,* **351**, 143–5.

Sokal, R. R. and Wartenberg, D. E. (1983). A test of spatial autocorrelation analysis using an isolation by distance model. *Genetics,* **105**, 219–37.

Szymura, J. M. and Barton, N. H. (1991). The genetic structure of the hybrid zone between the fire-bellied toads *Bombina bombina* and *B. variegata*: comparisons between transects and between loci. *Evolution,* **45**, 237–61.

Templeton, A. R. (1992). Human origins and analysis of mitochondrial DNA sequences. *Science,* **255**, 737–737.

Turelli, M. Hoffmann, A. A., and McKechnie, S. W. (1992). Dynamics of cytoplasmic incompatibility and mtDNA variation in natural Drosophila simulans populations. *Genetics,* **132**, 713–23.

Wade, M. J. and McCauley, D. E. (1988). Extinction and recolonisation: their effects on genetic differentiation of local populations. *Evolution,* **42**, 995–1005.

Weir, B. S. and Cockerham, C. C. (1984). Estimating *F* statistics for the analysis of population structure. *Evolution,* **38**, 1358–70.

Whitlock, M. C. and McCauley, D. E. (1990). Some population genetic consequences of colony formation and extinction: genetic correlations within founding groups. *Evolution,* **44**, 1717–24.

Wolfram, S. (1991). *Mathematica.* Addison Wesley, New York.

Wright, S. (1943a). An analysis of local variability in flower color in *Linanthus parryae. Genetics,* **28**, 139–56.

Wright, S. (1943*b*). Isolation by distance. *Genetics,* **28**, 114–38.

Wright, S., Dobzhansky, Th., and Hovanitz, W. (1942). Genetics of natural populations VII The allelism of lethals in the third chromosome of *Drosophila pseudoobscura. Genetics,* **27**, 363–94.

4

The coalescent process and background selection

R. R. Hudson and N. L. Kaplan

4.1 Introduction

Gene trees or gene genealogies, which represent the history of a sample of genes from a population are fascinating objects whose statistical properties have recently been investigated under a few models (for example Kingman (1982), Tajima (1983), Slatkin (1989), Hudson (1990) and Hey (1991)). Different models can lead to gene trees with very different shapes and sizes. Examples of gene trees under three different models are shown in Fig. 4.1. Though gene trees are not directly observable entities, important observable quantities are indirectly related to them. It frequently aids our intuition, as well as mathematical analysis, to think about genetic variation as a by-product of gene genealogical (or coalescent) processes which generate gene trees. In fact, consideration of gene genealogies often leads to simple results concerning directly observable quantities. Incidentally, the gene genealogical approach also leads to efficient computer simulation algorithms for studying sampling distributions (Hudson 1990).

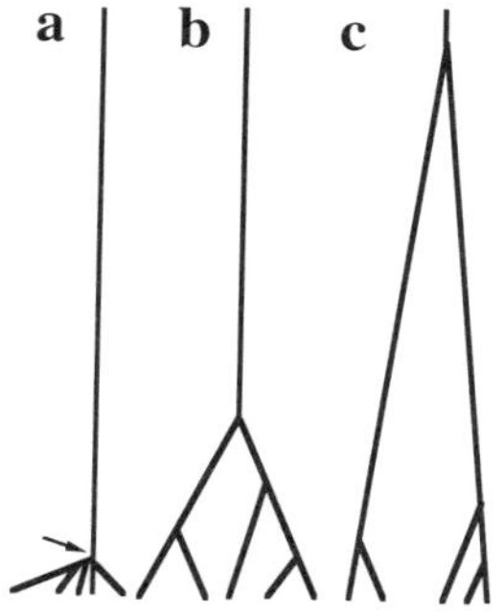

Fig. 4.1 Three sample gene genealogies that might arise under very different models. (a) A recent directional hitchhiking event at the time indicated by the arrow results in most coalescent events occurring recently and forming a star genealogy. (b) A typical genealogy under a strict neutral model. (c) A tightly linked balanced polymorphism can result in this type of genealogy in which two lineages extend deep into the past. (Redrawn from Aguadé and Langley (1994).)

Two examples will illustrate the connection between sample gene trees and observable population genetics quantities. First, consider nucleotide diversity, denoted π, which is a commonly employed measure of the amount of DNA variation in a population. (Nucleotide diversity is the average pairwise difference between sampled sequences, usually scaled by the length of the region sequenced.) The expected nucleotide diversity in a population is very simply related to a property of gene trees of samples of size two from the population. The gene tree for a sample of size two is extremely simple (Fig. 4.2), and can be characterized by a single number, T, the number of generations back to the most recent common ancestor (MRCA) of the two sampled genes. Expected nucleotide diversity can be expressed as simply the product of twice the neutral mutation rate per base pair and $E(T)$, the mean time back to the MRCA of a pair of genes. In other words, to calculate the expected nucleotide diversity (due to neutral mutations) under any model, one need only determine the mean time back to the MRCA of two sample genes under the model.

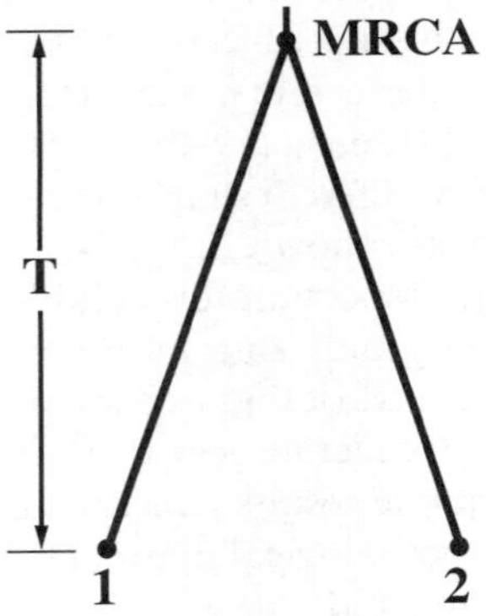

Fig. 4.2 The genealogy of a sample of size two. The most recent common ancestor of the two sampled alleles occurred T generations ago. The expected number of differences between the two sampled alleles per base pair, due to neutral mutations, is $2\upsilon T$, where μ is the neutral mutation rateper generation per base pair. (This assumes that multiple hits can be ignored.)

The second example concerns the frequency spectrum of variation. Each polymorphic site in a sample of genes corresponds to one or more mutations that occurred somewhere on the gene tree. When a polymorphic site is due to a single mutation on the gene tree it partitions the sample, say of size n, in two parts: those sampled genes descended from the mutation and those that retain the ancestral state. Typically, we cannot determine which state is ancestral and which is derived. Consider a particular mutation which produces a polymorphism in the sample. Let j denote the frequency in the sample of the less common nucleotide at the dimorphic site being considered. We designate such a site as a j-site. The frequency spectrum of a sample is simply a tabulation of the number of 1-sites, 2-sites, 3-sites, etc. The sample frequency spectrum is clearly related to the shape of the gene tree. If the gene tree for a sample is like that shown in Fig. 4.1a, then almost all polymorphic sites will be 1-sites. If the gene tree is like that in Fig. 4.1c, most polymorphic sites will be not be 1-sites or 2-sites but will have intermediate frequency variants. Consider all the branches on a tree with the property that if a mutation occurred on the branch the resulting polymorphism would create a j-site. We will call such branches j-branches. The number of j-sites in a sample has an expectation that is proportional to the expected sum of the lengths of the j-branches of the gene tree of the sample. The expected proportion of polymorphic sites that are j-sites is equal to the expected fraction of the total gene tree which is made up of j-branches. Thus the shape (including topology and relative lengths of branches) of a gene tree determines the sample frequency spectrum. The frequency spectrum under a strictly neutral model is known. Statistical tests of the neutral model are available that are based, at least in part, on the sample frequency spectrum (Tajima 1989). The shape of the gene tree under a strict neutral model is determined by the following two properties: (1) as we go back in time tracing the history of a sample of genes, each coalescent event (MRCA event) brings together a random pair of extant lineages (that is, if n ancestral lineages are being followed back in time, any of the $n(n-1)/2$ possible pairs of lineages are equally likely to coalesce at the next coalescent event), and (2) the time between such coalescent events is exponentially distributed with a mean which is proportional to $1/n\ (n-1)$, where n is the number of distinct ancestral lineages in the interval. Any model with these properties will have the neutral frequency spectrum.

In the following paragraphs, a method for calculating E-(T) and hence the expected nucleotide diversity under a deleterious background selection model will be sketched. This background selection model has recently aroused some interest because it may help explain the very low levels of variation observed in regions of the *Drosophila* genome that have low rates of recombination (Charlesworth *et al.* 1993). We will also argue that under a model with background selection the coalescent process has the two properties indicated above approximately, leading to gene trees with a shape like neutral trees, and thus to a sample frequency spectrum like the neutral one.

4.2 Deleterious background selection model

We consider a locus, denoted locus A, at which neutral mutations occur at a rate μ per generation. The goal is to calculate the expected nucleotide diversity at this locus, assuming that the locus is linked to loci at which deleterious mutations occur. We consider a deleterious mutation–selection balance model described in more detail in Hudson and Kaplan (1994). In the absence of background selection the expected nucleotide diversity would be equal to $2\mu\epsilon(T)=2\mu 2N$, the standard result for an equilibrium Wright–Fisher model with N diploids. We will see, however, that with linkage to loci that experience deleterious mutation $E(T)$ is less than $2N$, and consequently the expected nucleotide diversity is reduced (Charlesworth *et al.* 1993). We now describe very briefly how $E(T)$ can be calculated. A more detailed derivation is given by Hudson and Kaplan (1995).

To begin with we assume that locus A is linked to one other locus, locus 1, at which deleterious mutations occur at a rate μ_1. All deleterious mutations are assumed to have the same effect, *sh*, in the heterozygous state. Interaction is assumed to be multiplicative, so that an individual heterozygous for *i* such mutations has fitness $(1-sh)^{\mathrm{i}}$. It is assumed that no back mutation occurs. In a very large population, an equilibrium is reached in which the frequency of locus 1 genes with *j* deleterious mutations is approximately $f_j(u_1)=(u_1/2sh)\exp(-u_1/2sh)$. We refer to chromosomes that carry *i* deleterious mutations at locus 1 as the *i*-class. If the recombination rate between locus 1 and locus A is zero, the history of a sample of genes at locus A can be described as follows. As one traces the lineages of the sampled genes back in time, all lineages very quickly revert to the 0-class. For example the mean time for a single lineage in the 1-class to move over to the 0-class is $1/sh$ generations. If $sh = 0.02$, as has been suggested (Crow and Simmons 1983), then only 50 generations are required on average for a 0-class ancestor to be found for a sampled 1-class chromosome. Once all the lineages are in the 0-class, the coalescent process occurs as for the neutral model, with a population size of $f_0(u_0)2N$, instead of $2N$. Thus the time to the most recent common ancestor of two sampled genes will be approximately (ignoring the time to get to the 0-class) $f_0(u_1)2N$, and the expected nucleotide diversity is $4N\mu f_0\,(u_0)$, as found by Charlesworth *et al.* (1993). Ignoring the very short period when lineages are shifting over to the 0-class, the shape of the gene tree will be just like the shape of a neutral tree (except the time-scale will be different) and hence the frequency spectrum will be as expected under the neutral model.

Now consider the case where the recombination rate between locus A and locus 1 is non-zero. This two-locus model was analyzed by Hudson and Kaplan (1994), but the method of analysis outlined here, as well as the results for the multilocus models to be considered next, are described in greater detail in Hudson and Kaplan (1995). As one traces backward in time along the ancestral lineage of a sampled gene under this two-locus model one will encounter two kinds of events, mutation events and recombination events. At points in the history of a lineage when mutation events occur, the ancestor has fewer deleterious mutations than the immediately descendent chromosome. At recombination events, the ancestor has a random number of deleterious mutations, the number being drawn from the distribution, $f_i(u_0)$. In other words, recombination has the effect of randomizing the number of deleterious mutations at the locus linked to the ancestor of the sampled locus A gene. As one traces back the history of a lineage, there is a continuous 'pressure' toward the 0-class, due to the effects of mutation and selection, but this is balanced by recombination that at random time points associates with the ancestor a random locus 1 allele. The process is a standard continuous-time Markov chain. At stationarity it can be shown that the distribution of the number of deleterious mutations on the ancestor chromosome is a geometric mixture of Poisson distributions. If the mutation rate is small the distribution is simply

$$P(1)=\frac{u_1 R}{2sh(R+sh)} \text{ and } P(0)=1-P(1). \tag{4.1}$$

where $P(i)$ is the probability that the ancestor belongs to the i-class. In a large population the numbers of deleterious mutations linked to locus A genes of two distinct ancestral lineages are independent of each other. If two ancestral chromosomes have parents in different classes then these parents are necessarily distinct, that is the two lineages will not have their most recent common ancestor in the previous generation. If two individuals have parents in the same class, say the i-class, the probability that they have the same parent is $1/2Nf_i$. Thus, the probability, Λ that two lineages are coalescent in a particular generation is given by the following sum:

$$\Lambda=\sum_k \frac{P(k)^2}{2Nf_k(u_1)}. \tag{4.2}$$

If the Markov chain moves at a sufficiently rapid pace compared with the coalescent probabilities given by (4.2), (which requires that N be large) then the time back to the common ancestor of two lineages will be approximately exponentially distributed with a mean given by the inverse of Λ. That is, using eqns (4.1) and (4.2), which assume that u_1 is small,

$$E(T)\approx\Lambda^{-1} \approx 2Nf(u,R), \tag{4.3}$$

where

$$F(u,R)=\frac{ush}{2(R+sh)^2}. \tag{4.3}$$

In this case, the expected time back to the the MRCA of a pair of genes is reduced by the factor, $F(u_1, R)$.

Now consider a second linked locus, locus 2, with deleterious mutation rate $u2$. In this case we say that a chromosome belongs to the i,j-class if it has i deleterious mutations at locus 1 and j deleterious mutations at locus 2. It is easily shown that under the multiplicative model the equilibrium frequency of the i,j class is $f_i(u_1)f_k(u^2)$. Assume that locus 1 is to the right of locus A, and that locus 2 is to the right of locus 1. Let R_2 denote the recombination rate between locus 2 and locus A. The process is still a continuous-time Markov chain, but on a two-dimensional state space. The stationary distribution is specified by giving the joint probabilities, $P(i,j)$, of a lineage belonging to the i,j-class. By the same logic that leads to (4.2), we find that the probability of a common ancestor in a particular generation is

$$\Lambda = \sum_{i,j} \frac{P(i,j)^2}{2Nf_i(u_1)f_j(u_2)}. \tag{4.4}$$

It is easy to show by the same methods used to obtain (4.2) that if u_1 and u_2 are small then the numbers of deleterious mutations at the two loci on ancestral chromosomes are approximately independent, and thus that

$$P(i,j) \approx P_1(i)P_2(j). \tag{4.5}$$

where $P_1(i)$ is given by (4.1) and $P_2(j)$ is given by (4.1) with u_1 and R replaced by u_2 and R_2 respectively. Using (4.5) in (4.4), one finds that

$$E(T) \approx \Lambda^{-1} \approx 2NF(u_1,R)F(u_2,R_2). \tag{4.6}$$

This equation shows that the effects of different loci accumulate in a multiplicative fashion. For many linked loci we then have

$$E(T) \approx 2N \prod_i F(u_i,R_i). \tag{4.7}$$

Instead of considering that locus A is linked to a collection of discrete loci, we now suppose that locus A is located in a continuous region, for which $u(x)$ is the deleterious mutation rate per unit length of DNA at position x, and $R(x)$ is the recombination rate between position x and locus A. In this case, we can approximate the effect of the linked region by a dividing the region up into a large number of small intervals and apply (4.7). The product approaches an exponential function with the exponent being an integral. In this case, using the fact the $E(\pi) = 2\mu E(T)$, we find

$$E(\pi) \approx 4N\mu \exp\left[-\int_{L_1}^{L_2} \frac{u(x)\, sh\, \mathrm{d}x}{2(sh + R(x))^2} \right]. \tag{4.8}$$

If $u(x)$ is constant and $R(x)$ is linear (i.e the recombination rate is constant per base pair, and can be taken to be proportional to distance) the integral can be calculated explicitly. If the region with constant recombination is large, the result is

$$E(\pi) \approx 4N\mu \exp\left[-\frac{u}{r} \right]. \tag{4.9}$$

where u is the deleterious mutation rate per unit length of DNA and r is the recombination rate (in morgans) for the same unit length of DNA. Remarkably, the effect of background selection in this case is independent of the selection coefficient. For this result to hold, the recombination rate per unit length of DNA must be constant for a distance of say $2sh/r$ on both sides of locus A. In regions with very low r this is unlikely to be true.

Summarizing the theory of this section, we have found that if the product of selection coefficient and population size is large, then the history of a small sample is a quite accurately represented as a continuous-time Markov chain. The stationary distribution is quickly achieved and the coalescent events are widely spaced in time relative to the Markov chain transitions. In this case it not difficult to show that the two properties of gene genealogies mentioned in the introduction hold. Therefore the frequency spectrum of neutral variation under this background selection model will be approximately the same as under a strict neutral model. In addition, we have found that for low mutation rates the stationary distribution of the states for the multilocus model is approximately the product of the stationary probabilities for the individual loci. This leads to the effects of many loci being approximately the product of the effects of the individual loci. Simulations have verified that the results we have obtained are quite accurate over a wide range of plausible parameter values (Hudson and Kaplan 1995).

Equation (4.8) has been applied to predict levels of variation on each of the chromosomes of *Drosophila melanogaster*. The total genome deleterious mutation rate is taken to be 1.0. Three different selection coefficients were considered, $sh = 0.03$, 0.02, and 0.005. Recombination rates are estimated from published maps of the *Drosophila* genome. The level of polymorphism to be expected in the absence of background selection was chosen to fit the data ($4N\mu = 0.014$). The results for chromosome 3 are shown in Fig. 4.3, along with data from a number of studies. The fit is quite good except for the loci at tips of the chromosomes.

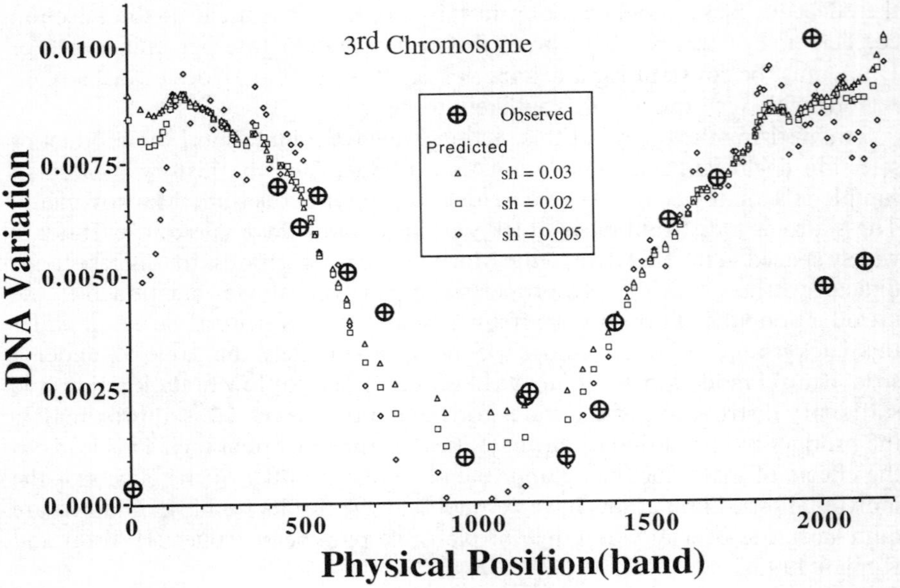

Fig. 4.3 Observed and predicted levels of DNA variation as a function of physical position on the third chromosome of *D. melanogaster*. The observed data, from left to right, are from the following loci: *Lspl-γ*, *Hsp26*, *Sod*, *Est6*, *fz*, *tra*, *Pc*, *Antp*, *Gld*, *MtnA*, *Hsp70A*, *ry*, *Ubx*, *Rh3*, *E(spl)*, *Tl*, and *Mlc2*. The DNA variation at *MtnA* is an estimate of π from Lange *et al.* (1990) as reported in Begun and Aquadro (1992). The observed level of DNA variation at *Hsp70A* is an estimate of π from Leigh Brown (1983) as reported in Begun and Aquadro (1992). The other observed values are estimates of θ from Kindahl and Aquadro (1995). The predicted values are based on eqn (4.8) and assume that $4N\mu = 0.014$ and u, the deleterious mutation rate per cytological band, is 0.0002. (Redrawn from Hudson and Kaplan (1995).)

References

Aguadé, M. and Langley, C. H. (1994). Polymorphism and divergence in regions of low recombination in *Drosophila*. In *Non-neutral evolution: theories and molecular data*, (ed. G. B. Golding), pp. 67–76. Chapman and Hall, New York.

Begun, D. J. and Aquadro, C. F. (1992). Levels of naturally occurring DNA polymorphism correlate with recombination rates in *D. melanogaster*. *Nature*, **356**, 519–20.

Charlesworth, C., Morgan, M. T. and Charlesworth, D. (1993). The effect of deleterious mutations on neutral molecular variation. *Genetics*, **134**, 1289–303.

Crow, J. F. and Simmons, M. J. (1983). The mutation load in *Drosophila*. In *The genetics and biology of drosophila*, (ed. M. Ashburner, H. L. Carson, and J. N. Thompson), pp. 1–35. Academic. London.

Hey, J. (1991). A multi-dimensional coalescent process applied to multi-allelic selection models and migration models. *Theor. Popul. Biol.*, **39**, 30–48.

Hudson, R. R. (1990). Gene genealogies and the coalescent process. In *Oxford Surveys in Evolutionary Biology*, (ed. D. Futuyma and J. Antonovics), pp. 1–44. Oxford University Press.

Hudson, R. R. and Kaplan, N. L. (1994). Gene trees with background selection. In *Non-neutral evolution: theories and molecular data*, (ed. G. B. Golding), pp. 140–53). Chapman and Hall, New York.

Hudson, R. R. and Kaplan, N. L. (1995). Deleterious background selection with recombination. *Genetics*, **141**, 1605–17.

Kingman, J. F. C. (1982). On the genealogy of large populations. *J. Appl. Probab.*, **19A**, 27–43.

Lange, B. W., Langley, C. H. and Stephen, W. H. (1990). Molecular evolution of *Drosophila* metallothionein genes. *Genetics*, **126**, 921–32.

Leigh Brown, A. J. (1983). Variation at the 87A heat shock locus in *Drosophila melanogaster*. *Proc. Nat. Acad. Sci. USA*, **80**, 5350–4.

Slatkin, M. (1989). Detecting small amounts of gene flow from phylogenies of alleles. *Genetics*, **121**, 609–12.

Tajima, F. (1983). Evolutionary relationship of DNA sequences in finite populations. *Genetics*, **105**, 437–60.

Tajima, F. (1989). Statistical method for testing the neutral mutation hypothesis by DNA polymorphism. *Genetics*, **123**, 585–95.

5

Inferring population history from molecular phylogenies

Sean Nee, Eddie C. Holmes, Andrew Rambaut, and Paul H. Harvey

5.1 Introduction

Elsewhere, we have presented and applied an exact theory for making inferences about the past history of a clade on the basis of internode distances in its molecular phylogeny (Nee *et al.* 1994*a,b,c*; Harvey *et al.* 1994*a,b*). Here we use coalescence theory from population genetics (as developed for non-recombining sequences), which has proved to be a powerful tool for studying small samples of sequences from populations (Hudson 1990), to make inferences about population dynamic history. Coalescence theory is particularly applicable to genealogical trees of mitochondrial and viral genome sequences because only a very small proportion of extant lineages is ever sequenced.

5.2 Background in population genetics

The frequency distribution of pairwise differences between sequences and the number of segregating or polymorphic sites in a sample from a population are statistics that have been used in the past to determine whether a population has been of constant size throughout its history. The theoretical approach has been to show that these statistics behave differently in equilibrium (constant size) and non-equilibrium populations. The models of non-equilibrium are varied and have been applied to human mitochondrial DNA sequence data to infer a past history of population growth. They include an exponentially growing population (Slatkin and Hudson 1991), a population going through a bottleneck (Tajima 1989), and a population of one size which suddenly expands to a larger, constant size (Harpending *et al.* 1993).

The theory for the distributions of the number of pairwise differences between sequences and the number of segregating sites in a sample was developed in the context of constructing estimates of the size of an equilibrium population (more precisely, the inbreeding effective population size, N_e, and, if the mutation rate μ is unknown, the composite parameter $N_e\mu$) (Watterson 1975; Tajima 1983; Fu and Li 1993). However, assuming infinitely long sequences, Felsenstein (1992) showed that methods of estimating population size based on these statistics are far less efficient than methods which use, as their raw information, the times between nodes (the coalescence events) in the genealogical tree of the sampled sequences (Fig. 5.1)—the case of finite sequences is discussed by Fu and Li (1993). Felsenstein notes that 'pairwise methods are attractive because they do not involve estimating the tree [genealogy] and have an aura of robustness' which, he claims, is misleading because pairwise estimates take most of their information from the earliest bifurcations in the genealogical tree (see legend to Fig. 5.1). Hence, they make inefficient use of the information contained in the data and any signal the data contain may be swamped by non-independence. Felsenstein (1992) concluded that 'there is much to gain by explicitly taking the tree structure' of the sampled sequences into consideration and alluded to the possibility of using the internode times to detect changes in population size.

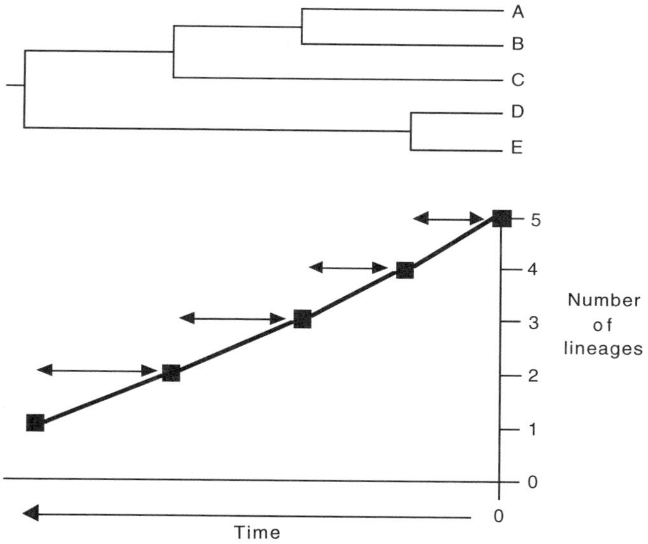

Fig. 5.1 A phylogenetic tree linking a sample of lineages (A to E). Time is measured from the viewpoint of 'how long ago?', so zero is the present, and time increases as we look backward into the past. Horizontal lines show the times between coalescent events, that is when 5, 4, 3, and 2 lineages are represented in our phylogeny. The bold line, which is the line joining the dots in subsequent lineages-through-time plots, links the points denoting the earliest historical times at which each number of lineages existed. The data provided by such phylogenies consist of the times between the nodes, and the lineage-through-time plots are the graphical representations of the data. Techniques which analyse all pairwise differences between extant lineages include many non-independent comparisons. For example, six of the ten pairwise comparisons that are possible between the five extant lineages represent estimates of genetic divergence between pairs of lineages that last shared common ancestry at the time of the earliest node.

5.3 Basic theory

Let the present be time zero and the size of the population from which we are sampling be $N(t)$ at time t in the past. Given a random sample of sequences from a population, we are interested in the properties of the internode distances in their bifurcating molecular phylogeny (Fig. 5.1). We assume that the phylogeny of the sampled sequences has a temporal dimension providing time intervals between the nodes and the time from the last node to the present. In this theoretical section we measure time in units of generations, for convenience. When we present examples, we measure time as per cent base substitutions derived from the molecular phylogeny.

As we look back in time, the lineages in the phylogeny 'coalesce' (a bifurcation seen in reverse) one by one until only a single lineage remains. If $n+1$ lineages have coalesced to n lineages at time t in the past, then the probability density for the additional amount of time, s, that will elapse before the coalescence of the n lineages to $n-1$, $p(s)$ ds, is given by

$$p(s)\mathrm{d}s = H'(s)\exp[-H(s)]\,\mathrm{d}s. \tag{5.1a}$$

where

$$H(s) = \binom{n}{2}\int_{t}^{t+s}\frac{1}{N(s')}\,\mathrm{d}s'. \tag{5.1b}$$

This way of expressing the density, in terms of an 'integrated hazard', $H(.)$, is a convenient stylistic device to extract our analysis from population genetics and insert it into the field of the statistical analysis of series of events. It also highlights the separate roles played by the changing number of lineages in the phylogeny, n, and the (possibly) changing population size from which the sample has been drawn, $N(t)$. An interpretation of the integrated hazard is that it specifies the piecewise transformation of the time-scale necessary to induce a unit exponential distribution of coalescence times (Cox and Oakes 1984).

Given a molecular phylogeny with a temporal dimension, the data we wish to analyse consist of the time intervals between nodes (coalescences). Our analytical weapon is simply probability expression (5.1a) for these intervals. Our task, which is to make inferences about the population dynamical history, can be expressed as a statistical problem: with what hypotheses of population dynamical histories, $N(t)$, are our data consistent?

5.4 Landmarks in hypothesis space

The problem we face is how to navigate through the enormous hypothesis space of all possible past population histories. As is usual with data analysis, we exploit graphical representations to guide us through this space. We need graphical representations which give the data an interpretable form. The basic representation of the data, the time intervals between nodes in the phylogeny, is illustrated in Fig. 5.1—the lineages-through-time plot of the phylogeny.

Consider the two distinct directions of concavity illustrated in Fig. 5.2, where we plot the the logarithmically scaled number of lineages against time. The direction of concavity of the lower curve is the easiest case to deal with, so we consider that first. It is convenient to reverse our temporal perspective and to think of the phylogeny as 'growing', like a tree. The slope of the plot at any point is the the rate at which a branch produces new branches, the per lineage rate of appearance of lineages in the phylogeny. The direction of concavity of the lower curve indicates an increasing rate of branching as we move toward the present. We recognize this as the qualitative signature of a phylogeny of sequences sampled from a population which has been of constant size over the period of time covered by the coalescences. This is because sequences from a population which has been of constant size are expected to be related by a genealogical tree that grows hyperbolically as it approaches the present. However, a population that was smoothly declining in size, or one that was slowly increasing in size, over this period would also display the same qualitative signature, so we need further analysis.

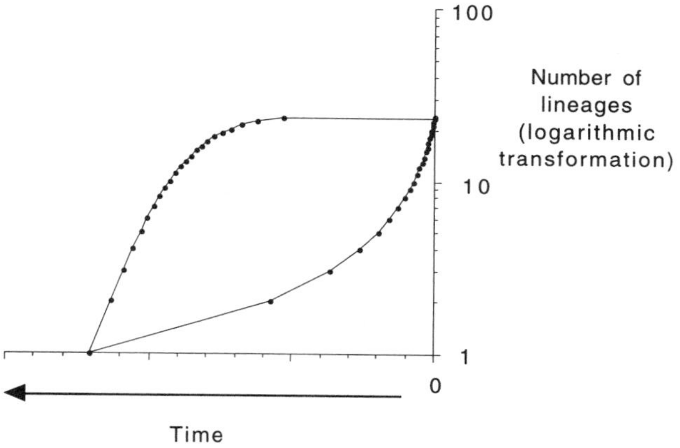

Fig. 5.2 When the lineages-through-time graph is given as a semilogarithmic plot, with the number of lineages axis being logarithmically transformed, then the direction of concavity of the upper line is the pattern expected for a sample from a population which is undergoing rapid exponential growth, over the time period covered by the coalescences in the phylogeny, while the lower line is expected from a population of constant size. To construct this figure, and Fig. 5.4, we concatenate the theoretical expectations of the internode intervals as experience with simulation shows this to provide a reliable visual guide to the behaviour of the processes. Simulation algorithms can be found in Hudson (1990) (constant-size population) and Slatkin and Hudson (1991) (exponential growth). The expected internode intervals for the top curve were calculated from expression (5.3), with $N(0)r = 100\,000$ (see legend to Fig. 5.4).

Now, the hypothesis of a constant-sized population is a very convenient landmark in hypothesis space because expression (1) assumes a particularly tractable form:

$$p(s)\mathrm{d}s = \frac{\binom{n}{2}}{N} \exp\left[-\frac{\binom{n}{2}}{N} s\right] \mathrm{d}s. \tag{5.2}$$

where N is the population size (see, for example, Hudson (1990) and Felsenstein (1992). To explore this hypothesis further, we simply multiply each internode interval by the appropriate value of $n(n-1)/2$. So, looking right to left in Fig. 5.1, from the tips of the tree to the root, the first interval, i.e. the divergence time of D and E, would be multiplied by $5 \times 4/2$ ($=10$), the second by 6, and so on. Under the hypothesis of size constancy, after the application of this transformation the data are all drawn from an exponential distribution with parameter $1/N$. If we do not know the mutation rate, the parameter is $1/N\mu$. Graphically, this transformation is a piecewise transformation of the time axis. Having applied the transformation, we can inspect a plot of the number of lineages in the phylogeny against this transformed time. If, indeed, the sample of sequences has come from a constant-sized population, then this plot should appear linear, in the sense that there will be no consistent trend in its slope (there will, of course, be stochastic wobbles). Figure 5.3 presents an actual example.

In the example of Fig. 5.3, the plot is indeed linear, so for these data we have finished our exploration of hypothesis space and conclude that the data are consistent with the sequences having been drawn from a constant-sized population. Now, if the plot became steeper toward the present (so it had the same direction of concavity as the lower line in Fig. 5.2) then this would suggest we entertain the hypothesis that the population from which the sequences had been drawn had been declining in size. The opposite direction of curvature would suggest a gently increasing population. These results can be readily derived by contemplating expression (5.2).

However, before rejecting the hypothesis of size constancy, we would want to perform a statistical analysis against the hypothesis that the curvature has just arisen by chance. This is very easy. The transformation we have applied to the data has, under the hypothesis of constancy, given us a realization of a Poisson process and the graphical representation is simply the cumulative plot of events through time, a very informative way of viewing a Poisson process. Over the years, numerous statistical tests have been developed for Poisson processes. Of particular relevance here are uniform conditional tests (Cox and Lewis 1966, Section 6.3), which are important because they are based on normalized times to events and thereby manage to eliminate the nuisance parameter which is the parameter of the exponential distribution, either $1/N$ or $1/N\mu$. (A computer package including some relevant tests is available on the Internet at http:/evolve.zps.ox.ac.uk/.)

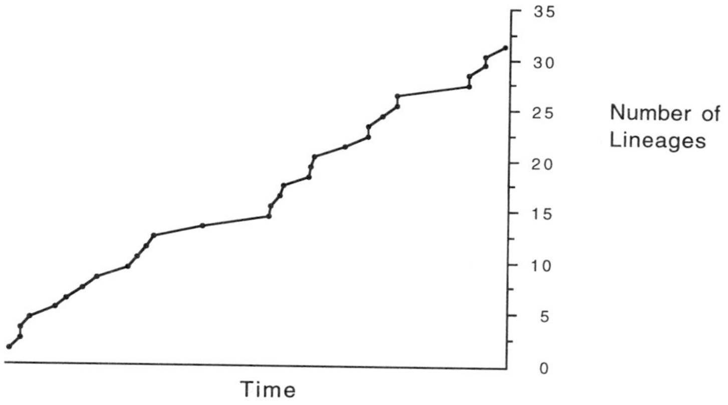

Fig. 5.3 Baker *et al.* (1993) report data on mitochondrial DNA control region sequences (283 base pairs) from 32 humpback whales (*Megaptera novaeangliae*) collected from around the world (North Pacific, North Atlantic, and Southern Ocean) that differ by more than a single base pair. The semilog representation of their phylogeny exhibits the direction of concavity of the lower curve in Fig. 5.2 (not shown). The internode intervals in their phylogeny exhibit linearity under the subsequent transformation described in the text, suggesting that they are consistent with having been drawn from a population of roughly constant size over the period of time covered by the coalescences. Note that with this number of sequences we do not expect the first coalescence to occur before, at the very least, about 10 generations ago (assuming that the whale population has been 5000 since the cessation of harvesting). Hence, the impact of hunting is not expected to be manifested in these data. Because the time axis is formed by concatenating transformed intervals, there is no obvious scale for the axis, so none is shown. The phylogenetic tree on which the plot is based was reconstructed using the ultrametric KITSCH clustering program from the PHYLIP package (Felsenstein 1993). Similar trees were obtained using methods which allow variable rates of substitution. Distances were corrected for multiple substitution using a model of molecular evolution which allows different base frequencies and different rates of transition and transversion (PHYLIP program DNADIST). The transition/transversion ratio was set to 20:1.

Much has been written in the population genetics literature on the estimation of this parameter which, here, is simply a nuisance which we are pleased to be able to eliminate. This is because, for our present purposes, we simply want to know whether or not the data are consistent with the hypothesis of size constancy, not what that size might actually be. Subsequently, the estimation of the parameter might be of interest. Logically, the sort of analysis described here must occur prior to any estimate of population size based on genealogical information: it makes no sense to estimate the size of a population on the basis of a constant-size model unless the data are consistent with that model.

We now consider how to proceed if the data exhibit the direction of concavity of the top line in Fig. 5.2, indicating a decrease in the rate of appearance of lineages in the phylogeny as we approach the present. We need a new landmark in hypothesis space, because we are now nowhere near the hypothesis of size constancy which need not be considered further: with a reasonable number of data no statistical analysis is required to reject this hypothesis. The landmark we choose is the hypothesis of rapid exponential growth, where, again, rapid refers to the time period covered by the coalescences in the phylogeny.

If the population has been growing exponentially at a rate r, so $N(t) = N(0) \exp(-rt)$, then expression (5.1) becomes

$$p(s)\mathrm{d}s = \frac{\binom{n}{2}\exp[r(t+s)]}{N} \exp\left[-\frac{\binom{n}{2}\exp(rt)}{N(0)r}[\exp(rs)-1]\right]\mathrm{d}s. \tag{5.3}$$

This expression is implicit in a simulation algorithm presented by Slatkin and Hudson (1991).

Expression (3) is very awkward: further progress is hindered by the fact that there is no informative transformation of the data (visually, no transformation of the time axis) that can be carried out without knowledge of the values of the nuisance parameters which are, in the best-case scenario of knowing the mutation rate, $N(0)$ and r.

To circumvent this difficulty, we have developed a reasonably robust transformation of the number-of-lineages axis for subsequent analysis. Let $n(t)$ be the number of lineages in the phylogeny at time t, where t is measured in whatever units are being employed. Under this transformation, the transformed data, $m(t)$, are given by

$$m(t) = \ln\left[\frac{n(0)-n(t)}{n(0)n(t)}\right]. \tag{5.4}$$

which we call the 'epidemic' transformation. If the sequences have been drawn from a population which has been growing exponentially at a constant rate, then the data will appear linear under this transformation, with the same stochastic wobbles and wanderings that one expects to see in any time-series analysis. The transformation is derived in the Appendix. Because of the effects of the nuisance parameters, the performance of the transformation in the recent past starts to decline when, simultaneously, both the product of the population growth rate and the current population size ($rN(0)$) become small and the number of lineages in the sample grows large. Figure 5.4 provides a sense of the meanings of the words 'large' and 'small'.

If the plot becomes less steep towards the present under the epidemic transformation, this suggest that the population has been growing more slowly than our landmark hypothesis—it may have been growing exponentially at a rate that has been decreasing through time. The opposite direction of concavity suggests the opposite. We consider two examples of the use of the epidemic transformation.

In what follows, we do not show the semilogarithmic lineages-through-time data but, having carried out that analysis and observed the decreasing rate of appearance of lineages in the phylogeny, go directly to the epidemic transformation. Figure 5.5 shows the analysis for two phylogenies which are consistent with

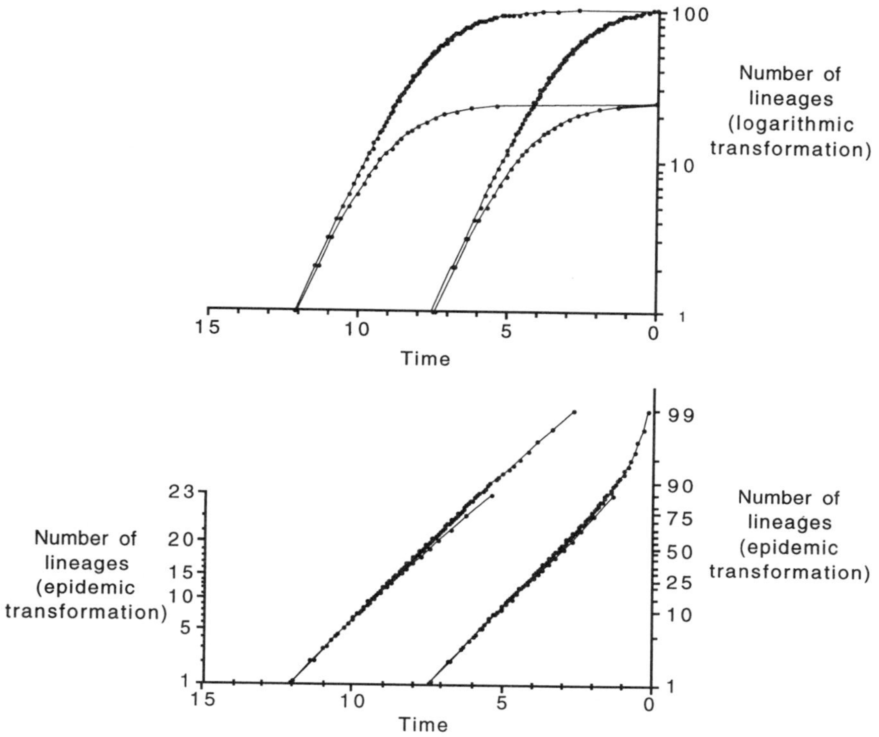

Fig. 5.4 To construct this figure, we have followed Slatkin and Hudson (1991) and transformed the time-scale of expression (5.3) to $\tau = rt$. On this scale (3.5) becomes a function of the composite parameter $N(0)r$. For the examples given, sample sizes are 100 or 24 (this latter number was chosen to compare with the analysis of 24 sequences of HIV-1 data given in Fig. 5.5), with $N(0)r = 1000$ and 100 000, making a total of four combinations. The top of the figure shows the lineages-through-time plots for the four samples and the bottom shows the corresponding transformed data plotted against time. As is evident from expression (5.4), the transformation is not defined for $n(t) = n(0)$, the total sample size, and values are therefore taken from the first coalescence (i.e. 99 and 23). The four samples are plotted on the same time-scale, thus forcing the interpretation that they come from populations with the same growth rate, r, but different values of $N(0)$.

the sequences having been drawn from exponentially growing populations: the HIV-1 *env* gene sequenced from 24 (mainly North American) patients infected with subtype B of the virus (figure 5.5a) and the human mitochondrial DNA control region from a large world-wide sample (figure 5.5b). Equivalent plots for the *gag* gene of HIV-1 (sequences from subtypes A and B of this virus taken separately), albeit from a different sample of patients, are also linear under the epidemic transformation providing additional support for an exponentially growing population. The HIV data have also been used to estimate the number of people infected at the time of their isolation (Nee 1994*a*), illustrating that there are more uses for such data than those we are describing here.

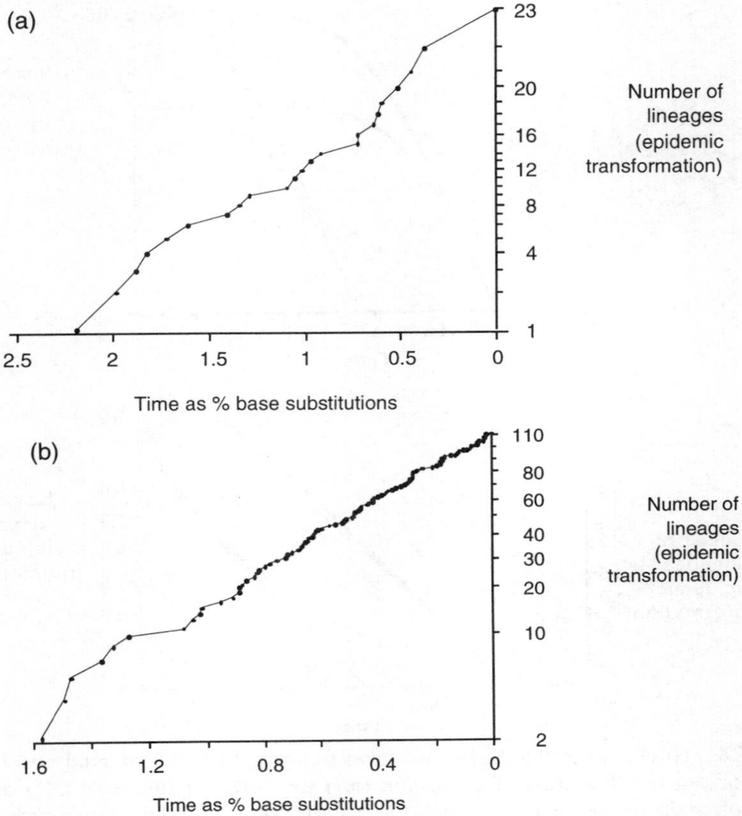

Fig. 5.5 Two population samples that provide good linear fits under the epidemic transformation. (a) This figure represents sequences taken from the *env* gene of 24 HIV infected individuals. All sequences were extracted from the Los Alamos AIDS database (Myers *et al.* 1993) and have been classified as belonging to HIV-1 subtype B. The sequences used were ADA, ALA1, BAL1, BRVA, CAM1, CDC4, D31, HAN, JFL, JH3, JRCSF, LAI, MN, NY5CG, OYI, RF, SC, SF2, SF33, SF162, SIMI84, TB132, WMJ22, YU2. Sequences were aligned using the ClustalV package (Higgins *et al.* 1992). Similar lineages-through-time plots were obtained when the highly variable regions were removed from the alignment. (b) This figure represents 110 sequences of mitochondrial DNA control region taken from a variety of individuals world-wide. These data derive from a number of published sources (Vigilant *et al.* 1989, 1991; Di Rienzo and Wilson 1991; Ward *et al.* 1991) and constitute a 338 base-pair region which could be successfully aligned. Only sequences differing by more than one nucleotide substitution are plotted. All sequences were downloaded from GenBank. The phylogenetic trees on which both plots (a) and (b) are based were reconstructed using the ultrametric KITSCH clustering program from the PHYLIP package (Felsenstein 1993). Similar trees were obtained using methods which allow variable rates of substitution. Distances were corrected for multiple substitution using a model of molecular evolution which allows different base frequencies and different rates of transition and transversion (PHYLIP program DNADIST). The transition/transversion ratio was set to 20:1 for the mitochondrial DNA data.

By way of contrast, the data from a molecular phylogeny of Dengue-3 virus isolates from 23 infected individuals exhibits curvature under the epidemic transformation, which suggests that the population has been growing at an accelerating rate (Fig. 5.6). This is consistent with the fact that the frequency of Dengue fever epidemics has been increasing and regions of hyperendemism have been expanding (Monath 1994).

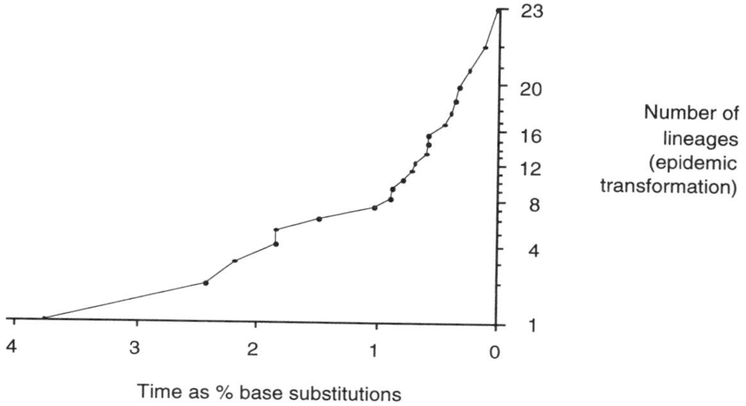

Fig. 5.6 One population that has been growing more rapidly than exponentially. Plotted under the epidemic transformation, 23 Dengue-3 virus lineages sampled from patients in 13 different countries (Lanciotti *et al.* 1994) show a curved line characteristic of a population increasing more rapidly than if it was expanding exponentially at a constant rate.

5.5 Problems and perspectives for the future

We are profoundly ignorant of hypothesis space and how to navigate it. We have identified two landmarks so far, but all we can say of the rest is, 'here be dragons'. Fortunately, anyone can go on a voyage of discovery on their own computer, which we hope will result in reports of discoveries and maps of how to get there.

The assumption of constant molecular clocks in the reconstruction of phylogenies may be a limiting factor for the examples given here, and maximum likelihood methods are being developed to detect rate variation, assign it to specific lineages, and correct for it. Furthermore, although corrections have been made for multiple substitutions at single nucleotide sites, all methods of phylogenetic reconstruction may be subject to error in this respect, especially when considering organisms with extreme mutation biases like HIV.

Another general class of problems arises from the fact that the only source of error acknowledged in expression (5.1a) comes from the realization of the stochastic process itself. A comprehensive analysis would also incorporate errors in the estimates of internode distances allowing for the fact that the clock does not tick like a metronome. Although most work on phylogeny reconstruction has focused on topology (Hillis *et al.* 1994), there has been progress in assigning confidence intervals to estimates of the dates of nodes (Hasegawa *et al.* 1993). To see what we would like to have in the future, consider the following two points. First, with reference to expression (5.1a), the information that the internode intervals provide about population processes consists of the changing relationships between these intervals that are not accounted for by changes in n as we look along the tree. Second, in the simple case of two sequences, it is known that the coefficient of variation of an estimate of the time the sequences shared a common ancestor, based on their per cent sequence difference, decreases the longer ago they actually did share a common ancestor (Tajima 1983). Hence, uncertainty in our estimates of coalescence times also changes as we look along the tree. We need a theory for generating confidence statements about the *pattern* of the relationships amongst the internode intervals, in the context of the sort of hypotheses that can be generated by the techniques developed in this chapter.

There is no general prescription for dealing, in practise, with the above problems. In specific analyses, unlike the illustrative examples we are presenting here, specific questions can be formulated by asking 'What is it I am particularly interested in inferring, and what particular weaknesses of my analysis do I consider most likely to lead to fallacious inference'. We imagine that further theoretical developments will come from attempts to answer such questions.

At the very least, it is desirable to test any analysis for internal consistency. So, for example, say that the inference is that a sample of sequences has been drawn from a constant-sized population. One can simulate such phylogenies (and suppose these are the 'true' phylogenies), simulate a mutational process on these phylogenies to generate hypothetical sequence data, and then analyse these data in the same way that the original data were analysed, using the same tree-building algorithms and so on.

As an example of progress that can be made, we have found a modification of the popular bootstrap procedure (Felsenstein 1985) useful for assessing confidence in the reconstructed population processes described in Figs. 5.5–5.7. Phylogenetic trees are reconstructed on a large number of replicate datasets. Lineages-through-time plots are then constructed on each replicate under the appropriate transformation. If significantly more than half of the second-order effects (the coefficient of the x^2 term in a polynomial regression) are either positive or negative, then there is a departure from linearity. This regression analysis is simply a convenient quantification of a visual impression. For example, when the Dengue-3 virus isolates plotted in Fig. 5.6 were subjected to this procedure, the second-order effects were positive in each of the 100 replicates. The use of this method will be described in detail elsewhere, together with the results from simulated datasets.

Two important issues concern the nature of the data themselves. In common with those analyses of human mitochondrial data which reject the hypothesis of an equilibrium population size (Harpending *et al.* 1993), we have assumed throughout that the sequences used are a phylogenetically random sample of extant lineages. This assumption is certainly not true for the human mitochondrial data, but may be closer to the truth for the HIV data, which was not collected as part of a single study with the intention of revealing the topological structure of a particular part of the tree. In general, we do not think that this non-randomness is a crippling problem, because it is easy to intuit the effect of various sampling biases on one's conclusions and take this into consideration at the interpretation stage.

The second issue concerning the nature of the data is the amount of variation present: is there enough information to estimate the genealogy? Compare, for example, the HIV envelope phylogeny data with the human mitochondria data used in Fig. 5.5. The HIV envelope sequence analysed is 2697 base pairs in length, of which about 120 substitutions are expected to have occurred since the common ancestor of the tree. The section of the human mitochondrial control region used for sequence comparison is 338 base pairs long, of which about four are expected to have changed since the root of the reconstructed phylogeny. Even in the best human mitochondrial datasets, which provide much smaller phylogenies than the one used in Fig. 5.5b, the control region examined is only about 670 base pairs long, of which about eight have changed since the root of the tree. Another reason the mitochondrial tree contains very little information is that mutations at some sites are much more frequent than at others (Wakeley 1993; Hasegawa *et al.* 1993) and, because of the sequence sampling biases (for example numerous Kung! Bushmen sequences), the majority of bifurcations will have occurred in the very recent past when the expected number of base substitutions is extremely low (between one and three). We are, therefore, not surprised that, although the available data accord with an exponential growth model, they might also fit other models of population growth, or even null models which scatter a small number of changes on to a star phylogeny. Indeed, the fact that there are so many equally parsimonious trees for the data warns against using the data to make other than very weak inferences. A more appropriate level of variation, which does reveal interpopulation relationships among humans, is provided by polymorphic microsatellites (Bowcock *et al.* 1994). In general, with too few data to resolve genealogical relationships, methods of analysis such as that of Templeton *et al.* (1992) are to be preferred.

Acknowledgements

We are grateful to Steve Palumbi for his helpful comments. This work was funded by grants from the AFRC (SN), the Wellcome Trust (PHH), and the BBSRC (PHH, SN).

Appendix

The steps which suggest the epidemic transformation (5.4) are as follows. We emphasize that their justification lies solely in the performance of the end product, studied on simulated data. It can be shown from (5.3) that the variable $\exp(rs)$-1 is exponentially distributed with mean $2N(0)r/\exp(rt)n(n-1)$. This inspires the following differential equation for the change in the number of lineages in the phylogeny, n, through time, z, where $z = \exp(rt)$:

$$\frac{dn}{dz} = \frac{-zn^2}{2N(0)r}.$$

Denoting the number of sequences in the sample by $n(0)$, this has the solution

$$n(z) = \left[\frac{z^2}{4N(0)r} - \left(\frac{1}{4N(0)r} - \frac{1}{n(0)}\right)\right]^{-1}.$$

Solving for t,

$$2rt = \ln\left(\frac{1}{4N(0)r}\right) + \ln\left(\frac{1}{4N(0)r} - \frac{n(0)-n(t)}{n(0)n(t)}\right);$$

so, for large $N(0)r$,

$$\ln\left(\frac{n(0)-n(t)}{n(0)n(t)}\right) \approx 2rt + \ln(4N(0)r),$$

suggesting the 'epidemic' transformation as a candidate for a linearizing transformation.

References

Baker, C. S., Perry, A., Bannister, J. L., Weinrich, M. T., Abernethy, R. B., Calambokidis, J. *et al.* (1993). Abundant mitochondrial DNA variation and world-wide population structure in humpback whales. *Proc. Natl. Acad. Sci. USA*, **90**, 8239–43.

Bowcock, A. M., Ruiz-Linares, A., Tomforhde J., Minch, E., Kidd, J. R., and Cavalli-Sforza L. L. (1994). High resolution of human evolutionary trees with polymorphic microsatellites. *Nature*, **368**, 455–7.

Cox, D. R. and Lewis, P. A. A. (1966). *The statistical analysis of series of events*. Methuen, London.

Cox, D. R. and Oakes, D. (1984). *Analysis of survival data*. Chapman and Hall, London.

Di Rienzo, A. and Wilson, A. C. (1991). Branching pattern in the evolutionary tree for human mitochondrial DNA. *Proc. Natl. Acad. Sci. USA*, **88**, 1597–601.

Felsenstein, J. (1985). Confidence limits on phylogenies: an approach utilizing the bootstrap. *Evolution*, **39**, 783–91.

Felsenstein, J. (1992). Estimating effective population size from samples of sequences: inefficiency of pairwise and segregating sites as compared to phylogenetic estimates. *Genet. Res. (Camb.)*, **59**, 139–47.

Felsenstein, J. (1993). *PHYLIP (Phylogeny Inference Package). Version 3.5c*. Distributed by Author at Department of Genetics, University of Washington, Seattle WA 98195, USA.

Fu, Y.-X. and Li, W.-H. S. (1993). Maximum likelihood estimation of population parameters. *Genetics*, **134**, 1261–70.

Harpending, H. C., Sherry, S. T., Rogers, A. R., and Stoneking, M. (1993). The genetic structure of ancient human populations. *Current Anthrop.*, **34**, 483–96.

Harvey, P. H., Holmes, E. C., Mooers, A. Ø., and Nee, S. (1994*a*). Inferring evolutionary processes from molecular phylogenies. In *Models in phylogeny reconstruction*, (ed. R. W. Scotland, D. J. Siebert, and D. M. Williams), pp. 313–33. Systematics Association, London.

Harvey, P. H., May, R. M., and Nee, S. (1994*b*). Phylogenies without fossils. *Evolution*, **48**, 523–9.

Hasegawa, M., Di Rienzo, A., Kocher, T. D., and Wilson, A. C. (1993). Toward a more accurate time scale for the human mitochondrial DNA tree. *J. Mol. Evol.*, **37**, 347–54.

Higgins, D. G., Bleasby, A. J., and Fuchs, R. (1992). CLUSTALV: improved software for multiple sequence alignment. *CABIOS*, **8**, 189–91.

Hillis, D. M., Huelsenbeck, J. P., and Cunningham, C. W. (1994). Application and accuracy of molecular phylogenies. *Science*, **264**, 671–7.

Hudson, R. R. (1990). Gene genealogies and the coalescent process. *Oxf. Surv. Evol. Biol.*, **7**, 1–44.

Lanciotti, R. S., Lewis, J. G., Gibler, D. J., and Trent, D. W. (1994). Molecular evolution and epidemiology of Dengue-3 viruses. *J. Gen. Virol.*, **75**, 65–75.

Monath, T. P. (1994). Dengue: the risk to developed and developing countries. *Proc. Natl. Acad. Sci. USA*, **91**, 2395–400.

Myers, G., Korber, B. T. M., Wain-Hobson, S., Smith, R. F., and Pavlakis, G. N., (1993). *Human retroviruses and AIDS*. Mexico: Los Alamos National Laboratory.

Nee, S., Holmes, E. C., May, R. M., and Harvey, P. H. (1994*a*). Estimating extinction from molecular phylogenies. In *Estimating extinction rates*, (ed. J. L. Lawton and R. M. May), pp. 164–82. Oxford University Press.

Nee, S., Holmes, E. C., May, R. M., and Harvey, P. H. (1994*b*). Extinction rates can be estimated from molecular phylogenies. *Phil. Trans. R. Soc.*, **B344**, 77–82.

Nee, S., May, R. M., and Harvey, P. H. (1994*c*). The reconstructed evolutionary process. *Phil. Trans. R. Soc.*, **B344**, 305–11.

Nee, S., Mooers, A. O. and Harvey, P. H. (1992). The tempo and mode of evolution revealed from molecular phylogenies. *Proc. Natl. Acad. Sci. USA*, **89**, 8322–26.

Purcell, R. H. (1994). Hepatitis viruses: changing patterns of human disease. *Proc. Natl. Acad. Sci. USA*, **91**, 2401–6.

Simmonds, P., Holmes, E. C., Cha, T. A., Chan, S.-W., McOmish, F., Irvine, B., *et al.* (1993). Classification of hepatitis C virus into 6 major genotypes and a series of subtypes by phylogenetic analysis of the NS-5 region. *J. Gen. Virol.*, **74**, 2391–9.

Slatkin, M. and Hudson, R. R. (1991). Pairwise comparison of mitochondrial DNA sequences in stable and exponentially growing populations. *Genetics*, **129**, 555–62.

Tajima, F. (1983). Evolutionary relationship of DNA sequences in finite populations. *Genetics*, **105**, 437–60.

Tajima, F. (1989). The effect of change in population size on DNA polymorphism. *Genetics*, **123**, 585–95.

Templeton, A., Crandall K. A., and Sing, C. F. (1992). A cladistic analysis of phenotypic associations with haplotypes inferred from restriction endonuclease mapping and DNA sequence data. III. Cladogram estimation. *Genetics*, **132**, 619–33.

Vigilant, L., Pennington, R., Harpending, H., Kocher, T. D., and Wilson, A. C. (1989). Mitochondrial DNA sequences in single hairs from a southern African population. *Proc. Natl. Acad. Sci. USA*, **86**, 9350–4.

Vigilant, L., Stoneking, M., Harpending, H., Hawkes, K., and Wilson, A. C. (1991). African populations and the evolution of human mitochondrial DNA. *Science*, **253**, 1503–7.

Wakeley, J. (1993). Substitution rate variation among sites in hypervariable region 1 of human mitochondrial DNA. *J. Mol. Evol.*, **37**, 613–23.

Ward, R. H., Frazier, B. L., Dew-Jager, K., and Paabo, S. (1991). Extensive mitochondrial diversity within a single Amerindian tribe. *Proc. Natl. Acad. Sci. USA*, **88**, 8720–4.

Watterson, G. A. (1975). On the number of segregating sites in genetical models without recombination. *Theor. Popul. Biol.*, **7**, 256–76.

6

Applications of intraspecific phylogenetics

Keith A. Crandall and Alan R. Templeton

6.1 Introduction

Recent advances in population genetics theory, especially coalescent theory (Tavaré 1984; Watterson 1984; Ewens 1990; Hudson 1990), coupled with an explosion of molecular techniques have allowed detailed phylogenetic information at the population level. Such genealogical relationships are termed gene trees, allele trees, or haplotype trees (we use haplotype throughout this chapter) in which different haplotypes are the operational taxonomic units. Avise and colleagues were the first to utilize this powerful molecular phylogenetic approach for studies of population genetics, especially relating to biogeographic patterns (Avise *et al.* 1987; Avise 1989, 1994). Their approach explores qualitatively the relationship between lineage divergence and the extent of geographical partitioning among haplotypes.

Slatkin and Maddison (Slatkin 1989; Slatkin and Maddison 1989) have extended this type of analysis in an attempt to quantify gene flow or migration among populations at equilibrium. Given a phylogeny, their method computes the minimum number of migration events between pairs of populations sampled. This value is then used to estimate the effective migration rate among populations. Hudson *et al.* (1992) demonstrated the effectiveness of this method, relative to estimating F_{st} based on frequencies at polymorphic sites, when there is no recombination. The method has been expanded to detect certain types of population structure (Slatkin and Maddison 1990).

Intraspecific phylogenetic approaches to the estimation of effective population size or mutation rate have also been developed recently. The parameter $\theta = 4N_{ei}\mu$ plays an important role in population genetics as a measure of nucleotide diversity; where N_{ei} is the inbreeding effective population size and μ is the mutation rate. Felsenstein (1992*a*, *b*) described a procedure that utilizes a maximum likelihood approach to estimate nucleotide diversity and, given a mutation rate, inbreeding effective population size. Felsenstein (1992*b*) demonstrated analytically and verified through computer simulation that this maximum likelihood estimate is more efficient than pairwise or segregating sites estimates because of the additional information inherent in the tree structure of the genealogies. Similarly, Fu (1994*a*) developed a phylogenetic approach to estimating nucleotide diversity based on a distance approach. Fu's study also showed that the estimated genealogies provide substantially more information for estimating θ than do non-phylogenetic methods (Fu 1994*a*). He recently extended this estimator over a variety of population genetic models (Fu 1994*b*). An advantage to the phylogenetic estimation of θ, besides it being more efficient, is that one can estimate the parameter for subclades of the phylogeny. This allows for the statistical testing of changes in effective population sizes, which play important roles in bottleneck and speciation theory.

Golding (1987) initiated the study of within-species selection using a phylogenetic framework by exploring deleterious selection. Individuals containing deleterious characters leave few or no descendants and thus should be localized at the tips of haplotype networks. Using this phylogenetic information, Golding (1987) developed a statistical test for the detection of deleterious selection. Additional models have been developed to examine other forms of selection (Hudson and Kaplan 1988, 1994; Golding and Felsenstein 1990; Golding 1993) including purifying selection (Golding 1994), overdominant and frequency-dependent selection (Takahata and Nei 1990), and balancing selection (Takahata *et al.* 1992; Satta 1993).

Finally, tree shape has been used to differentiate alternative models of speciation (Hey 1992; Kirkpatrick and Slatkin 1993). Likewise, the partitioning of nucleotide variation both within and between species relative to phylogenetic relationships has been used to test alternative hypotheses of speciation (Takahata 1993; Hey and Kliman 1994). Thus intraspecific phylogenies have proved useful in estimating a variety of parameters and testing a variety of hypotheses at the population level, from estimating nucleotide diversity to testing hypotheses of the speciation process itself.

6.2 Intraspecific cladogram estimation

While many applications utilize intraspecific phylogenies, these methods depend upon an accurate estimate of phylogenetic relationships. Most methods of phylogeny reconstruction were developed to estimate interspecific relationships. Many of the assumptions of these methods are violated by intraspecific datasets. In this chapter we outline a method developed specifically to estimate intraspecific gene genealogies and its associated hypothesis testing framework. We then demonstrate the method's utility for hypothesis testing in a number of areas in population biology; from human genetics of disease to species concepts and speciation.

Difficulties of interspecific methods at the intraspecific level

A problem common to many of the applications of intraspecific phylogenetics described above is that they depend heavily on a reliable estimate of phylogenetic relationships. There are many phenomena that exist at the population level that lead to a lack of resolution of phylogenetic relationships when traditional interspecific methods of phylogeny reconstruction (for example maximum likelihood, parsimony, and distance methods) are applied at this level. For example, populations typically have lower levels of variation over a given gene region relative to higher taxonomic levels, resulting in fewer characters for phylogenetic analyses. Another difference concerns the treatment of ancestral types. In higher-level systematics ancestors are generally assumed to be extinct. In populations, however, most haplotypes in the gene pool exist as sets of multiple, identical copies because of past DNA replication. When one copy mutates to form a new haplotype, it would be extremely unlikely for all the identical copies of the ancestral haplotype to also mutate. Thus, as mutations occur to create new haplotypes, they rarely result in the extinction of the ancestral haplotype. Ancestral haplotypes are thereby expected to persist in the population. Indeed, coalescent theory predicts that the most common haplotypes in a gene pool will tend to be the oldest (Watterson and Guess 1977; Donnelly and Tavaré 1986), and most of these old haplotypes will be interior nodes of the haplotype tree (Crandall and Templeton 1993; Castelloe and Templeton 1994). A third difference between haplotype and species trees also arises from the fact that gene pools often contain multiple, identical copies of a given haplotype. Species trees are traditionally regarded as strictly bifurcating. In haplotype trees, each gene lineage of the identical copies of a single haplotype is at risk of independent mutation. Hence, a single ancestral haplotype will often give rise to multiple, descendant haplotypes, thereby yielding a haplotype tree with multifurcations. Finally, gene regions examined in populations can undergo recombination. Traditional methods assume recombination does not occur in the region under examination. Furthermore, recombination is an additional reason why the assumption of a strictly bifurcating tree topology is likely to be violated. Thus, a method for reconstructing genealogical relationships is needed that takes into account these population genetic phenomena.

Estimating intraspecific gene genealogies

Templeton *et al.* (1992) have developed such a method based on the probability of multiple mutations at a specific site difference between a pair of haplotypes. This method is compatible with either nucleotide sequence or restriction site data. The estimation procedure first checks for recombination using the algorithm of Hein (1990, 1993). The method next seeks to estimate a haplotype tree either for the entire DNA region or for subregions defined by recombination events. By calculating the probability of multiple mutations at a site difference, the method estimates haplotype relationships with accompanying confidence bounds on the pairwise connections. The method has its greatest statistical power when there are few differences and many similarities between a pair of haplotypes. Crandall (1994) has shown that the method outperforms maximum parsimony when few characters are available to differentiate haplotypes.

Figure 6.1 presents the results of applying this method to the human mitochondrial DNA (mtDNA) restriction site data of Excoffier and Langaney (1989). There is no recombination for mtDNA, nor were there any probable deviations from parsimony in this case (Templeton 1996). Nevertheless, there was much ambiguity in the haplotype tree because of multiple, equally parsimonious alternatives. Connections between haplotypes that have a unique parsimonious solution are indicated by full lines; whereas those with equally parsimonious alternatives are indicated by broken lines. The broken lines define many closed loops within the haplotype tree. Obviously, a true evolutionary tree cannot have loops. This figure actually represents a 95 per cent plausible set of haplotype trees, with the members of this set representing the approximately 2700 ways in which the loops can be broken. This is a more precise way of representing tree ambiguity than the more standard consensus tree representation because it identifies precisely which connections are ambiguous and all the probable alternatives.

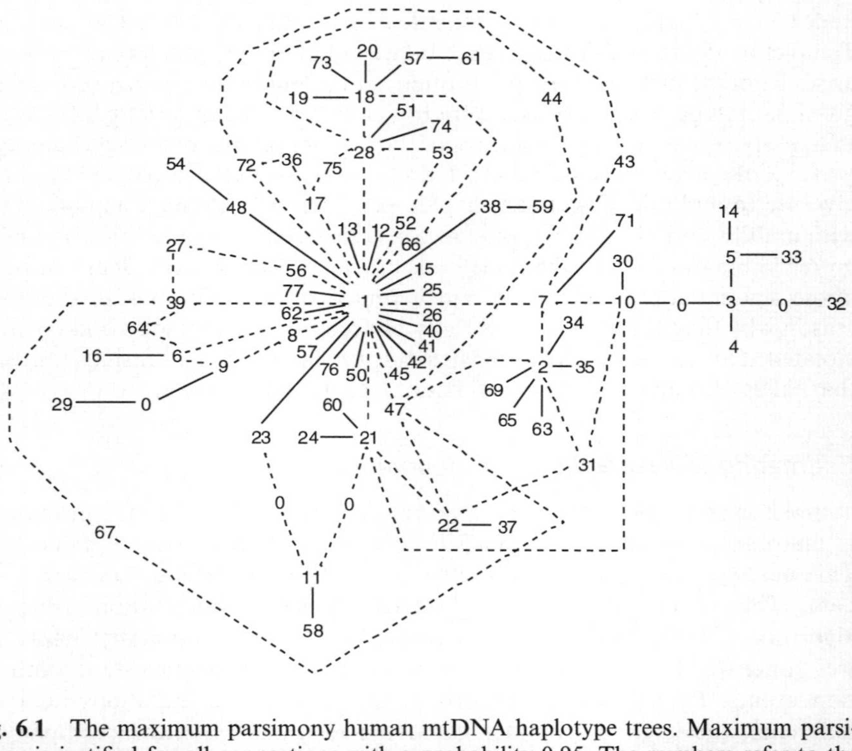

Fig. 6.1 The maximum parsimony human mtDNA haplotype trees. Maximum parsimony is justified for all connections with a probability 0.95. The numbers refer to the haplotype designations given in Excoffier and Langaney (1989). Zeros indicate an intermediate haplotype state under maximum parsimony that was not found in the sample. Full lines indicate a single mutational change that is unambiguous under the principle of maximum parsimony. Broken lines indicate single mutational changes that may or may not have occurred during the evolution of mtDNA, and thus indicate the positions of ambiguity under maximum parsimony.

Even though there are high levels of ambiguity in this plausible set of haplotype trees, this set of trees nevertheless is sufficiently resolved to illustrate many of the expected differences between haplotype trees and species trees. In Fig. 6.1., a '0' represents an interior node haplotype that was not actually present in the sample. These missing nodes may simply be absent because of insufficient sampling, or they may represent actual extinct ancestral haplotypes. Note also that multifurcations are common throughout this tree, and the single most common haplotype in the gene pool (haplotype 1) is at the centre of an extreme multifurcation. These topological properties, so different from species trees, are exactly what we expect from coalescent theory.

These tree connection probabilities calculated from the original Templeton *et al.* (1992) algorithm evaluate only the probability of parsimonious connections and the probabilities of deviations from parsimony by a specified number of additional steps. However, this algorithm cannot discriminate between equally parsimonious alternatives, such as those illustrated in Fig. 6.1 for human mtDNA. This failure to discriminate between equally parsimonious alternatives is also true for the estimation of species trees using cladistic principles. However, coalescent theory predicts a relationship between haplotype frequencies and haplotype tree topology. This means that information about the haplotype tree topology is contained in the haplotype frequency array. This basic prediction from coalescent theory was empirically confirmed by Crandall and Templeton (1993), who provide tables and equations for converting haplotype frequency data into probabilities of topological alternatives. Moreover, when dealing with restriction site data, the theory of restriction site evolution predicts some topological asymmetries (Templeton 1983) that can also be used as a source of information to refine haplotype tree estimation (Crandall *et al.* 1994). All of these represent sources of information that are available for haplotype tree estimation but that often are not applicable for species tree estimation. Figure 6.2 shows the impact of using these new sources of information on refining the estimate of the human mtDNA haplotype tree (from Templeton 1996). All broken connections that have a probability less than 0.05 were eliminated, broken connections having a probability greater than 0.95 were converted to full lines, and the remaining alternative connections were assigned a probability. As is evident from comparing Figs 6.1 and 6.2, this additional information had a great impact on the size of the 95 per cent plausible set of haplotype trees, reducing this set from about 2700 equally parsimonious alternatives to only 200 likely alternatives. Moreover, these remaining 200 alternatives are not equally likely (see Templeton (1996) for the actual probabilities for the alternative ways of breaking the loops rather than marginal probabilities assigned to a particular connection, as given in Fig. 6.2). As this example illustrates, tree estimation algorithms that do not use the topological information unique to intraspecific datasets are poor and inefficient estimators of haplotype trees regardless of their appropriateness for estimating species trees.

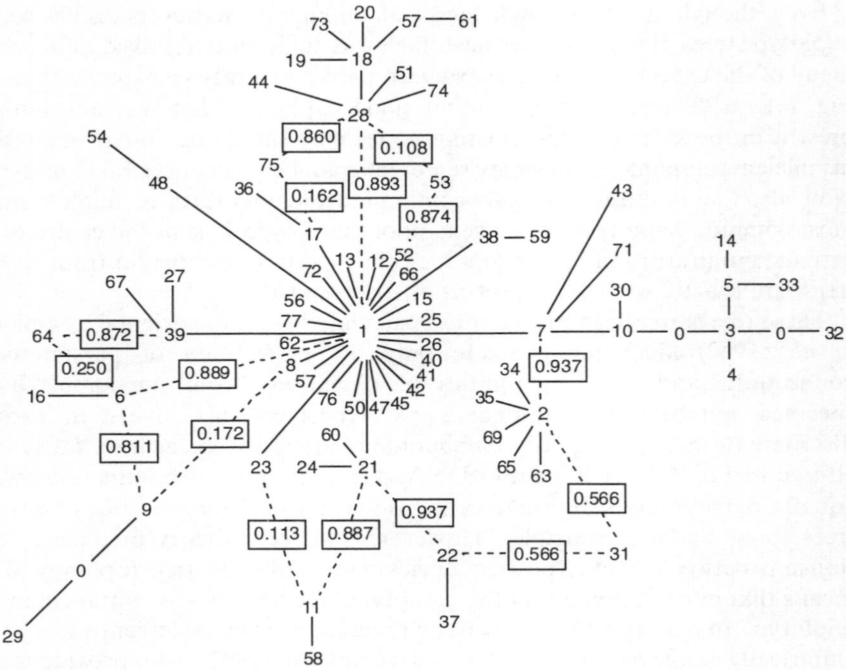

Fig. 6.2 The final 95 per cent plausible set of mtDNA cladograms. Full lines indicate mutational changes that have a probability ≥ 0.95 using information on the probability of parsimony, restriction site gain/loss asymmetries, and haplotype frequencies in a coalescent context. Broken lines indicate possible mutational changes that have a probability < 0.95, with the boxed number superimposed upon the broken line being the estimated probability that that particular mutational event actually occurred.

Often the question of interest in population biology depends on the directionality of mutational change or relative ages of haplotypes. Castelloe and Templeton (1994) have provided an approach for assigning root probabilities based on the cladogram estimation procedure of Templeton *et al.* (1992) and results from coalescent theory. For example, when this algorithm is applied to the human mtDNA data, the root probabilities (which explicitly incorporate tree uncertainty shown in Fig. 6.2) vary from 0.125 57 for haplotype 1 to 0.000 01 in a triple tie for haplotypes 60, 61, and 62. This ability to calculate root probabilities is particularly important, because the more traditional ways of rooting a tree, such as the outgroup method, often fail to statistically resolve the root of an intraspecific haplotype tree (for example the failure of chimp mtDNA resolving the root of the human mtDNA haplotype tree (Templeton 1992, 1993). Thus, a complete phylogenetic estimation framework has been developed to estimate within-species gene genealogies using this statistical parsimony/coalescent approach.

Hypothesis testing

Once the genealogical relationships between haplotypes have been estimated using the above statistical parsimony approach, the mutational connections underlying established relationships define a nested statistical design for testing hypotheses of associations between haplotypes and phenotypes (Templeton *et al.* 1987; Templeton 1993; Templeton and Sing 1993). The nesting procedure consists of nesting n-step clades within $(n+1)$ -step clades, where n refers to the number of transitional steps used to define the clade. By definition, each haplotype is a 0-step clade. The $(n+1)$ -step clades are formed by the union of all n-step clades that can be joined together by $n+1$ mutational steps. The nesting procedure begins with tip clades, i.e. those clades with a single mutational connection, and proceeds to interior clades. Nesting continues until all the data would become nested into a single clade at the $n+1$ level. The nesting procedure results in hierarchical nests with the nesting level directly correlated with evolutionary time, i.e. the lower the nesting level the more recent the evolutionary events relative to higher nesting levels.

The nesting design is then used to test for significant associations of haplotypes with phenotypes by either a nested analysis of variance (NANOVA) for continuous data (Templeton *et al.* 1987) or a permutation chi-squared contingency test for categorical data (Roff and Bentzen 1989; Templeton and Sing 1993). Because the nesting levels are direct correlates of evolutionary time, these nesting levels can also be used to statistically partition effects of current-day population structure from those of population history. There are three distinct advantages to this approach over those previously described: (1) a more accurate estimation of phylogenetic relationships for data with low levels of divergence (Crandall 1994); (2) a rigorous hypothesis testing framework which provides a quantitative partitioning of population phenomena across evolutionary time (Templeton 1993; Templeton and Sing 1993); and (3) this approach allows for uncertainty in the cladogram estimation, therefore, it does not rely on a single estimate of phylogenetic relationships, but is robust over a set of plausible alternative phylogenies (Templeton *et al.* 1992; Templeton and Sing 1993).

As an example of this nesting procedure, recall that 200 alternative haplotype trees reside in our final 95 per cent plausible set for human mtDNA (Fig. 6.2). Figure 6.3 shows the 1-step nesting categories that result when the nesting rules are applied only to the unambiguous connections in the 95 per cent plausible set. Note that even though 200 alternative trees exist in this set, only four haplotypes have ambiguous nesting relationships. Thus, there is little nesting ambiguity even when there are large amounts of tree ambiguity. The ambiguity that does remain can either be dealt with by additional nesting rules that incorporate the ambiguity (Templeton and Sing 1993) or by exhaustively analysing all alternative nesting categories (Templeton 1996). We will now demonstrate the value of this approach for addressing a variety of issues in population biology.

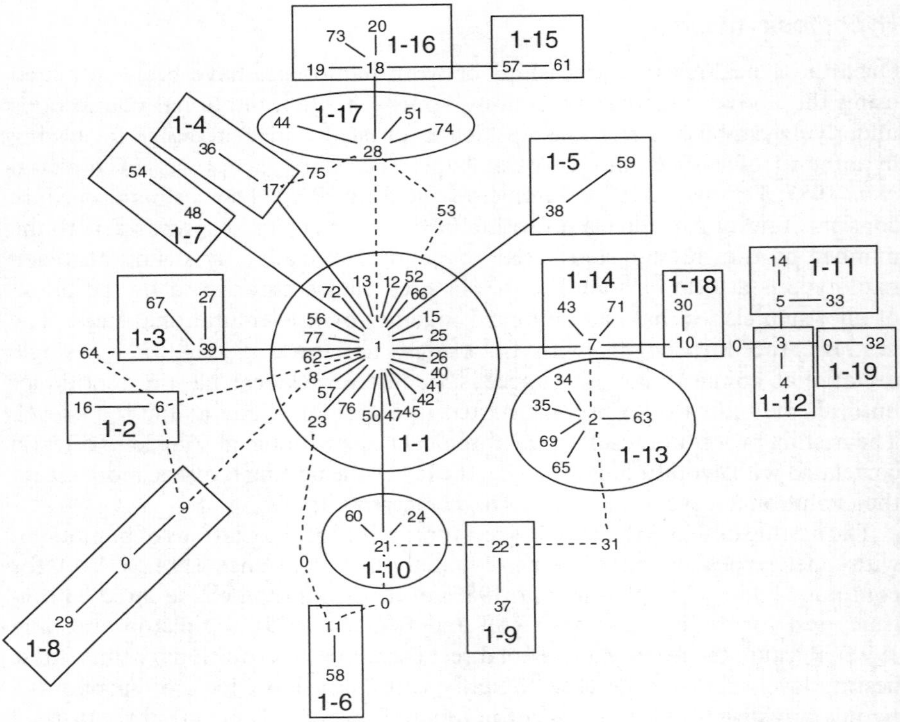

Fig. 6.3 The haplotypes nested into 1-step clades using the 95 per cent plausible set of haplotype trees given in Fig. 6.2. Each 1-step clade is indicated by an enclosed oval or box and designated by '1-#' where '1' refers to the clade level and '#' represents the number assigned to a particular 1-step clade.

6.3 Applications of the intraspecific comparative method

Using the cladogram estimation procedure and the resulting nested analysis, significant associations between phenotype and genotype can be detected. The detection of a significant association indicates either a correlated change in the underlying genotype with the phenotype of interest or, in the case of candidate genes, perhaps an underlying mutation responsible for the shift in phenotype. Thus, a phenotype can be used in the traditional, quantitative genetic sense, for example height. Alternatively, phenotypes can be geographic locations, species names, etc. This broad spectrum of possibilities has led to numerous, diverse applications of this method, from quantitative genetics to speciation theory.

Quantitative genetics and human epidemiology

The most extensive use of this cladistic analysis has been in the area of quantitative genetics relating to questions in human epidemiology, specifically with the candidate gene approach. The aetiology of complex disease involves the identification of underlying genetic architecture responsible for associated phenotypes. The genetic architecture of a quantitative trait refers to the number of genes involved in the manifestation of a given phenotype, the arrangement of these alleles into genotypes, and the impact of alleles and genotypes on the trait of interest and other traits (Boerwinkle *et al.* 1986; Sing *et al.* 1988). The candidate gene approach identifies a gene region of functional significance (via knowledge of the biochemical pathway or from marker locus studies) in the manifestation of a phenotype of interest. The candidate gene can then be surveyed for genetic variation using restriction site or nucleotide sequence data. Using the cladistic approach, a gene tree is estimated from these data and associations between the phenotype of interest and the underlying genetic causes are identified using the nested analysis. The fundamental premise of this approach is that any phenotypically important mutations at the candidate gene occurred at some point during the evolutionary process that created the current array of genetic variation. By reconstructing the evolutionary history of this genetic variation, the associated nested statistical design can utilize the evolutionary history of the candidate gene to focus statistical power (Templeton *et al.* 1987, 1988). This is accomplished by using the nested design to pool individuals by genotype rather than by phenotype, allowing an unbiased (relative to the phenotype) increase in sample size for statistical testing. The use of the cladistic analysis of candidate genes in quantitative genetic studies has been recently reviewed by Weiss (1993), Routman and Cheverud (1994), and Crandall (1996*a*). The implications of these studies for evolutionary biology have been explored by Cheverud and Routman (1993).

This approach has been used extensively to identify DNA regions responsible for variation in the risk of coronary artery disease (Sing *et al.* 1992*a, b*). For example, Haviland *et al.* (1995) explored genetic variation at the apolipoprotein AI-CIII-AIV gene cluster and its effects on plasma lipid, lipoprotein, and apolipoprotein levels. Their analysis indicated three haplotypes associated with significant differences in adjusted plasma lipid, lipoprotein, and apolipoprotein levels. They also identified gender-specific pleiotropic effects on other measures of lipid metabolism (for example, adjusted levels of total cholesterol) (Haviland *et al.* 1995).

The application of the cladistic approach in quantitative genetics has recently been expanded in two ways. First, Templeton (1995) has extended the method to include the analysis of case/control data, a common sampling design in genetic/disease association studies. He demonstrates the method by showing associations between the candidate locus for apoproteins E, CI, and CII and the phenotype of sporadic early and late-onset forms of Alzheimer's disease. Second, Hallman *et al.* (1994) have developed a likelihood-based approach to the cladistic analysis for employing family data in the test for association between candidate genes and quantitative phenotypes. This method is demonstrated by identifying effects of the apolipoprotein B (Apo B) gene on total-lipoprotein, low-density lipoprotein, and high-density-lipoprotein (HDL)–cholesterol, triglyceride, and Apo B levels using haplotypes from 121 French nuclear families. They concluded that 10 per cent of the genetic variance and 5 per cent of the total variance in the HDL–cholesterol and triglyceride levels was associated with haplotype effects at the Apo B locus (Hallman *et al.* 1994).

Population genetics

A principal aim of population genetics is to identify and quantify population substructure and levels of gene flow among subdivided populations. The traditional approach to quantifying population substructure is with the use of Wright's (1969) *F*-statistics or some variant of them (for example Lynch and Crease 1990). These analyses typically assume a model of gene flow (for example one-dimensional stepping stone or two-dimensional isolation by distance) to calculate the *F*-statistics then use the result to estimate various population parameters (for example inbreeding effective size, mutation rate, gene flow). Ideally, one would want to test the assumed model of population structure used to estimate these population level parameters. Templeton and colleagues (Templeton 1993; Templeton *et al.* 1995) utilized the cladistic approach to partition population historical factors from population structure.

To achieve the partitioning of population history from population structure, two measures of geographic distance are used in the context of the nested, cladistic design. The first distance is the geographic centre of all individuals bearing haplotypes that fall within a given clade. This distance is the clade distance and measures how geographically widespread the individuals are from that clade. The second distance is the nested clade distance which measures the distance of clades from the geographic centre of their nesting category. Thus, the nested clade distance is a measure of how far individuals bearing evolutionarily closely related clades are located from one another. These distances are then used to test alternative hypotheses of patterns of gene flow and population structure based on predictions from computer simulation models and coalescent theory (Takahata 1988; Neigel *et al.* 1991; Crandall and Templeton 1993; Nath and Griffiths 1993; Neigel and Avise 1993; Slatkin 1993; Castelloe and Templeton 1994). The resulting predictions expected under three distinct patterns of population subdivision are given in Table 6.1. When this type of analysis is applied to human mtDNA data, all significant rejections of the null hypothesis of no haplotype tree/geographic association at the intercontinental level fit the pattern of restricted gene flow (Templeton 1993), and all these inferences were completely robust to the uncertainty in the estimated haplotype tree shown in Fig. 6.2 (Templeton 1996). Within continents, most associations also reflected restricted gene flow, but two indicated population range expansion: one within Europe and the other within Africa (Templeton 1993), although the intra-Africa range expansion was *not* robust to haplotype tree uncertainty (Templeton 1996). These results clearly indicate that this methodology does not regard restricted gene flow and historical events as mutually exclusive alternatives, but rather identifies the mixture of restricted gene flow patterns and historical events that have interacted to yield the current spatial distributions of haplotype variation.

Table 6.1 Expected patterns under the different models of population structure and historical events (modified from Templeton *et al.* 1995)

Pattern I: **Restricted gene flow**

(a) Tip clades narrowly geographically distributed. Some interior clades broadly distributed.

(b) Average clade distance of interior clades—average clade distance of tip clades is significantly large.

(c) Average clade distances should increase with increasing clade level in a nested series of clades.

(d) The same patterns hold for nested clade distances unless some gene flow is due to long-distance dispersal events, then significant reversals of the above patterns can occur with the nested clade distances.

Pattern II: **Range expansion**

(a) Significantly large clade distances and nested clade distances for tip clades, and sometimes significantly small for interior clades under contiguous range expansion, but some tip clades should show significantly small clade distances under long-distance colonization.

(b) Interior clade distance—tip clade distance significantly small for contiguous range expansion; interior nested clade distance—tip nested clade distance significantly small for long-distance colonization.

(c) The above patterns are not recurrent in the cladogram or are geographically congruent

Pattern III: **Allopatric fragmentation**

(a) Significantly small clade distances at higher clade levels.

(b) The pattern of distances in (a) should represent a break or reversal of the distance pattern established by the lower level nested clades.

(c) Clades showing patterns (a and b) should be connected to the remainder of the cladogram by a larger than average number of mutational steps.

(d) The above patterns are not recurrent in the cladogram or are geographically congruent.

This approach is complementary to other cladistic approaches to the study of gene flow (for example Slatkin 1989; Slatkin and Maddison 1989, 1990; Hudson *et al.* 1992) because it is designed to identify and localize the effects of restricted gene flow and historical events on geographic associations of haplotypes. The other approaches assume that geographic associations are due to restricted gene flow and attempt to measure the amount of gene flow given this assumption. Thus, the method presented in Templeton *et al.* (1995) offers a test of the basic assumptions of the gene flow estimation procedures based on intraspecific phylogenies. For example, when this method was applied to the mtDNA haplotype tree for tiger salamanders (*Ambystoma tigrinum*), a fragmentation event was inferred between eastern and western subspecies (probably due to Pleistocene glaciation) followed by range expansions into formerly glaciated areas overlaid with isolation by distance within each subspecies (Templeton 1994; Templeton *et al.* 1995). Hence, it would only be appropriate to use the cladistic gene flow estimation procedures within the more southerly populations within each subspecies. If these gene flow estimation procedures were applied to populations living in formerly glaciated areas or across subspecies, they would yield biologically meaningless estimators because the spatial patterns are primarily due to historical events in those samples and not to restricted gene flow.

In addition to testing hypotheses of population substructure, the cladistic analysis has also been used to detect the transmission of viral sequences (Crandall 1996*b*) and hybridization events (Matos and Schaal 1996). In testing either viral transmission or hybridization events, the null hypotheses of no transmission or no hybridization give the same expected relationships among haplotypes, i.e. the nesting of a single species (or host species) before that species is nested with an additional species (Fig. 6.4a). This hypothesis is tested against the alterative of the nesting of different species at lower clade levels before the entire species is nested together (Fig. 6.4b). In addition to testing this basic hypothesis, the nesting structure allows for an historical quantification of the number of relative amounts of transmission (or cross-species gene flow) based on the topological positioning of cross-species nests and nesting level. For example, in an analysis of cross-species transmission of primate T-cell leukaemia/lymphoma virus type I, Crandall (1996*b*) showed that multiple cross-species transmissions had occurred at low levels throughout the phylogenetic history of the viral sequences. Likewise, Matos and Schaal (1996) were able to identify the presence of historical hybridization events between two species of pine tree within the *Pinus montezumae* complex and reject the alternative hypothesis of association due to ancestral polymorphism using the nested analysis of haplotypes.

Conservation biology

While the analysis of population structure, levels of gene flow, and hybridization events are clearly important aspects of conservation biology, cladistic analysis has been expanded to have even greater utility in the area of conservation biology through the analysis of ecosystems or landscapes (Templeton and Georgiadis 1996). By utilizing the cladisitic analysis of population structure described above, it is possible to separate the roles of restricted gene flow, colonization, and population fragmentation in determining a species' population genetics pattern over evolutionary time. These results can be useful in the management decisions of single species (for example, Georgiadis *et al.* 1994) and, with information from additional species, can be used to infer landscape-level processes rather than mere patterns that influence the current distribution of genetic diversity across many species (Templeton and Georgiadis 1996).

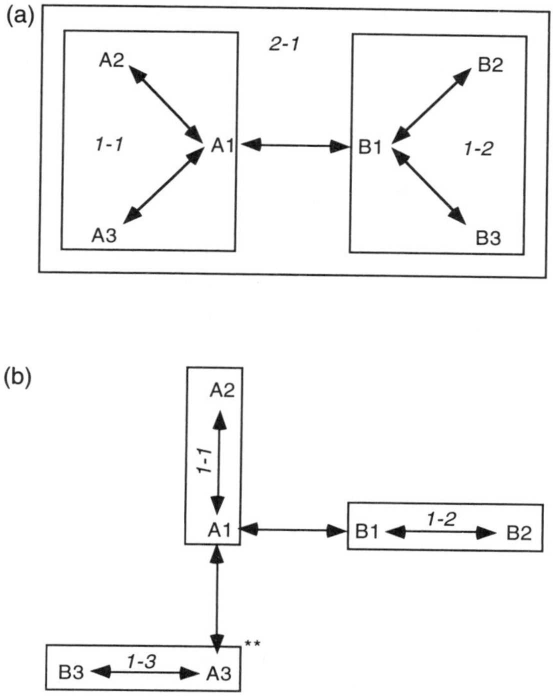

Fig. 6.4 (a) The biological null hypothesis of no transmission among species is accepted when all haplotypes from one species (A) are nested together before they are nested with haplotypes from another species (B). (b) The null hypothesis of no transimission is rejected when nesting of two species occurs before the nesting of all haplotypes from a single species (A3 and B3).

Speciation theory

Recently, Templeton (1994) has extended cladistic analysis upward to the micro/macroevolutionary interface in the definition of species and analysis of the speciation process. Templeton rephrased the cohesion concept of species into a series of testable hypotheses within the framework outlined above. The cohesion concept of species is 'the most inclusive population of individuals having the potential for phenotypic cohesion through intrinsic cohesion mechanisms' (Templeton 1989). These intrinsic cohesion mechanisms are classified broadly into two categories: *genetic exchangeability*—the factors that define the limits of spread of new genetic variants through gene flow with impact on the limits of action of drift and selection—and *demographic exchangeability*—organisms are ecologically or demographically constrained such that their descendants or genes can either replace (through drift) or displace (through selection) the descendants or genes of other individuals in the lineage, even if the lineage is not reproducing sexually. Fundamental to this species concept, and many others including the evolutionary species concept (Wiley 1981) and the phylogenetic species concept (Cracraft 1989), is the establishment of an independent evolutionary lineage. Under the cohesion concept, this hypothesis is the first null hypothesis to be statistically tested using the nested analysis framework. For example, in the analysis of human mtDNA, it was concluded that Africans, Europeans, and Asians have had recurrent but restricted gene flow throughout the entire time period marked by the coalescence of mtDNA (Templeton 1993, 1996). Hence, there is no evidence to reject the null hypothesis of a single evolutionary lineage for humans (Templeton 1994, 1996). By contrast, the analysis of tiger salamander mtDNA revealed a significant genetic fragmentation, so that the eastern and western populations do indeed constitute distinct evolutionary lineages. Thus, each subspecies is a potential cohesion species. The next step is to perform a nested analysis on data relating to genetic exchangeability and/or ecological interchangeability. Overlays of ecological parameters and traits upon the salamander haplotype tree revealed highly significant associations that were exactly concordant with the lineages defined by the geographic overlay analysis (Templeton 1994). This two-tiered system of hypothesis testing resulted in the inference that the eastern and western lineages of tiger salamanders do indeed constitute distinct cohesion species. This approach to species inference has the advantages that it makes the criteria and data used to identify a population as a species completely explicit and it quantifies rigorously the degree to which the data support the inference.

6.4 Summary and future directions

The statistical parsimony approach outlined above provides a statistical framework for researchers studying diverse population level phenomena. The main advantages to this approach are a robust cladogram estimation procedure and a robust framework for hypothesis testing given a set of plausible cladograms. The estimation procedure takes into account many of the difficulties associated with gene genealogies within populations. It then provides a set of plausible relationships upon which hypothesis testing can be achieved via a nested statistical analysis. This nested analysis detects significant associations between haplotypes and phenotypes and is powerful even when many alternative trees exist within the plausible set. This design allows for the identification of causal links between phenotype and genotype and the partitioning of historical population genetic structure from current patterns of gene flow. We have demonstrated the broad application of such information in the studies of quantitative genetics, population biology, conservation biology, and speciation theory.

Improvement of this method will be achieved by further incorporation of coalescent theory information into the cladogram estimation algorithm. Because many of these procedures rely on the predictions of coalescent theory, future work will focus on the robustness of these methods to violations of the assumptions of coalescent theory and of the procedures themselves. Part of such a study includes an analysis of statistical power and the statistical properties of both the tree estimation algorithm and the nested analyses. Finally, through applications of these methods to numerous datasets and through computer simulation, the robustness of the predictions of population structure (Table 6.1) will be more thoroughly examined. Such studies will also be informative for questions of speciation and gene genealogies at the tokogenetic/phylogenetic interface.

Acknowledgements

This work was supported by grants from the National Science Foundation (DEB-9303258) and the Alfred P. Sloan Foundation to K. A. C. and the National Institutes of Health (R01 GM31571) to A. R. T.

References

Avise, J. C. (1989). Gene trees and organismal histories: a phylogenetic approach to population biology. *Evolution*, **43**, 1192–208.

Avise, J. C. (1994). *Molecular markers, natural history and evolution*. Chapman and Hall, New York.

Avise, J. C., Arnold, J., Ball, R. M., Bermingham, E., Lamb, T., Neigel, J. E., *et al.* (1987). Intraspecific phylogeography: the mitochondrial DNA bridge between population genetics and systematics. *Ann. Rev. Ecol. Sys.*, **18**., 489–522

Boerwinkle, E., Chakraborty, R., and Sing, C. F. (1986). The use of measured genotype information in the analysis of quantitative phenotypes in man. *Ann. Hum. Gene.*, **50**, 181–94.

Castelloe, J. and Templeton, A. R. (1994). Root probabilities for intraspecific gene trees under neutral coalescent theory. *Mol. Phylogenet. Evol.*, **3**, 102–13.

Cheverud, J. M. and Routman, E. (1993). Quantitative trait loci: individual gene effects on quantitative characters. *J. Evol. Biol.*, **6**, 463–80.

Cracraft, J. (1989). Species as entities of biological theory. In *What the philosophy of biology is*, (ed. M. Ruse, pp. 31–52, Kluwer Dordrecht.

Crandall, K. A. (1994). Intraspecific cladogram estimation: accuracy at higher levels of divergence. *Syst. Biol.*, **43**, 222–35.

Crandall, K. A. (1995). Intraspecific phylogenetics: support for dental HIV transmission. *J. Virol.*, **69**, 2351–6.

Crandall, K. A. (1996*a*). Identifying links between genotype and phenotype using marker loci and candidate genes. In *The impact of plant molecular genetics*, (ed. B. W. S. Sobral, pp. 137–57. Birkhauser, Berlin.

Crandall, K. A. (1996*b*). Multiple interspecies transmissions of human and simian T-cell leukemia/lymphoma virus type I sequences. *Mol. Biol. Evol.*, **13**, 115–31.

Crandall, K. A. and Templeton, A. R. (1993). Empirical tests of some predictions from coalescent theory with applications to intraspecific phylogeny reconstruction. *Genetics*, **134**, 959–69.

Crandall, K. A., Templeton, A. R., and Sing, C. F. (1994). Intraspecific phylogenetics: problems and solutions. In *Models in phylogeny reconstruction*, (ed. R. W. Scotland, D. J. Siebert, and D. M. Williams), pp. 273–97. Clarendon, Oxford.

Donnelly, P. and Tavaré, S. (1986). The ages of alleles and a coalescent. *Adv. Appl. Prob.*, **18**, 1–19.

Ewens, W. J. (1990). Population genetics theory-the past and the future. In *Mathematical and statistical developments of evolutionary theory*, (ed. S. Lessard), pp. 177–227. Kluwerd New York.

Excoffier, L. and Langaney, A. (1989). Origin and differentiation of human mitochondrial DNA. *Am. J. Hum. Genet.*, **44**, 73–85.

Felsenstein, J. (1992*a*). Estimating effective population size from samples of sequences: a bootstrap Monte Carlo integration method. *Genet. Res., (Camb.)*, **60**, 209–20.

Felsenstein, J. (1992*b*). Estimating effective population size from samples of sequences: inefficiency of pairwise and segregating sites as compared to phylogenetic estimates. *Genet. Res. (Camb.)*, **59**, 139–147.

Fu, Y.-X. (1994*a*). A phylogenetic estimator of effective population size or mutation rate. *Genetics*, **136**, 685–92.

Fu, Y.-X. (1994*b*). Estimating effective population size or mutation rate using the frequencies of mutations of various classes in a sample of DNA sequences. *Genetics*, **138**, 1375–86.

Georgiadis, N., Bischof, L., Templeton, A., Patton, J., Karesh, W., and Western, D. (1994). Structure and history of African elephant populations: I. Eastern and Southern Africa. *J. Hered.*, **85**, 100–4.

Golding, B. (1993). Maximum-likelihood estimates of selection coefficients from DNA sequence data. *Evolution*, **47**, 1420–31.

Golding, B. (1994). Using maximum likelihood to infer selection from phylogenies. In *Non-neutral evolution: theories and molecular data*, (ed. B. Golding), pp. 126–39. Chapman and Hall, New York.

Golding, G. B. (1987). The detection of deleterious selection using ancestors inferred from a phylogenetic history. *Genet. Res. (Camb.)*, **49**, 71–82.

Golding, G. B. and Felsenstein, J. F. (1990). A maximum likelihood approach to the detection of selection from a phylogeny. *J. Mol. Evol.*, **31**, 511–23.

Hallman, D. M., Visvikis, S., Steinmetz, J., and Boerwinkle, E. (1994). The effect of variation in the apolipoprotein B gene on plasma lipid and apolipoprotein B levels. I. A likelihood-based approach to cladistic analysis. *Ann. Hum. Genet.*, **58**, 35–64.

Haviland, M. B., Kessling, A. M., Davignon, J., and Sing, C. F. (1995). Cladistic analysis of the apolipoprotein AI-CIII-AIV gene cluster using a healthy French Canadian sample. I. Haploid analysis. *Ann. Hum. Genet.*, **59**, 211–31.

Hein, J. (1990). Reconstructing evolution of sequences subject to recombination using parsimony. *Math. Biosci.*, **98**, 185–200.

Hein, J. (1993). A heuristic method to reconstruct the history of sequences subject to recombination. *J. Mol. Evol.*, **36**, 396–405.

Hey, J. (1992). Using phylogenetic trees to study speciation and extinction. *Evolution*, **46**, 627–40.

Hey, J. and Kliman, R. M. (1994). Genealogical portraits of speciation in the *Drosophila melanogaster* species complex. In *Non-neutral evolution: theories and molecular data*, (ed. B. Golding), pp. 208–16. Chapman and Hall, New York.

Hudson, R. R. (1990). Gene genealogies and the coalescent process. *Oxford Surv. Evol. Biol.*, **7**, 1–44.

Hudson, R. R. and Kaplan, N. L. (1988). The coalescent process in models with selection and recombination. *Genetics*, **120**, 831–40.

Hudson, R. R. and Kaplan, N. L. (1994). Gene trees with background selection. In *Non-neutral evolution: theories and molecular data*, (ed. B. Golding), pp. 140–53. Chapman and Hall, New York.

Hudson, R. R., Slatkin, M., and Maddison, W. P. (1992). Estimation of levels of gene flow from DNA sequence data. *Genetics*, **132**, 583–9.

Huelsenbeck, J. P. and Hillis, D. M. (1993). Success of phylogenetic methods in the four-taxon case. *Syst. Biol.*, **42**, 247–64.

Kirkpatrick, M. and Slatkin, M. (1993). Searching for evolutionary patterns in the shape of a phylogenetic tree. *Evolution*, **47**, 1171–81.

Lynch, M. and Crease, T. J. (1990). The analysis of population survey data on DNA sequence variation. *Mol. Biol. Evol.*, **7**, 377–94.

Matos, J. A. and Schaal, B. A. (1996). Chloroplast evolution in the *Pinus montezumae* complex: II. A coalescent approach to hybridization. *Evolution*. (In press.)

Nath, H. B. and Griffiths, R. C. (1993). The coalescent in two colonies with symmetric migration. *J. Math. Biol.*, **31**, 841–52.

Neigel, J. E. and Avise, J. C. (1993). Application of a random-walk model to geographic distributions of animal mitochondrial DNA variation. *Genetics*, **135**, 1209–20.

Neigel, J. E., Ball, R. M., and Avise, J. C. (1991). Estimation of single generation migration distances from geographic variation in animal mitochondrial DNA. *Evolution*, **45**, 423–32.

Roff, D. A. and Bentzen, P. (1989). The statistical analysis of mitochondrial DNA polymorphisms: chi-square and the problem of small samples. *Mol. Biol. Evol.*, **6**, 539–45.

Routman, E. and Cheverud, J. M. (1994). Individual genes underlying quantitative traits: molecular and analytical methods. In *Molecular ecology and evolution: approaches and applications*, (ed. B. Schierwater, B. Strait, G. P. Wagner, and R. DeSalle), pp. 593–606, Birkhauser, Basel.

Satta, Y. (1993). Balancing selection at *HLA* loci. In *Mechanisms of molecular evolution*, (ed. N. Takahata and A. G. Clark), pp. 129–50. Sinauer, Sunderland, MA.

Sing, C. F., Boerwinkle, E., Moll, P. P., and Templeton, A. R. (1988). Characterization of genes affecting quantitative traits in humans. In *Proceedings of the second international conference on quantitative genetics*, (ed. B. S. Weir, E. J. Eisen, M. M. Goodman, and G. Namkoong), pp. 250–69. Sinauer, Sunderland, MA.

Sing, C. F., Haviland, M. B., Templeton, A. R., Zerba, K. E., and Reilly, S. L. (1992*a*). Biological complexity and strategies for finding DNA variations responsible for inter-individual variation in risk of a common chronic disease, coronary artery disease. *Ann. Med.*, **24**, 539–47.

Sing, C. F., Haviland, M. B., Zerba, K. E., and Templeton, A. R. (1992*b*). Application of cladistics to the analysis of genotype–phenotype relationships. *Eur. J. Epidemiol.*, **8**, 3–9.

Slatkin, M. (1989). Detecting small amounts of gene flow from phylogenies of alleles. *Genetics*, **121**, 609–12.

Slatkin, M. (1993). Isolation by distance in equilibrium and nonequilibrium populations. *Evolution*, **47**, 264–79.

Slatkin, M. and Maddison, W. P. (1989). A cladistic measure of gene flow from the phylogenies of alleles. *Genetics*, **123**, 603–13.

Slatkin, M. and Maddison, W. P. (1990). Detecting isolation by distance using phylogenies of genes. *Genetics*, **126**, 249–60.

Takahata, N. (1988). The coalescent in two partially isolated diffusion populations. *Genet. Res. (Camb.)*, **52**, 213–22.

Takahata, N. (1993). Allelic genealogy and human evolution. *Mol. Bio. Evol.*, **10**, 2–22.

Takahata, N. and Nei, M. (1990). Allelic genealogy under overdominant and frequency-dependent selection and polymorphism of major histocompatibility complex loci. *Genetics*, **124**, 967–78.

Takahata, N., Satta, Y., and Klein, J. (1992). Polymorphism and balancing selection at major histocompatibility complex loci. *Genetics*, **130**, 925–38.

Tavaré, S. (1984). Line-of-descent and genealogical processes, and their applications in population genetics models. *Theor. Pop. Bio.*, **26**, 119–64.

Templeton, A. R. (1983). Convergent evolution and nonparametric inferences from restriction data and DNA sequences. In *Statistical analysis of DNA sequence data*, (ed. B. S. Weir), pp. 151–79. Dekker, New York.

Templeton, A. R. (1989). The meaning of species and speciation: a genetic perspective. In *Speciation and its consequences*, (ed. D. Otte and J. A. Endler), pp. 3–27. Sinauer, Sunderland, MA.

Templeton, A. R. (1992). Human origins and analysis of mitochondrial DNA sequences. *Science*, **255**, 737.

Templeton, A. R. (1993). The 'Eve' hypothesis: a genetic critique and reanalysis. *Am. Anthropol.*, **95**, 51–72.

Templeton, A. R. (1994). The role of molecular genetics in speciation studies. In *Molecular ecology and evolution: approaches and applications*, (ed. B. Schierwater, B. Streit, G. P. Wagner, and R. Desalle), pp. 455–77. Birkhauser, Basel.

Templeton, A. R. (1995). A cladistic analysis of phenotypic associations with haplotypes inferred from restriction endonuclease mapping or DNA sequencing. V. Analysis of case/control sampling designs: Alzheimer's disease and the apoprotein E locus. *Genetics*, **140**, 403–9.

Templeton, A. R. (1996). Testing the out-of-Africa replacement hypothesis with mitochondrial DNA data. In *Conceptual issues in modern human origins research*, (ed. G. A. Clark and C. Willermet), in press. de Gruyter, Chicago, IL.

Templeton, A. R. and Georgiadis, N. J. (1996). A landscape approach to conservation genetics: conserving evolutionary processes in African bovids. In *Conservation genetics: case histories from nature*, (ed. J. Avise and J. Hamrick), in press. Chapman and Hall, New York.

Templeton, A. R. and Sing, C. F. (1993). A cladistic analysis of phenotypic associations with haplotypes inferred from restriction endonuclease mapping. IV. Nested analyses with cladogram uncertainty and recombination. *Genetics*, **134**, 659–69.

Templeton, A. R., Boerwinkle, E., and Sing, C. F. (1987). A cladistic analysis of phenotypic associations with haplotypes inferred from restriction endonuclease mapping. I. Basic theory and an analysis of alcohol dehydrogenase activity in *Drosophila*. *Genetics*, **117**, 343–51.

Templeton, A. R., Sing, C. F., Kessling, A., and Humphries, S. (1988). A cladistic analysis of phenotypic associations with haplotypes inferred from restriction-endonuclease mapping. II. The analysis of natural populations. *Genetics*, **120**, 1145–54.

Templeton, A. R., Crandall, K. A., and Sing, C. F. (1992). A cladistic analysis of phenotypic associations with haplotypes inferred from restriction endonuclease mapping and DNA sequence data. III. Cladogram estimation. *Genetics*, **132**, 619–33.

Templeton, A. R., Routman, E., and Phillips, C. A. (1995). Separating population structure from population history: a cladistic analysis of geographical distribution of mitochondrial DNA haplotypes in the tiger salamander, *Ambystoma tigrinum*. *Genetics*, **140**, 767–82.

Watterson, G. A. (1984). Lines of descent and the coalescent. *Theor. Pop. Biol.*, **26**, 77–92.

Watterson, G. A. and Guess, H. A. (1977). Is the most frequent allele the oldest? *Theor. Pop. Biol.*, **11**, 141–60.
Weiss, K. M. (1993). *Genetic variation and human disease: principles and evolutionary approaches*. Wiley, New York.
Wiley, E. O. (1981). *Phylogenetics: the theory and practice of Phylogenetic systematics*. Wiley, New York.
Wright, S. (1969). *Evolution and the genetics of populations*. University of Chicago Press.

Inferences from trees

7

Inferring phylogenies from DNA sequence data: the effects of sampling

Sarah P. Otto, Michael P. Cummings, and John Wakeley

7.1 Introduction

The various chapters in this volume illustrate the utility of a phylogenetic approach to a wide variety of questions. For example, Nee *et al.* (Chapter 5) describe methods of determining whether a clade is species-rich and other methods of detecting patterns of population growth from a tree topology (see the application to hepatitis C virus by Holmes *et al.* (Chapter 11)). Other authors use phylogenetic trees to study biogeography, to examine the cross-species transmission of viruses, and to determine correlations between evolutionary events. We must remember, however, that inferences made from a tree about the evolution of a group depend on the accuracy of the tree. In this chapter we describe some of the difficulties involved in reconstructing phylogenetic trees. We focus on our recent work investigating the sampling properties of DNA sequences in phylogenetic analyses (Cummings *et al.* 1995).

Sampling sequence data

A phylogenetic tree describes the pattern of historical relationships among a group of individuals or species. As long as the genomes of interest have remained intact and unmixed, the same pattern of branching will recount the ancestry of every region of the genome. For asexual lineages or non-recombining genomes, every sample of the genome is expected to depict the same tree. Even with recombination within a species, different genes will lead to the same tree as long as the species are distantly related relative to the length of time that sites remain polymorphic for particular alleles (see Takahata 1990). Due to varying patterns of mutation and selection, however, different regions contain different records of this common phylogeny. For example, DNA sequences within the stems of stem-loop structures may be highly conserved, with low rates of molecular evolutionary change (Kumazawa and Nishida 1993). Sites in other regions, such as the displacement-loop of the mitochondrial genome, evolve remarkably rapidly. Even coding regions vary greatly in their rate of evolution, some exhibiting high (for example cytochrome oxidase 1) and others low (for example NADH6) rates of change (Gadaleta *et al.* 1989; Cummings *et al.* 1995).

Despite the widespread recognition that the underlying processes of molecular evolution vary across the genome, the extent to which phylogenetic inference depends on the particular sample of DNA sequenced has not been adequately examined. We undertook a systematic study of the sampling properties of DNA sequences drawn from a genome using the entire mitochondrial genomes of 10 vertebrate taxa as a database (Cummings *et al.* 1995). We evaluated whether genes are adequate samples of the genomes from which they are drawn and, more generally, what kinds of samples provide the most representative picture of the whole genome phylogeny. Three types of samples were drawn: genes, contiguous segments starting at a random position, and sequences composed of sites drawn at random from the genome. By comparing these different sampling regimes, we were able to quantify the negative impact on phylogenetic inference that comes from gathering sites in a single contiguous block rather than from dispersed positions within the genome. The number of sites within the samples was also varied to determine the sequence length required to get a good estimate of the mitochondrial genome tree. To ensure that the sampling properties were not dependent on a particular method, three methods of phylogenetic inference were used: maximum parsimony, neighbour joining, and maximum likelihood.

7.2 Methods

Data

Complete mitochondrial genomes of ten vertebrate species, including one bird, one amphibian, two fish, and six mammalian species were obtained from GenBank (see Table 7.1). Alignment of the genome was done by aligning each gene on the basis of amino acid sequence or secondary structure, as described in Cummings *et al.* (1995), and then assembling the genes to conform to the order found in the majority of the taxa. The entire genome was analysed except for the control region, for which homology of sites is unclear. The aligned genomes resulted in a data matrix of 16 075 sites (including gaps) for each of the ten species.

Phylogenies were reconstructed from each gene and from the entire genome using maximum parsimony, maximum likelihood, and neighbour-joining methods (see Swofford and Olsen (1990) for a description of these methods and Cummings *et al.* (1995) for further details). To estimate the amount of support given by a gene for the tree estimated from it, a bootstrap analysis was performed (Felsenstein 1985; Efron and Tibshirani 1993). 1024 resampled sequences of the same length as the original gene were formed by sampling sites with replacement. Bootstrap trees were then inferred from each of these resampled sequences, and these trees were used to construct a majority-rule consensus tree. The number of times that any particular clade appeared among these 1024 trees was the bootstrap support for that clade.

Table 7.1 Mitochondrial genomes from 10 vertebrate taxa

Species name	Common name	Accession number	Source
Bos taurus	Cow	V00654	Anderson *et al.* (1982)
Cyprinus carpio	Carp	X61010	Chang *et al.* (1994)
Gallus gallus	Chicken	X52392	Desjardins and Morais (1990)
Homo sapiens	Human	V00662	Anderson *et al.* (1981)
Crossostoma lacustre	Loach	M91245	Tzeng *et al.* (1992)
Mus musculus	Mouse	V00711	Bibb *et al.* (1981)
Rattus norvegicus	Rat	X14848	Gadaleta *et al.* (1989)
Phoca vitulina	Harbor seal	X63726	Arnason and Johnson (1992)
Balaenoptera physalus	Fin whale	X61145	Arnason *et al.* (1991)
Xenopus laevis	Frog	M10217, X01600 X01601, X02890	Roe *et al.* (1985)

7.3 Results

Genome tree

The phylogeny estimated from the entire dataset was the same for all three methods of phylogeny reconstruction (the 'whole genome tree' – Fig. 7.1). The bootstrap support for each clade was extremely high (>99.6 per cent) except for the whale–cow–seal–human clade which nevertheless had high bootstrap support (parsimony: 95.2 per cent, maximum likelihood: 99.6 per cent, and neighbour-joining: 87.9 per cent). Therefore, when using all the data available within the mitochondrial genome, a strongly supported phylogeny can be obtained.

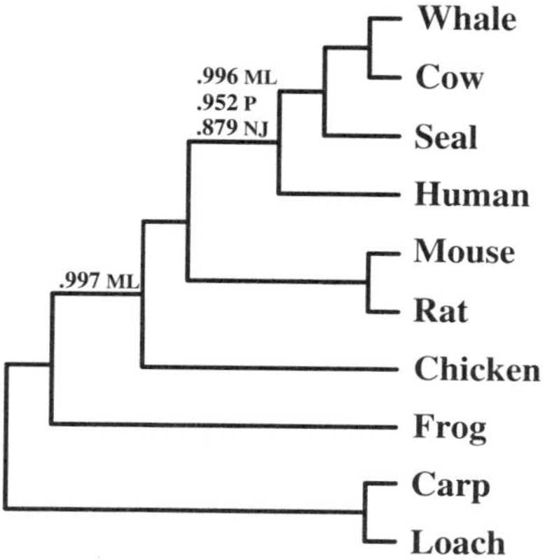

Fig. 7.1 The tree inferred from the entire mitochondrial genome from Cummings *et al.* (1995). Bootstrap values are given for each method of analysis; where no value is given all 1024 replicates supported that node.

This tree may or may not reflect the actual phylogeny of the group of species. However, the fact that the same tree was inferred from different methods of phylogenetic reconstruction each based on fairly different sets of assumptions lends support to the hypothesis that the tree in Fig. 7.1 reflects the actual ancestry of the group of species. Whether or not the the phylogeny shown in Fig. 7.1 is the actual tree for these organisms, it *is* the phylogeny for their mitochondrial genomes as inferred from the methods used. We therefore asked whether a subset of the entire dataset (for example, a gene) was sufficient to represent the entire genome and specifically whether it would lead to the whole genome tree.

Sampling genes from the genome

Different genes are more likely to provide the same phylogenetic information if mutational and evolutionary processes are uniform or homogeneous across the genome. However, base composition is extremely non-uniform across the mitochondrial genome (Churchill 1989, 1992). Similarly, the distribution of nucleotide positions that vary among the 10 species is also highly heterogeneous. The degree of heterogeneity across a sequence can be assessed using tests that are sensitive to departures from a random distribution of points along a line or a circle (Pearson 1963); results for the distribution of variable sites across the genome are highly significant (Kuiper's $V = 0.039$, $p < 10^{-11}$; Watson's $U^2 = 0.898$, $p < 10^{-8}$). Figure 7.2 depicts this heterogeneity by showing the number

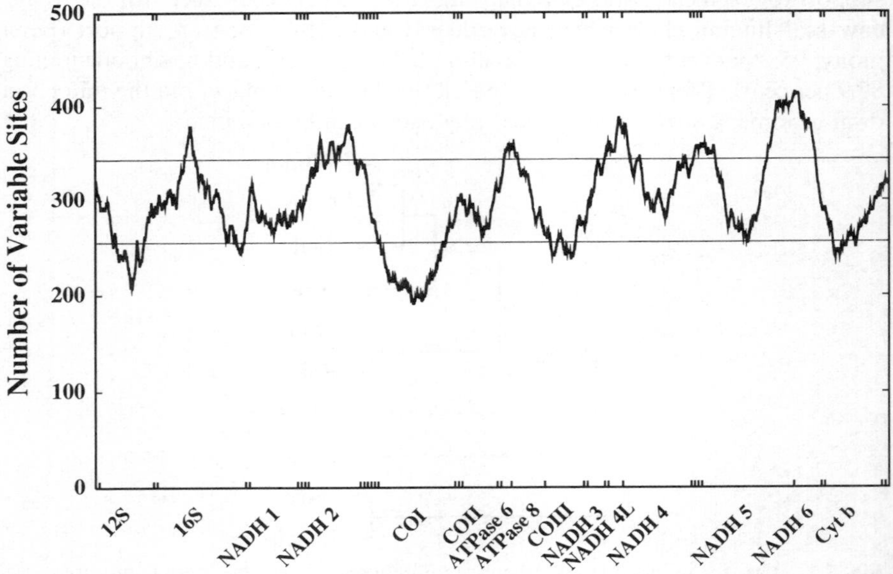

Fig. 7.2 The number of variable sites within a sliding window of 500 sites (step size = 1), adapted from Cummings *et al.* (1995). The full lines represent the range of values observed across 20 randomly shuffled genomes. Twenty shuffled genomes were created by random permutation of the aligned genomes. The sliding-window approach was used on each of these genomes and every value fell within the two lines.

of variable sites within a sliding window of 500 sites. The straight lines on the graph indicate the maximum peak and minimum trough observed over a total of 20 shuffled genomes. These 20 genomes were created by random permutation of the aligned genomes; each was then analysed using the same sliding-window approach. Clearly, some regions of the genome are much more variable and some much less variable than expected.

Given the extreme heterogeneity in base composition and variability across the mitochondrial genome, we tested whether there was also heterogeneity in the phylogenetic signal by comparing trees estimated from genes ('gene trees' – Table 7.2). The 13 protein-coding genes and 2 rRNAs gave a combined total of 12 different trees for parsimony and 9 different trees each for maximum likelihood and neighbour-joining analyses. The whole genome tree was inferred by maximum likelihood more often than the other methods, but even then only one-third of the genes gave the genome tree. Nearly every gene favoured a different tree, and no alternative tree was represented many times. Every clade that was not on the genome tree had bootstrap support under 95 per cent, with one exception. That exception was NADH4L, which placed *Xenopus* as a sister taxon to mammals with 95.7 per cent bootstrap support under parsimony. While the ordering of the mammalian taxa is under some debate (Penny and Hendy 1986; Adachi *et al.* 1993), it is accepted that birds are more closely related to mammals than amphibians, and we may regard the amphibian–mammal sister group relationship as an incorrect phylogenetic inference. This example serves as a reminder that all statistical tests are expected to give occasional errors.

Table 7.2 Genes for which the whole genome tree was inferred. '×' denotes the gene phylogenies with the same topology as the genome tree (see Cummings *et al.* (1995) for the alternative gene trees).

Gene	Length	Parsimony	Likelihood	Neighbour-Joining
NADH5	1860			×
16S rRNA	1786		×	×
COI	1560		×	
NADH4	1387	×		×
CYTB	1149		×	
12S rRNA	1111	×	×	
NADH2	1047			
NADH1	981			
COIII	785	×	×	
COII	705			
ATPase6	687			
NADH6	561			
NADH3	350			
NADH4L	297			
ATPase8	207			
Number of different trees		12	9	9
Trees identical to genome tree		3/15	5/15	3/15

If the bootstrap procedure were perfectly accurate then we would expect that 5 per cent of incorrect clades would have bootstrap values greater than 95 per cent, as in the case of NADH4L. In these cases, we would falsely reject the null hypothesis that the clade is incorrect (a type I error). In fact, recent analyses indicate that the bootstrap as applied to phylogenies is not perfectly accurate (Zharkikh and Li 1992; Felsenstein and Kishino 1993; Li and Zharkikh 1994). High bootstrap probabilities tend to be conservative (fewer than 5 per cent of incorrect clades have >95 per cent support), but even high values may not be conservative if the tree is not specified in advance. Figure 7.3 illustrates the distribution of bootstrap values under neighbour-joining using each clade on each gene tree as a data point (a similar figure for parsimony is given in Cummings *et al.* (1995)). The conservative nature of high bootstrap values is illustrated in Fig. 7.3, since the number of incorrect clades with high bootstrap proportions is lower than expected (if the test were accurate, we would expect to have observed the same number for each bootstrap proportion). However, clades not on the whole genome tree were frequently observed with bootstrap support within the range of 50–90 per cent. Therefore, evolutionary conclusions based on low bootstrap values must be suspect.

Sampling random sequences from the genome

Individual genes clearly are not sufficient to estimate the phylogeny of these taxa. By analysing sequences of increasing length we were able to show that many more sites than contained in a single gene were needed to infer, reliably, the whole genome tree (Fig. 7.4). For the same number of sites, sequences drawn from the genome in contiguous segments (squares) inferred the whole genome tree significantly less often than sequences drawn from randomly chosen sites from throughout the genome (triangles). Using maximum likelihood, 8000 sites (nearly half of the genome) were needed before contiguous sequences inferred the genome tree 95 per cent of the time. Only about 5500 sites were necessary, however, if the sites were drawn without replacement from throughout the genome. Even longer sequences were necessary using parsimony or neighbour-joining methods. Similar conclusions are reached if one measures the average distance between an inferred tree and the genome tree using a metric that measures the number of branches that need to be contracted and decontracted to make the two trees identical (described in Fig. 7.5; see Robinson and Foulds (1981) and Penny and Hendy (1985)). Trees inferred from longer sequences and sequences drawn from random positions are less distant, on average, from the whole genome tree (Fig. 7.6). Figure 7.6 also indicates that while we are unlikely to obtain the whole genome tree using short sequences, the tree obtained will, on average, be fairly similar to the genome tree.

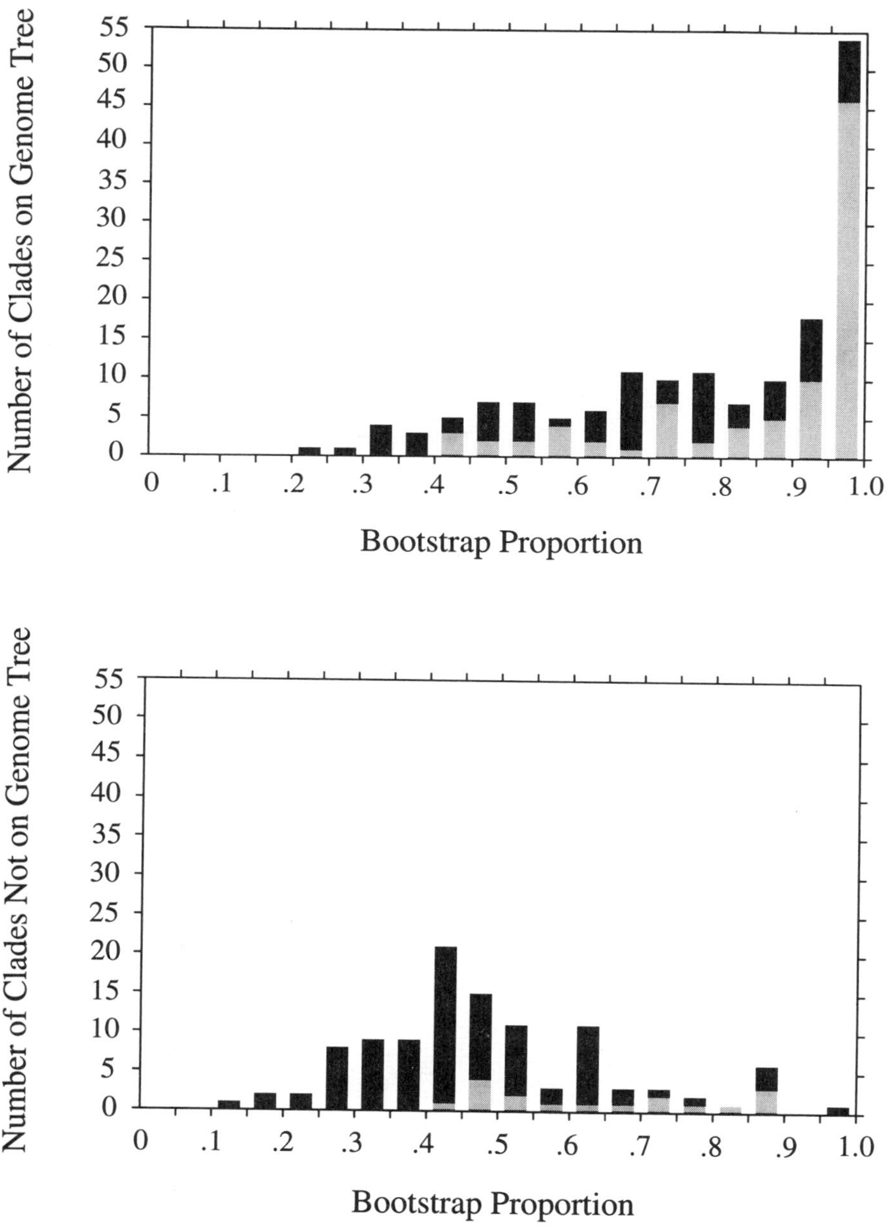

Fig. 7.3 Histograms of bootstrap proportions on clades contained within the whole genome tree and clades not on that tree. Each clade from each gene (rRNA and protein-coding genes—grey bars) and each tRNA (black bars) contributes one data point. The distribution shows the results under neighbour-joining; similar distributions are observed for parsimony and maximum likelihood methods.

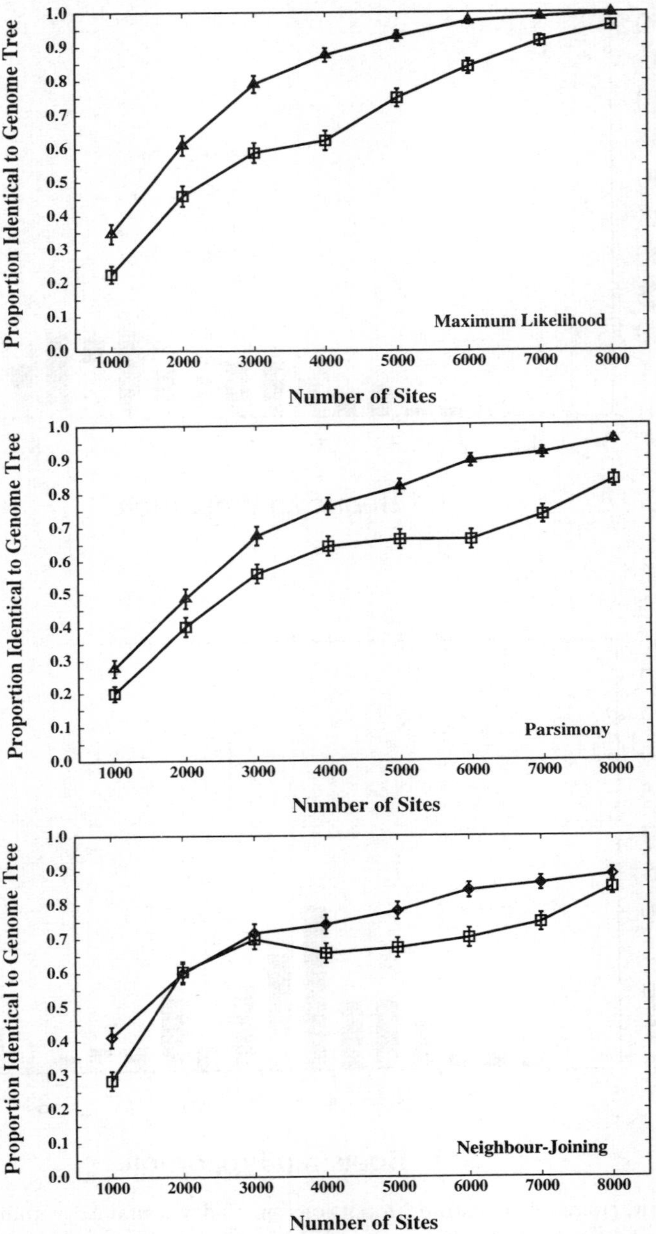

Fig. 7.4 Proportion of trees that are identical to the whole genome tree, from Cummings *et al.* (1995). 1024 sequences for each length were drawn either by choosing a contiguous block starting from a randomly chosen position (squares) or by drawing sites randomly without replacement from throughout the genome (triangles). Bars denote the 95 per cent confidence interval for the mean.

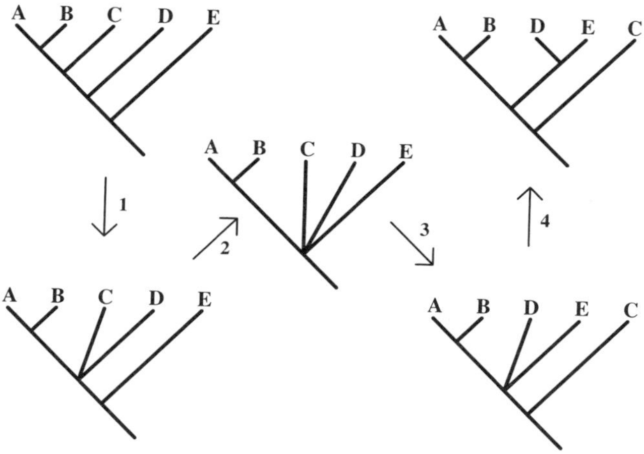

Fig. 7.5 An example demonstrating the contraction/decontraction metric of Robinson and Foulds (1981). The distance between the top-left tree and the top-right tree is 4, since two branch contractions and two branch decontractions are necessary to convert one tree topology into the other. The tree in the centre represents the most resolved tree that is consistent with both the top-left tree and the top-right tree. The maximum value of the metric is equal to 2 $(N - 3)$, where N is the number of taxa.

7.4 Conclusions

In the light of the above results, we must be fairly cautious when we draw conclusions that rest upon the accuracy of a particular tree topology. In the study of Cummings *et al.* (1995), analysis of individual genes infrequently gave the whole genome tree, and sequences much longer than contained in a gene (approximately half of the mitochondrial genome) were needed to obtain, reliably, a tree identical to the whole genome tree. This work strongly suggests that thousands, rather than hundreds, of base pairs may be necessary to draw accurate phylogenies and to draw appropriate conclusions from them.

The particular group of organisms included in our study was chosen simply on the basis of available mitochondrial genomes. As is typical, these taxa include some groupings that should be easy to resolve (placing the fish together and the rodents together) as well as other groups whose relationship is in question (the mammalian orders). In other groups, firm phylogenetic inferences may be possible with fewer data. However, if one works with groups that have rapidly expanded in number or if one chooses taxa to be distinct from one another, the tree topology will tend to be star-like with very short internal branches separating the clades (Slatkin and Hudson 1991). To resolve such trees, even more sites may be necessary than we needed in our study.

Low bootstrap values (50–90 per cent) were often observed supporting clades, both for clades found on and absent from the whole genome tree. Low bootstrap values need to be taken seriously. They mean that there is a good

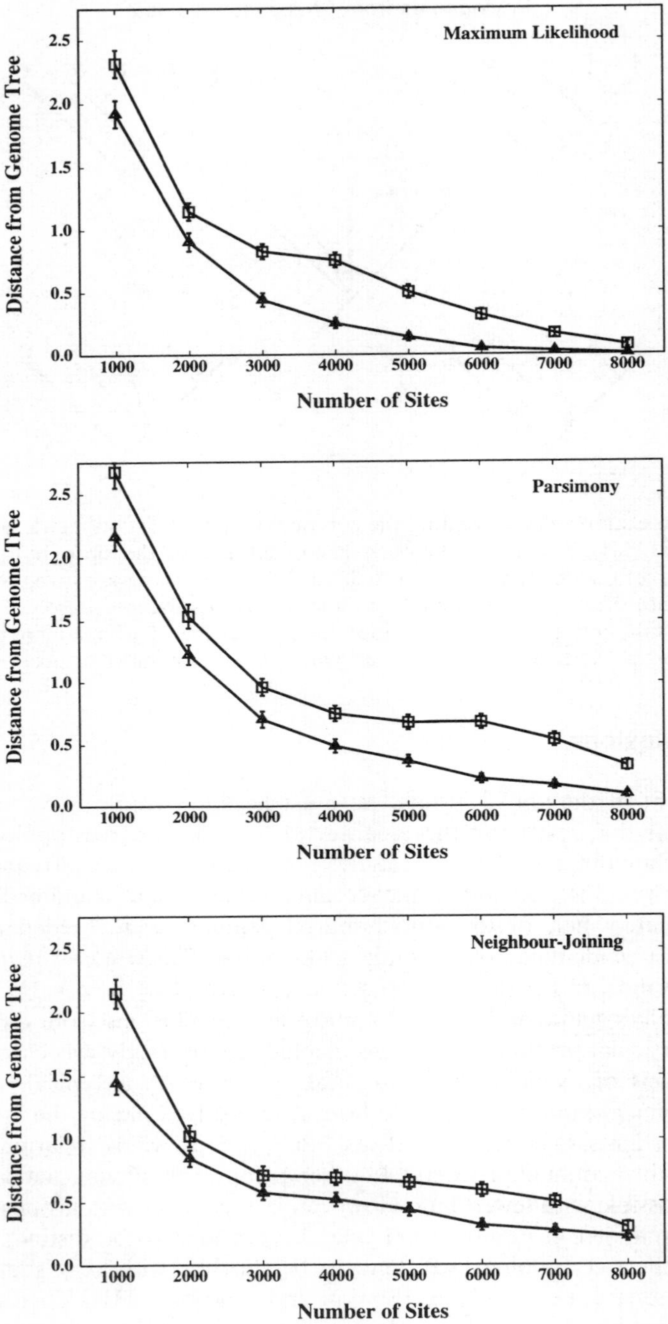

Fig. 7.6 The distance between a sample tree and the whole genome tree, from Cummings *et al.* (1995). The same samples as in Fig. 7.4 were compared with the whole genome tree using the contraction/decontraction metric (see Fig. 7.5). Symbols as in Fig. 7.4. The maximum value of the metric for these data is 14.

chance of getting a bootstrap value that high or higher *even if the clade is wrong*. What is worse is that low bootstrap values actually tend to underestimate this chance (Zharkikh and Li 1992). Furthermore, since bootstrap values do not estimate the probability that a clade is correct, they should not be used as relative measures of support for a clade. The null hypothesis that the clade is incorrect cannot be rejected whether the bootstrap value is 25, 50, or 75 per cent.

A further issue that must be remembered is that the bootstrap procedure assumes that every position evolves independently and that the distribution of sampled sites reflects the distribution of all the sites. This distribution need not be simple; some sites can evolve slower and others faster, some may tend to have a high purine content and others a low one, but the sampled sites must have the same probability of being in any particular class as the unsampled sites. Statistical tests indicate, however, that the four nucleotide types and variable sites are not independent and identically distributed across the genome (Cummings *et al.* 1995). The fact that regions of the genome differ in nucleotide distribution means that bootstrap resampling of a sequence from one region does not mimic sequencing new sites from other regions but only new sites that happen to follow the same distribution. A more problematic issue arises from sites that do not evolve independently. Lack of independent evolution is likely, for example, whenever secondary structure is critically dependent on weak bonds between nucleotides. In such cases, mutations at one site are generally followed by compensatory mutations at another site. As long as sites do not evolve independently, the assumptions of statistical tests will be violated. Studies investigating the sensitivity of the bootstrap to lack of independence among sites are needed.

Nevertheless, there are some routes that can be taken to make the most of a limited dataset. One can determine the sensitivity to resampling of a particular evolutionary or ecological conclusion by testing the hypothesis in each bootstrap sample. For example, one might test whether a tree topology is consistent with an exponentially increasing number of lineages on each bootstrap tree (see Nee *et al.*, Chapter 5, and Holmes *et al.*, Chapter 11, in this volume). Maximum likelihood methods may also be used to assess the strength of a conclusion across alternative trees (for example Pagel 1994).

Alternatively, one can attempt to increase the accuracy of a phylogeny through appropriate data sampling. Regions of the genome can be chosen that evolve at a rate that is high enough that the taxa will differ in sequence but not so high that the signal to noise ratio becomes unfavourable (Graybeal 1994). Furthermore, one can draw sequences from dispersed sites across a genome to get a more representative picture of the whole genome phylogeny (Cummings *et al.* 1995). We found that a tree was much more likely to match the whole genome tree if sites were drawn from throughout the genome; with 1000 sites, for example, matching occurred 39–54 per cent more often. For the same total number of base pairs sequenced, one can gain a more accurate picture of genomic evolution by sequencing blocks of sites from throughout a genome.

Acknowledgements

The subject matter of this chapter comes almost entirely from Cummings *et al.* (1995). The authors wish to thank A. Graybeal and M. Whitlock for helpful comments on the manuscript. S.P.O. was supported by an NSERC grant to N. Barton, M.P.C. was supported by an Alfred P. Sloan Fellowship in Molecular Studies of Evolution, and J. W. was supported by NIH Post-Graduate Training Program in Genetics grant GM07127 at UC Berkeley. The work reviewed was supported by NIH grant GM40282 to M. Slatkin.

References

Adachi, J., Cao, Y., and Hasegawa, M. (1993). Tempo and mode of mitochondrial DNA evolution in vertebrates at the amino acid sequence level: rapid evolution in warm-blooded vertebrates. *J. Mol. Evol.*, **36**, 270–81.

Anderson, S., Bankier, A. T., Barrell, B. G., De Bruijn, M. H., Coulson, A. R., Drouin, J., *et al.* (1981). Sequence and organization of the human mitochondrial genome. *Nature*, **290**, 457–65.

Anderson, S., De Bruijn, M. H. L., Coulson, A. R., Eperon, I. C., Sanger, F., and Young, I. G. (1982). Complete sequence of bovine mitochondrial DNA, conserved features of the mammalian mitochondrial genome. *J. Mol. Biol.*, **156**, 683–717.

Arnason, U., Gullberg, A., and Widegren, B. (1991). The complete nucleotide sequence of the mitochondrial DNA of the fin whale, *Balaenoptera physalus*. *J. Mol. Evol.*, **33**, 556–68.

Arnason, U. and Johnson, E. (1992). The complete nucleotide sequence of the mitochondrial DNA of the harbor seal, *Phoca vitulina*. *J. Mol. Evol.*, **34**, 493–505.

Bibb, M. J., Van Etten, R. A., Wright, C. T., Walberg, M. W., and Clayton, D. A. (1981). Sequence and gene organization of mouse mitochondrial DNA. *Cell*, **26**, 167–80.

Chang, Y.-S., Huang, F.-L., and Lo, T.-B. (1994). The complete nucleotide sequence and gene organization of carp (*Cyprinus carpio*) mitochondrial genome. *J. Mol. Evol.*, **38**, 138–55.

Churchill, G. A. (1989). Stochastic models for heterogeneous DNA sequences. *Bull. Math. Biol.*, **51**, 79–94.

Churchill, G. A. (1992). Hidden Markov chains and the analysis of genome structure. *Comput. Chem.*, **16**, 107–15.

Cummings, M., Otto, S., and Wakeley, J. (1995). Sampling properties of DNA sequence data in phylogenetic analysis. *Mol. Biol. Evol.*, **12**, 814–22.

Desjardins, P. and Morais, R. (1990). Sequence and gene organization of the chicken mitochondrial genome, a novel gene order in higher vertebrates. *J. Mol. Biol.*, **121**, 599–634.

Efron, B. and Tibshirani, J. (1993). *An introduction to the bootstrap*. Chapman and Hall, New York.

Felsenstein, J. (1985). Confidence limits on phylogenies: an approach using the bootstrap. *Evolution*, **39**, 783–91.

Felsenstein, J. and Kishino, H. (1993). Is there something wrong with the bootstrap on phylogenies? A reply to Hillis and Bull. *Syst Biol.*, **42**, 193–9.

Gadaleta, G., Pepe, G., De Candia, G., Quagliariello, C., Sbisa, E., and Saccone, C.

(1989). The complete nucleotide sequence of the *Rattus norvegicus* mitochondrial genome: signals revealed by comparative analysis between vertebrates. *J. Mol. Evol.*, **28**, 497–516.

Graybeal, A. (1994). Evaluating the phylogenetic utility of genes: a search for genes informative about deep divergences among vertebrates. *Syst. Biol.*, **43**, 174–93.

Kumazawa, Y. and Nishida, M. (1993). Sequence evolution of mitochondrial tRNA genes and deep-branch animal phylogenetics. *J. Mol. Evol.*, **37**, 380–98.

Li, W.-H. and Zharkikh, A. (1994). What is the bootstrap technique? *Syst. Biol.*, **43**, 424–30.

Pagel, M. (1994). Detecting correlated evolution on phylogenies—a general method for the comparative analysis of discrete characters. *Proc. R. Soc.*, **B255**, 37–45.

Pearson, E. S. (1963). Comparison of tests for randomness of points on a line. *Biometrika*, **50**, 315–25.

Penny, D. and Hendy, M. D. (1985). The use of tree comparison metrics. *Syst. Zool.*, **34**, 75–82.

Penny, D. and Hendy, M. D. (1986). Estimating the reliability of evolutionary trees. *Mol. Biol. Evol.*, **3**, 403–17.

Robinson, D. F. and Foulds, L. (1981). Comparison of phylogenetic trees. *Math. Biosci.*, **53**, 131–47.

Roe, B. A., Ma, D.-P., Wilson, R. K., and Wong, J. F.-H. (1985). The complete nucleotide sequence of the *Xenopus laevis* mitochondrial genome. *J. Biol. Chem.*, **260**, 9759–74.

Slatkin, M. and Hudson, R. R. (1991). Pairwise comparisons of mitochondrial DNA sequences in stable and exponentially growing populations. *Genetics*, **129**, 555–62.

Swofford, D. L. and Olsen, G. J. (1990). Phylogenetic reconstruction. In *Molecular systematics*, (ed. D. M. Hillis and C. Moritz), pp. 411–501. Sinauer, Sunderland, MA.

Takahata, N. (1990). A simple genealogical structure of strongly balanced allelic lines and trans-species evolution of polymorphism. *Proc. Natl. Acad. Sci. USA*, **87**, 2419–23.

Tzeng, C.-S., Hui, C.-F., Shen, S.-C., and Huang, P. C. (1992). The complete nucleotide sequence of the *Crossostoma lacustre* mitochondrial genome: conservation and variations among vertebrates. *Nucl. Acids Res.*, **20**, 4853–8.

Zharkikh, A. and Li, W.-H. (1992). Statistical properties of bootstrap estimation of phylogenetic variability from nucleotide sequences. I. Four taxa with a molecular clock. *Mol. Biol. Evol.*, **9**, 1119–47.

8

Uses for evolutionary trees

Walter M. Fitch

The general impression of molecular evolution is often that one sequences a gene from a number of organisms and infers the evolutionary relationships of the organisms. Indeed, if the sequences turn out to be orthologous and the data robust, one will get a phylogeny (tree) depicting those historical relations. But what one really obtains is a gene tree (I will henceforth assume that the data are robust; that is another problem) and the biological messages implicit in that tree can be quite various. This chapter enumerates a number of those messages which you may have or may wish to look for.

8.1 Evolutionary trees and genetic processes

Species trees

Once a species divides into two non-interbreeding groups, the mutational process, nucleotide substitutions in a given gene, cause the genes of the two incipient species to diverge. This divergence following each speciation creates a record that permits one to recover the cladogenic (speciation) process so long as one uses only genes (orthologous genes!) reflective of that process. In the spirit of new uses for new phylogenies, this is the oldest use. The other uses below are newer uses, perhaps new to some readers.

Gene duplications

Perhaps the most important genetic process to evolution, after nucleotide substitution, is the duplication of a gene to give the gene two non-allelic copies. This permits the organism to modify one (or both) of the copies either to create a new function or differentiate two functions with greater specificity. While speciation allows an organism to adapt preferentially to each of two different niches, gene duplications allow an organism to adapt better to both of two different niches. Although the protease study of Kurosky *et al.* (1980) is a good example of studying such paralogous genes, perhaps the earliest example is that of Zuckerkandl and Pauling (1962) who estimated the times of gene duplication producing the beta and gamma lineages of the haemoglobins.

Partial internal duplications

Gene duplications arise by unequal crossings over outside the gene. If the crossing over occurs within the gene, it gives rise to an internal repeat. An early example of this is apolipoprotein A which has 14, 11-amino-acid repeats (Fitch 1977). These have the interesting property that, because the repeats are ordered, not all possible trees of the repeating units are acceptable in explaining the history of the repeats. This is to say that all partial duplications must give rise to tandem elements.

Recombination

I restrict the use of recombination to cross-over events to distinguish it from reassortment which involves the choice among alternatives whenever the cell has two (or more) different chromosomes that it can pass on to the progeny. Trees may sometimes clearly indicate the presence of recombination. For example, the clupeine tree (Fig. 8.1) has one branch of the tree with no change upon it, indicating that the molecule at its tip, clupeine Z, is an intermediate between clupeines YI and YII (Fitch 1971). The conclusion is verified by noting that all the differences between Z and YI are proximal to the differences between Z and YII. An even earlier example is that of haemoglobin Lepore which is a recombination of the human beta and delta hemoglobins (Baglioni 1962).

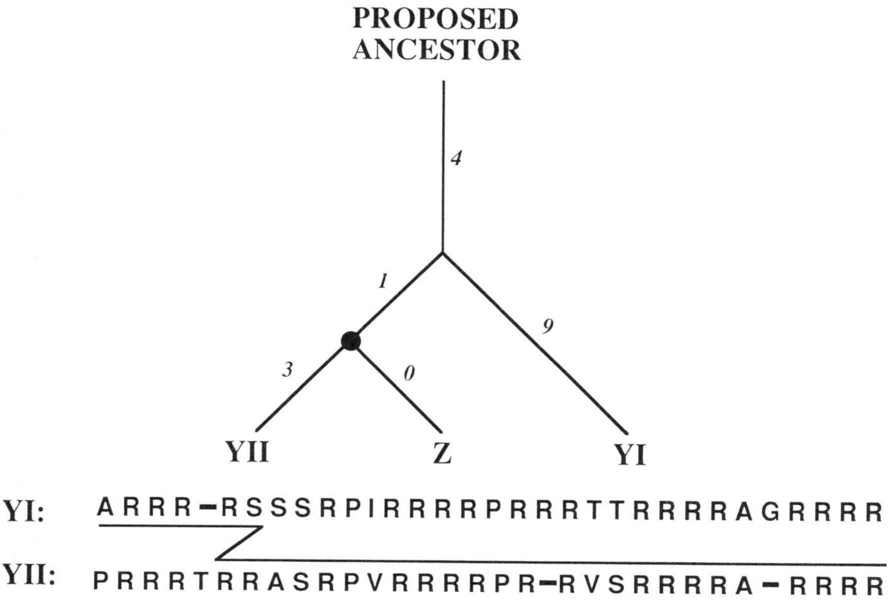

Fig. 8.1 Clupeine evolution by recombination. At the time this figure was drawn there were three known clupeine sequences, YI, YII, and Z. A tree based on differences is shown at the top. The sequences of YI and YII in the single letter code are shown at the bottom with the recombinant Z shown by the zig-zag line between the two. (From Fitch 1971.)

Reassortment

Reassortment can occur not only at meiosis but also in viruses whenever two or more viruses infect the same cell. This and the next three subsections deal with matters that arise from the problem that different data may give rise to trees that conflict with each other. Such conflicts most commonly arise because the data are not robust enough to resolve the conflict. There are occasions, however, when both trees are correct and reassortment is illustrated here by influenza type C where the tree for the *NS* (non-structural) gene was identical to that of the *HA* (haemagglutinin) gene provided one did not include the isolate from England, *Eng83*. (*Buonagurio et al.* 1986). *Eng83* clustered with *Cal78* or with *Ya81* depending upon whether one examined the *HA* gene or the *NS* gene but the most recent common ancestor of *Cal78* and *Ya81* is deep in the tree so that the disagreement is enormous. The resolution of the conflict (see Fig. 8.2) is simply that *Eng83* arose when recent ancestors of *Cal78* and *Ya81* coinfected a cell and produced progeny with some genes, from one lineage and some from the other.

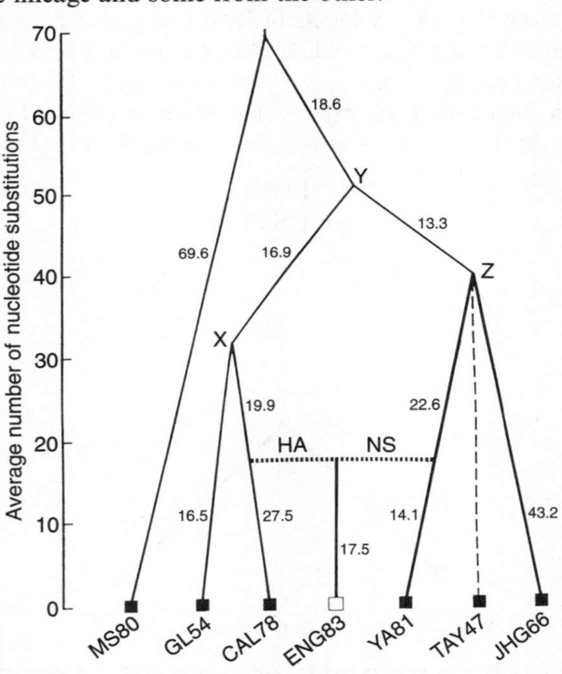

Fig. 8.2 Evolution of human influenza C genes, haemagglutinin (*HA*) and non-structural (*NS*). The trees for each gene alone are identical except for where the isolate *ENG83* attaches. It is shown here as a hybrid reassortant. (From Buonagurio *et al.* 1986.)

Gene conversion

Gene conversion is an old genetic concept but was first shown at the molecular level by Slightom *et al.* (1980) for a region of the human gamma haemoglobins which have two loci. I will illustrate the idea with an example from the beta region of two paralogous T-cell receptors in three species of wild mice (Rudikoff *et al.* 1992). If one examines the sequences other that those in exon 1, the tree (see top left side of Fig. 8.3) shows an ancient gene duplication, giving rise to the paralogues, followed by speciation events that are topologically identical for both paralogues and confirming the order of the speciation events. If, however, one examines the tree for exon 1 (see bottom left of Fig. 8.3), the tree drastically conflicts with the previous tree by having the two speciation events precede the gene duplications which now must occur three times, once in each of the three species. The conflict is resolved by explaining the exon 1 tree as the result of gene conversion (see right side of Fig. 8.3). It seemed remarkable that gene conversion would occur in what is apparently the same position in each of three different species.

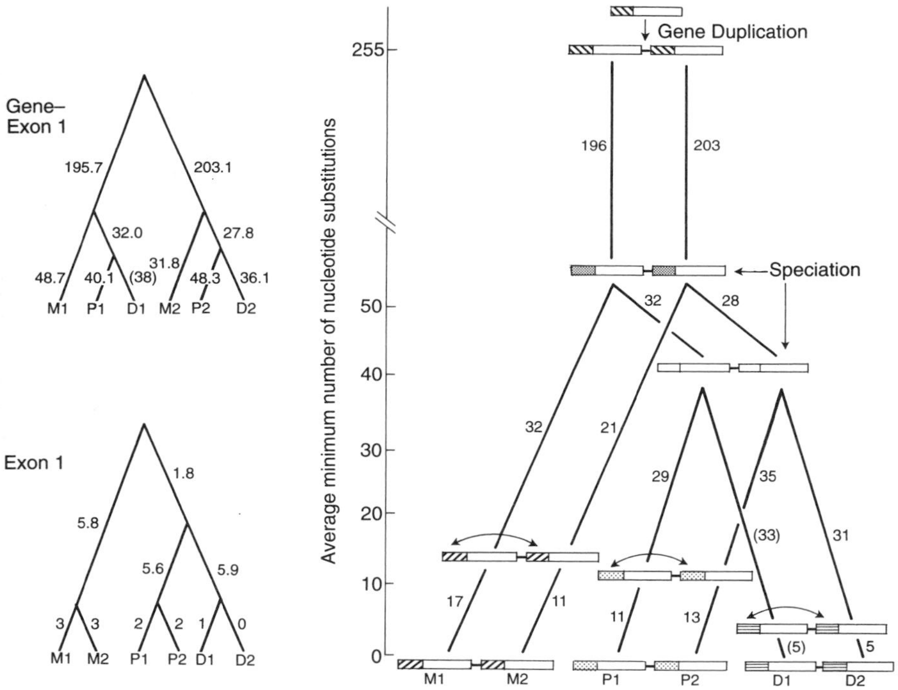

Fig. 8.3 Evolution of mouse T-cell receptor. On the left are two trees, the upper one being for all of the gene except exon 1, the lower tree for exon 1 only. They grossly disagree but the tree on the right that depicts the joint information and the resolution of the conflict by gene convergence in exon 1. The left-most part of each gene represents exon 1 and is shown in different shades to represent accumulated changes. The positions of those conversions are proportional to the substitutions from the tips. Numbers are the number of nucleotide substitutions. (From Rudikoff *et al.* 1992.)

Partial duplication with translocations

Kurosky *et al.* (1980) studied the evolution of proteases which may have a set of one or more motifs preceding the active site region. These motifs are called kringles in the case of plasminogen, which has five of them. The tree of the active site region (see Fig. 8.4) is largely congruent to that of the motifs but there is one gross exception. The first kringel of prothrombin appears to have been obtained as copy of a recent ancestor of the third kringel of plasminogen.

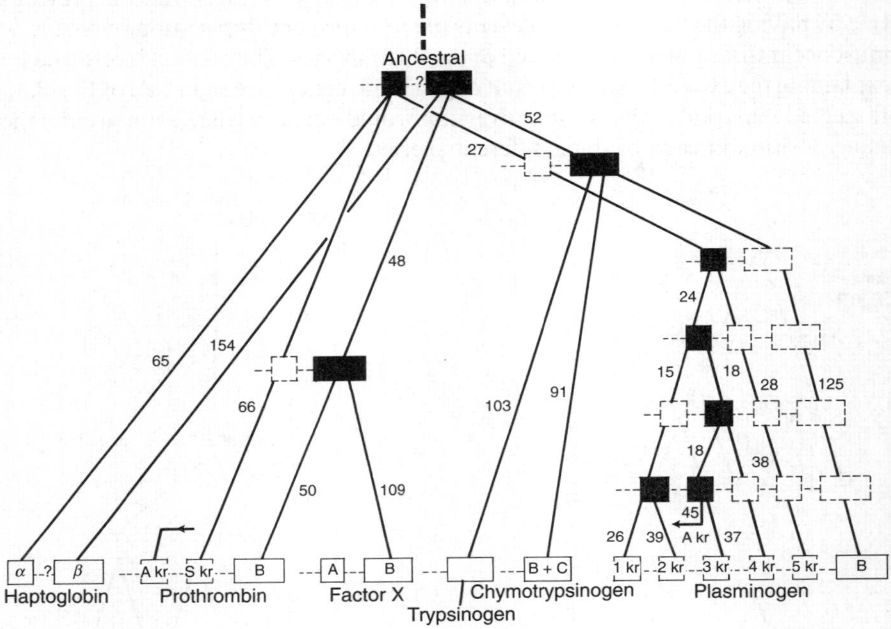

Fig. 8.4 Evolution of proteases. The tree depicts relationships for six different proteases. Open boxes are sequences known at the time, closed boxes are for sequences inferrable from parsimony procedures, dashed boxes are regions inferred to exist because they are present in various relatives. Dotted lines connect regions assumed to belong to a single gene. Numbers are the number of substitutions required on the tree. Note that the A kringle of prothrombin appears to descend via a copy from an ancestral form of kringle 3 of plasminogen. (From Kurosky *et al.* 1980.)

Xenology

This is the last of four ways in which trees apparently conflict, yet both trees are in fact correct. In this instance the gene tree (for phosphoaminotransferase, an anti-antibiotic) conflicts with the species tree because the gene conferring antibiotic resistance was obtained by way of a plasmid that permitted a trans-species movement of the gene (Gray and Fitch 1983). Symbiosis can lead to the same situation, mitochondria and chloroplasts being particularly good examples. The *NP* (nucleoprotein) gene of influenza A virus from human and birds has been isolated from swine and swine genes have been isolated from humans and birds, indicating substantial xenologous trafficking (Scholtissek *et al.* 1993*a*).

8.2 Evolutionary trees and rates and dates

Greatest rate

In Fig. 8.5 one can see a tree for human *HA* evolution with palaeontological dates (when the patient's throat was swabbed) that impel one to plot the substitutions accumulated since the root against the year of isolation (Fitch *et al.* 1991). The result is a remarkably high rate of 10^{-2} substitutions per site per year or more than a million times than seen in mammals.

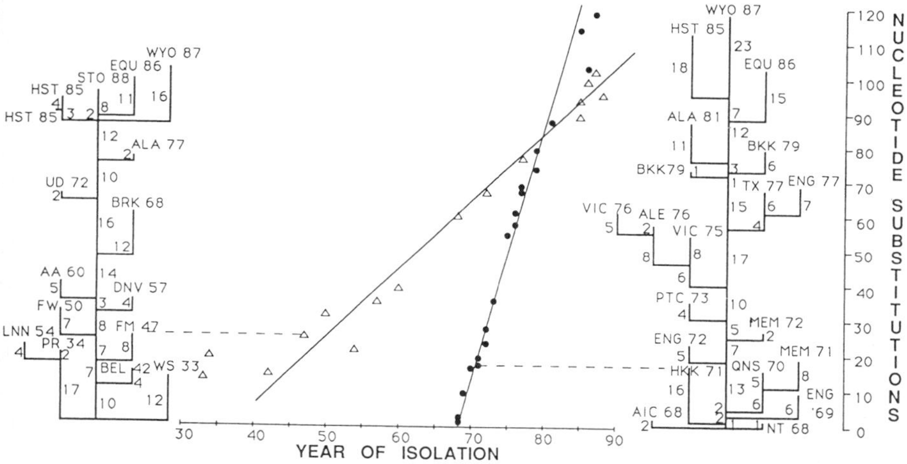

Fig. 8.5 Evolution of two human influenza genes. On the left of the figure is the tree for the non-structural gene and on the right is that for haemagglutinin. Numbers on tree are the number of substitutions required and the branches are drawn to scale. Isolate names indicate where they were isolated and the year of their isolation. When the distance from the root is plotted against the year of isolation, the result is the two lines shown whose slope is the rate of evolution, showing that the clock is quite accurate, especially for haemagglutinin. (From Fitch *et al.* 1991.)

Greatest change of rate

The *NP* gene has been isolated from many humans and birds (see Fig. 8.6). If one focuses on the evolution of the protein so that we are measuring only those nucleotide substitutions that produce amino acid replacements, one gets a very fast rate of evolution in the human lineage (10^{-3} replacement substitutions per year per site; Fig. 8.6) whereas the rate in birds is not significantly different from zero (Fig. 8.6) (Scholtissek *et al.* 1993*b*). The difference is not a property of the gene itself but of the host species as the bird gene, upon transfer into pigs (* in Fig. 8.6), shows an immediate increase in rate equal to that of the classical swine gene in pigs and of the human lineage shown in the figure.

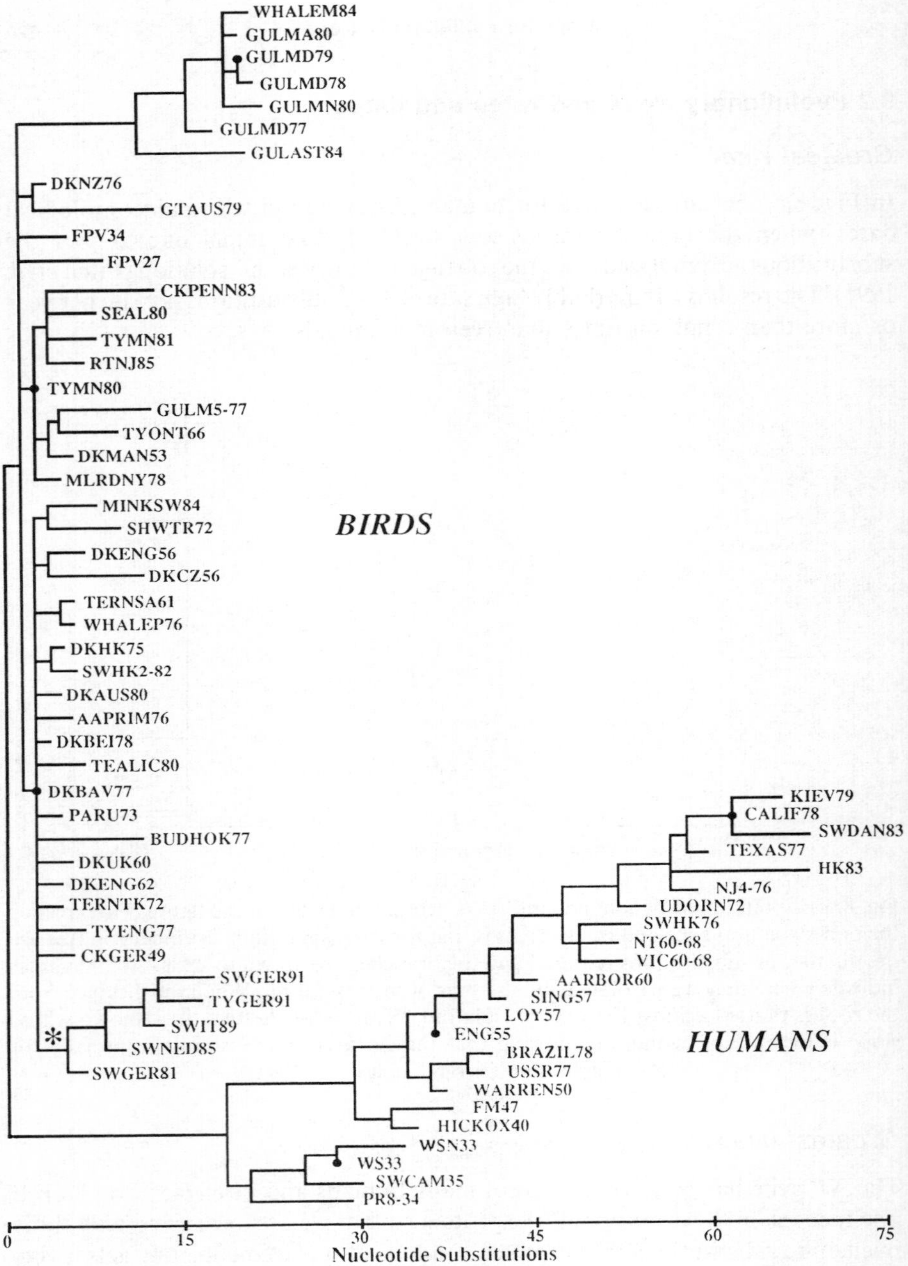

Fig. 8.6 Evolution of the *NP* gene of influenza A of birds and mammals. The tree is based on the amino acid sequences of this nucleoprotein with the amino acids back-translated into ambiguous nucleotides so that silent substitutions are not present. The sequences are clearly separated by the major forms of swine, and humans and birds, although any one form is occasionally found in other species. An especially notable case are the several swine species (SW. . .) descending from the asterisk which denotes the point at which the gene entered the swine population from the birds. (From Scholtissek *et al.* 1993*b*).

Covarions

If one wishes to know whether replacements are occurring randomly over the protein one can count how many times each amino acid site has had a replacement in its history and see if a Poisson distribution is obtained. When this is done, one finds an enormous non-randomness due largely, but not wholly, to an excess of unvaried positions (Fitch and Markowitz 1970). Their number is the sum of the sites that are invariable (cannot vary) and those that are variable but by chance did not vary (have not varied) and the best Poisson fit to the varied positions permits an estimate of the number of invariable sites. Such computations and their revelations cannot be done without an inferred tree. The presence of a sizeable number of invariable sites is worrisome because all standard methods of correcting for multiple substitutions at a single site assume there are no invariable sites and hence such correlations underestimate, sometimes grossly, the total change. Superoxide dismutase is thought to be a bad clock (Ayala 1986) but given an approximate number of COncomitantly VARIable codONS (covarions) and some turnover rates, it could in fact have a good clock that was not seen as such because the corrections for multiple replacements were very far off (Fitch and Ayala 1994).

Dating events

One often wishes to date past events. If molecules evolve at a constant rate, then the amount of change between two representatives of that molecule can be used to infer when the common ancestor of those molecules occurred. This process is rather old in molecular evolution (see Zuckerkandl and Pauling 1962), but a particularly good example of this occurs with the influenza A *NP* protein whose rate of evolution is demonstrably regular and for which the dates of isolates are accurate (Scholtissek *et al.* 1993*c*). Thus we may ask, 'When did the pig and human *NP* genes have their cenancestor (most recent common ancestor)?' Plotting nucleotide substitutions since the cenancestor versus year of isolation (see Fig. 8.7), the date of their cenancestor is where the best fitting line crosses the *x*-axis. These dates differ for the pig (1914) and human (1920) lineages but their average is remarkably close to 1918, the year of the great (Spanish) flu pandemic. Gorman *et al.* (1991) had earlier made an estimate of 1912 for that date.

Variation in rates

Actual estimation of rate variation among lineages has not been done often. Perhaps the first was by Langley and Fitch (1974) who examined seven proteins on a given tree for 17 mammalian taxa, fitting a maximum likelihood expression to the data where each protein was allowed its own rate but chosen to optimize expectations to observations. Time was measured in replacements as if the clock were operating (null hypothesis) The results, shown in Table 8.1, clearly reject the null hypothesis in two different ways. Do *total* rates (one lineage to another) remain

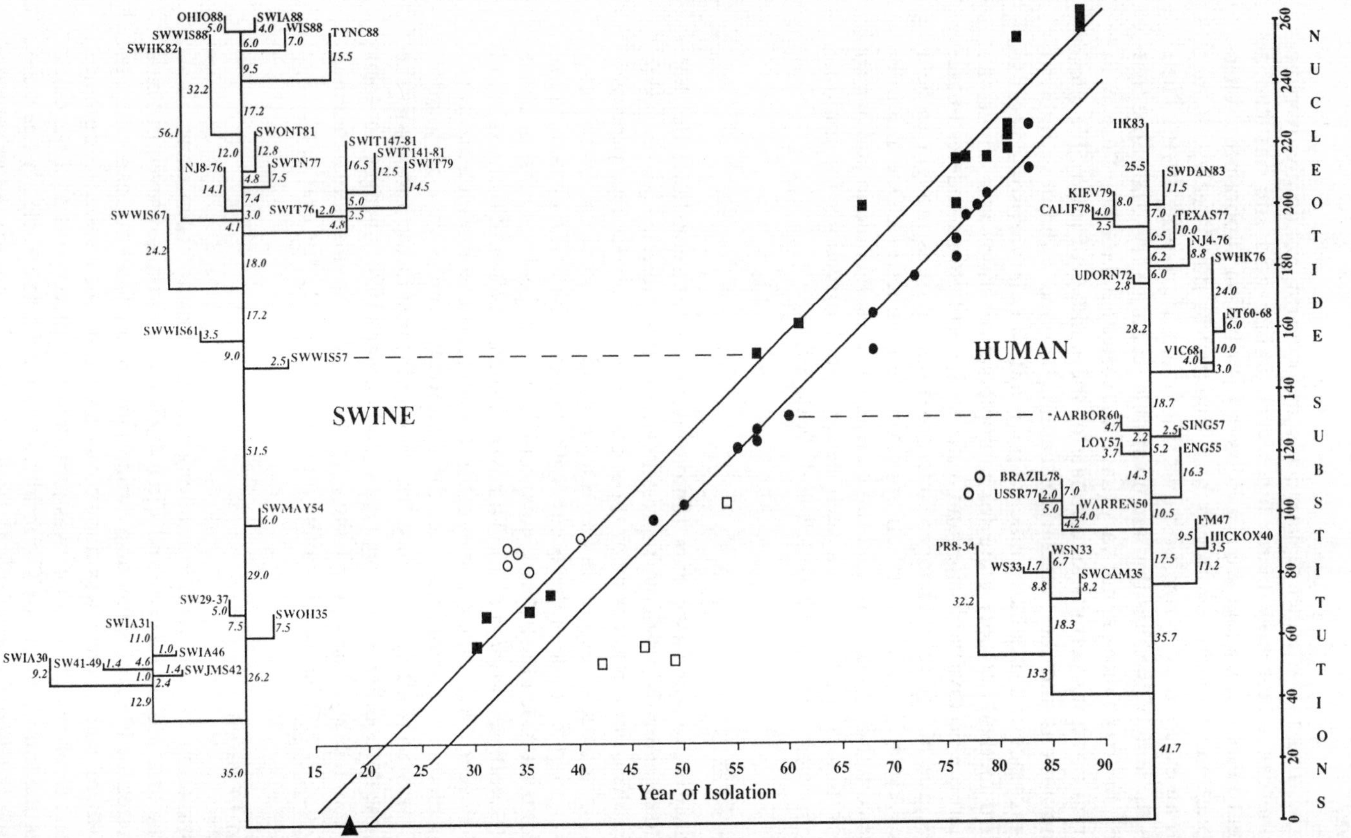

Fig. 8.7 Evolution of human and swine *NP* gene of influenza A. On the left-hand side is the tree for the swine and on the right that of humans. Symbols and plot as for Fig. 8.5. The major difference is that the bottom of the figure represents the common ancestor of the human and pig lineages. If they had their common origin during the great pandemic of 1918, the lines through the data points should intersect the line at the

constant regardless of rate deviations by individual proteins? No, with $p=0.0004$. Do *relative* rates (one protein to another) remain constant? No, with $p=0.006$. Thus, there is demonstrable rate heterogeneity. Nevertheless, averaged over all seven proteins, the amount of change relative to the palaeontological dates provided by Van Valen is remarkably linear (see Fig.6 of Langley and Fitch (1974)).

Table 8.1 Relative rates

	Corrected		
	χ^2	*d.f.*	*P*
Among proteins within legs (relative rates)	166.3	123	6×10^{-3}
Among legs over proteins (total rates)	82.4	31	4×10^{-6}
Total	248.7	154	6×10^{-6}

d.f. = degrees of probability; *P* = probability.

8.3 Evolutionary trees and theories

Neutrality

Kimura (1968) introduced the concept and maintained that the vast majority of amino acid replacements were neutral, that is, the fitnesses were sufficiently close to 1.0 that stochastic rather than selective processes dominated their fate. The study in the previous subsection was designed to test that hypothesis which, if correct, should find that chi-squared was approximately equal to the degrees of freedom. It is in fact about twice that expected so that neutralism could apply to, say, 95 per cent of the replacements only if the variance of the remaining 5 per cent were a huge 21 $(0.95\times1+0.05\times21=2)$.

Positive Darwinian evolution and convergence

This is hard to prove at the molecular level. Perhaps the first to do so was Stewart and Wilson (1987) who showed that the lysozymes of artiodactyls and of the primate langurs were more closely related than their non-ruminant ancestors. More statistically robust, Hughes and Nei (1989) demonstrated that at the mouse histocompatibility locus, the variable region was fixing replacements faster than the neutral substitution rate, which cannot happen except if there is positive Darwinian evolution. At the nucleotide level, but statistically weak, Smith *et al.* (1981) showed that two *E. coli* promoters were analogous only because the inactive homologue of one was identifiable.

A different method used data from the *HA* tree shown in Fig. 8.5. It was assumed that the trunk of the tree represented successful viruses while the branches, which went extinct, harboured the losers. If the difference between winners and losers were related to immune surveillance, one might expect that the successful viruses had more antigenic site replacements than the losers. The null hypothesis is that whether a replacement occurs in an antigenic site is independent of whether it occurred on the trunk or a branch. This gives rise to the 2×2 contingency matrix in Table 8.2 where the null hypothesis is rejected, implying that the successful virus is outrunning the immune system.

Table 8.2 Distribution of amino acid replacements in antigenic and non-antigenic sites as a function of their occurrence on the trunk or branches of the haemagglutinin tree

		Ag sites	Non-Ag sites	Total
Trunk	Obs.	33	10	43
	Exp.	25	18	
Branch	Obs.	31	35	66
	Exp.	39	27	
Total		64	45	109

$X^2{}_1 = 9.7$, $P < 0.001$ (one-tailed test).

Coevolution

It is common enough for people to make trees for hosts and parasites to see if they agree. They usually don't, suggesting that either the parasites have multiple hosts or that they can switch hosts. There have been at least five co-evolutionary hypotheses put forward for chlamydia relating to: 1, host; 2, infected cell type; 3, organ infected; 4, route of transmission; and 5, geography. The phylogeny of chlamydia as inferred from the tree for its outer-membrane protein is not congruent with any of those five hypotheses (Fitch *et al.* 1993). Such non-congruence is so often the case that it is obligatory to point out a case where there is congruence. Hafner *et al.* (1994) showed that the tree for gophers and chewing lice agreed for all but one taxon. Moreover, this very congruence meant that the determination of rates for the lice could be calculated because of known palaeontological dates for the host. Interestingly, the rates for the lice were 10 times that of the gophers, which is the inverse ratio of their generation times (but see Page and Hafner, Chapter 16 in this volume).

Origin of the genetic code

If all the tRNAs come from one common ancestral tRNA, then the order of the gene duplications might provide information on the code's origin and, in particular, test the ambiguity reduction theory which asserts that, in the earliest stages, any amino acid might be charged to any tRNA. But there would be an enormous selective advantage if hydrophobic amino acids were charged to *NYN* tRNAs while hydrophilic amino acids were charged to *NRN* tRNAs. This alone would permit coding for alpha helices and beta sheets. Such a study was performed (Fitch and Upper 1988) for seven tRNAs and those trees consistent with the ambiguity reduction theory were far out on the tail of the distribution, which was encouraging.

8.4 Evolutionary trees and biology

The cenancestor of everything alive today

If the cenancestor is the most recent common ancestor, then what is the cenancestor for all we now survey? It is far more recent than the progenote, which is the first cell with a genetic apparatus. The problem arises because there is no out-group to use to locate the root. Schwartz and Dayhoff (1978) were the first to note that a partial gene duplication common to all living organisms must have occurred prior to the cenancestor. Ferredoxin seemed to be such a molecule but the result was ambiguous. The tRNAs mentioned in the previous subsection should also be such molecules and provide seven tests of root location, but again the result was ambiguous (Fitch and Upper 1988). The cenancestor of us all is still to be located.

Tissue tropisms

Why should HIV have such apparent difficulty infecting the brain? The assumption is that certain changes in the envelope protein are required to permit access to receptors of brain cells. This is unfinished business, although Korber and Myers (1992) have made an excellent start with their signature analysis. A study using trees which possess pairs of nodes where the tissue specificity changes between them permits one to focus on just those changes that occur between them. A collection of such pairs should be very revealing. An example of brain and blood isolates where this could be done is that of Korber *et al.* (1994).

Geographic correlates

In a study of the vesicular stomatitis viruses (VSV), one of the two major branches could be plotted easily on the map of Central America and it is clear that the viruses moved from Mexico (perhaps near Chiappas) southward first into Guatamala and then successively into El Salvador, Honduras, Nicaragua, Costa Rica, and Panama (Nichol *et al.* 1993). (Note that due to a misrooting of the tree, the geographic movement of the virus presented there is northward rather than southward. Although the polarity is changed, the branching structure is not.)

Punctuated molecular equilibrium

In the study in the previous subsection, we observed an unusual phenomenon in which some branches seemed unusually long while at the ends of such branches there was a lot of very short branches (Fig. 8.8). This is not explainable by a mixture of long and short geographic distances. It is as if, as the virus moved south, it kept encountering new niches to which it had to adapt causing the most southerly forms to have evolved far more extensively than their northern forms which changed relatively little upon settling down into their new niche. Taking the long branches to represent spurts of change and the many short ones as relative stasis suggests a non-temporal analogy to Eldridge and Gould's (1972) punctuated equilibrium. We do not know the cause of the bursts but we suggest it could be the adaptation to a new arthropod vector.

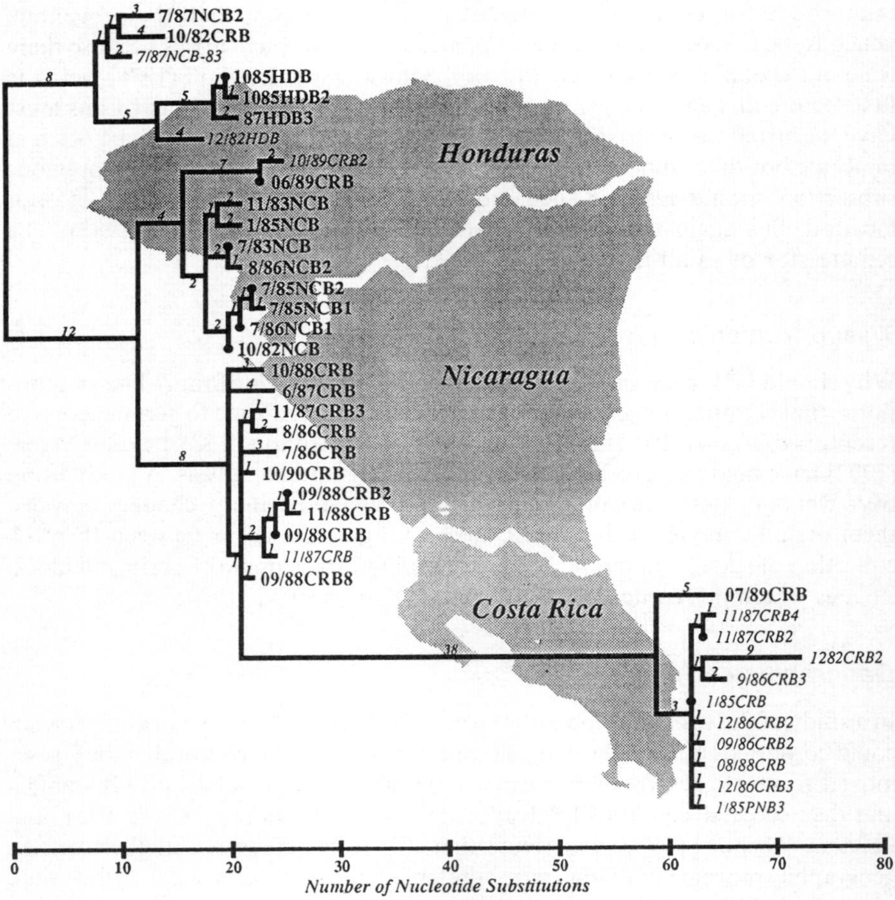

Fig. 8.8 Evolution of the vesicular stomatitis virus *P* gene. The tree shows only that part involving the lower part of Central America. It derives from portions from the north and shows a general trend of isolates south of the state of Chiappas, namely evolving southward. Isolate designations show the time of isolation and the country (HD = Honduras; NC = Nicaragua; CR = Costa Rica; PN = Panama). Note the large separation between two largely Costa Rican groups. (From Nichol *et al.* 1993).

Climatological correlates

Dr Luiz Rodriguez (personal communication) has examined where the cows from Costa Rica shown in Fig. 8.8 reside. There are two groups which are separated by 38 substitutions, but within the two groups the separations are very small. He has further shown that all but one of the 11 cows in the left group were from elevations below 600 m, while all but one of the cows in the right group came from elevations above 600 m. Given the frequent movement of cows, this is astonishing. Is there a change of arthropod vectors above and below 600 m?

Stress effects

Naas *et al.* (1994) examined the maps of insertion sequences from 118 clones from an *E. coli* stab made 30 years earlier and discovered there were 68 different haplotypes. Examining clones from stabs that were not as old (Naas *et al.* 1995), they discovered that the younger the stabs the fewer the changes observed (see Fig. 8.9). This sitting around with nothing to do (no food, no oxygen) is dangerous. Perhaps the cells are trying to escape their dead end by mutating. Perhaps mutation rates often increase in a hostile environment. That could be adaptive on average.

8.5 Evolutionary trees and non-trees

Network alternatives

The analysis of the data in the previous subsection led to the observation that among the isolates were seven that differed in only three map positions. As these data are for presence or absence of an insertion element at a site, there are only eight possible haplotypes and we have seven of them. If arranged according to which are one step away from each other, they appear at the vertices of a cube. There are over 50 equally most parsimonious trees and the usual tree algorithms would resolve this confusion with a concensus tree with a heptachotomy which is resolvable into 945 unrooted trees. I suggest that the network shown in Fig. 8.10 is the better representation, for several reasons: (1) it contains all of the most parsimonious trees (obtainable by removing branches that are in loops, thereby leaving a tree); (2) it contains no trees that are not most parsimonious; and (3) there have been so many mutations that it is reasonably likely that haplotype 201 may well have arisen twice, once from 128 and once from 242 (the difference lies only in the order of the two events that have given rise to haplotypes 128 and 242).

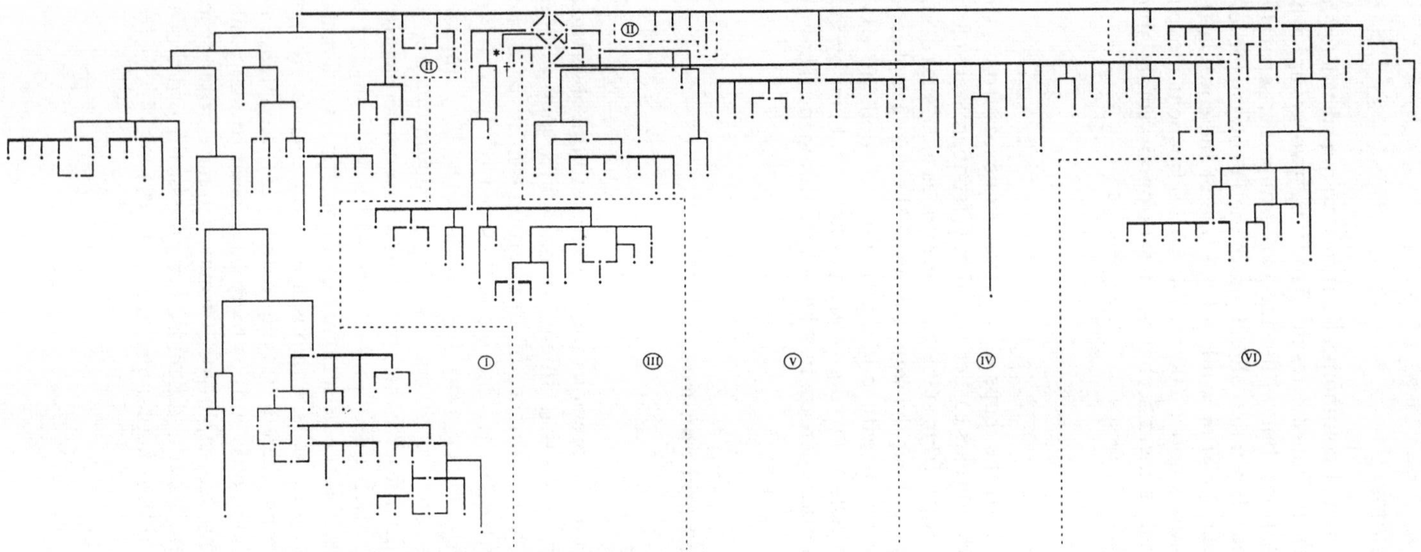

Fig. 8.9 Evolution of *E. coli* insertion sequences. Data were presence–absence of mapped insertion sequence elements. There were eight different insertion sequences used as probes and there was a total of 245 locations at which an element was observed in at least one isolate. The isolates all came from samples (stabs) stored for various numbers of years as follows: I, 30 years; II, 13 years, III, 12 years; IV, 10 years; V, 9 years; VI, 11 years. In addition, the dagger denotes a small group of isolates that are very similar (1 year) and the asterisk denotes 12 identical clones from the currently growing culture in the lab (zero time of storage). The short vertical segments are all one unit differences. (From Naas *et al.* 1994, 1995).

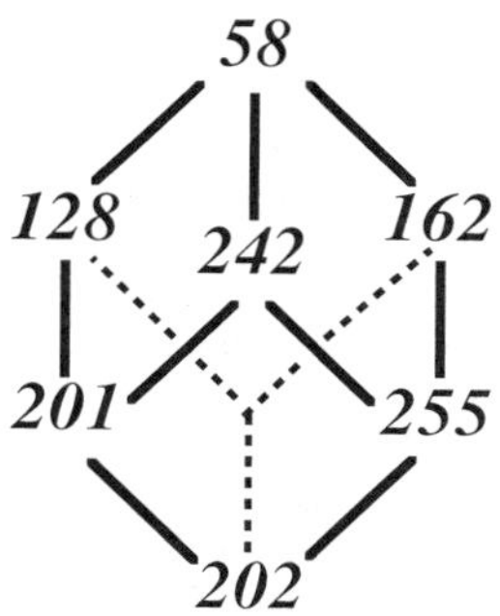

Fig. 8.10 Relationship of seven insertion sequences in Fig. 8.9. Isolate numbers are shown. The one numbered 58 is believed to be the ancestral form and was found in four different older stabs. Each branch is one step, the two isolates separated by the branch differing in only one of the 245 map positions. All seven differ in only three such positions and are shown as occupying the vertices of a cube. There are two cubed (equals eight) possible ways in which three positions can be one of two forms (present/absent); seven of them have been isolated as shown. The eighth would occupy the unlabelled vertex.

8.6 Conclusion

This chapter lists those uses that I have made of trees and is certainly far less than exhaustive. While a nimble mind might well discover the results inferred from these tree studies without recourse to a tree, the use of a tree simplifies and speeds the road to understanding evolutionary processes.

References

Ayala, F. J. (1986). On the virtues and pitfalls of the molecular evolutionary clock. *J. Heredity*, **77**, 226–35.

Baglioni, C. (1962). The fusion of two peptide chains in hemoglobin Lepore and its interpretation as a genetic deletion. *Proc. Natl. Acad. Sci. USA*, **48**, 1880.

Buonagurio, D. A., Nakada, S., Fitch, W. M., and Palese, P. (1986). Epidemiology of influenza C virus in man: multiple lineages and low rate of change. *Virology*, **153**, 12–21.

Eldredge, N. and Gould, S. J. (1972). Punctuated equilibria: an alternative to phyletic gradualism. *Models in paleobiology*. Freeman, San Francisco.

Fitch, W. M. (1971). Evolution of clupeine Z, a probable crossover product. *Nature, New Biol.*, **229**, 245–7.

Fitch, W. M. (1977). Phylogenies constrained by the cross-over process as illustrated by human hemoglobins and a thirteen cycle, eleven amino acid repeat in human apolipoprotein A1. *Genetics*, **86**, 623–44.

Fitch, W. M. and Ayala, F. J. (1994). The superoxide dismutase molecular clock revisited. *Proc. Natl. Acad. Sci. USA*, **91**, 6802–7.

Fitch, W. M. and Markowitz, E. (1970). An improved method for determining codon variability in a gene and its application to the rate of fixations of mutations in evolution. *Biochem. Genet.*, **4**, 579–93.

Fitch, W. M. and Upper, K. (1988). The phylogeny of tRNA sequences provides evidence for ambiguity reduction in the origin of the genetic code. *Cold Spring Harbor Symp. Quant. Biol.*, **LII**, 759–67.

Fitch, W. M., Leiter, J. M. E., Li, X., and Palese, P. (1991). Positive Darwinian evolution in human influenza A viruses. *Proc. Natl. Acad. Sci. USA*, **88**, 4270–4.

Fitch, W. M. Peterson, E. M., and de la Maza, L. M. (1993). Phylogenetic analysis of the outer membrane protein genes of chlamydiae and it's implication for vaccine development. *Mol. Biol Evol.*, **10**, 892–913.

Gorman, O. T., Bean, W. J. Kawaoka, Y., Donatelli, I., Guo, Y., and Webster, R. G. (1991). Evolution of influenza A virus nucleoprotein genes: Implications for the origins of H1N1 human and classical swine viruses. *J. Virol.*, **65**, 3704–14.

Gray, G. and Fitch, W. M. (1983) Evolution of Antibiotic resistance genes: the DNA sequences of a kanamycin resistance gene from *Staphylococcus aureus*. *Mol. Biol. Evol.*, **1**, 57–66.

Hafner, M. S., Sudman, P. D., Villablanca, F. X., Spradling, T. A., Demastes, J. W., and Nadler, S. A. (1994). Disparates rates of molecular evolution in cospeciating hosts and parasites. *Science*, **265**, 1087–90.

Hughes, A. L. and Nei, M. (1989). Nucleotide substitution at major histocompatibility complex class II loci: evidence for overdominant selection. *Proc. Natl. Acad. Sci. USA*, **86**, 958–62.

Kimura, M. (1968). Evolutionary rate at the molecular level. *Nature*, **217**, 624–6.

Korber, B. and Myers, G. (1992). Signature pattern analysis: a method for assessing viral sequence relatedness. *Aids Res. Human Retrovirus.*, **8**, 1549–60.

Korber, B. T. M., Kunstman, K. J., Patterson, B. K., Furtado, M., McEvilly, M. M., Levy, R., and Wolinsky, S. M. (1994). Genetic differences between blood- and brain-derived viral sequences from human immunodeficiency virus type 1-infected patients: evidence of conserved elements in the V3 region of the envelope protein of brain-derived sequences. *J. Virol.*, **68**, 7467–81.

Kurosky, A., Barnett, D. R., Lee, T.-H., Touchstone, B., Hay, R. E., Arnott, M. S., Bowman, B., and Fitch, W. M. (1980). Covalent structure of human haptoglobin: A serine protease homolog. *Proc. Natl. Acad. Sci. USA*, **77**, 3388–92.

Langley, C. H. and Fitch, W. M. (1974). An examination of the constancy of the rate of molecular evolution. *J. Mol. Evol.*, **3**, 161–77.

Naas, T., Blot, M., Fitch, W. M., and Arber, W. (1994). Insertion Sequence-related genetic variation in resting *E. coli* K-12. *Genetics*, **136**, 721–30.

Naas, T., Blot, M., Fitch, W. M., and Arber, W. (1995). Dynamics of IS-related genetic rearrangements in resting *Escherichia coli* K-12. *Mol. Biol. Evol.*, **12**, 198–207.

Nichol, S. T., Rowe, J. E., and Fitch, W. M. (1993). Punctuated equilibrium and positive Darwinian evolution in vesicular stomatitis virus. *Proc. Natl. Acad. Sci. USA*, **90**, 1042–8.

Rudikoff, S., Fitch, W. M., and Heller, M. (1992). Exon-specific gene correction (conversion) during short evolutionary periods: homogenization in a two-gene family encoding the b chain constant region of the T-lymphocyte antigen receptor. *Mol. Biol. Evol.*, **91**, 14–26.

Scholtissek, C., Ludwig, S., and Fitch, W. M. (1993*a*). Analysis of influenza A virus nucleoproteins for the assessment of molecular genetic mechanisms leading to new phylogenetic virus lineages. *Arch. Virol.*, **131**, 237–50.

Scholtissek, C., Altmueller, A., Schultz, U., Ludwig, S., Mandler, J., and Fitch, W. M. (1993*b*). Molecular epidemiology of influenza. In *Concepts in virology from Ivonovsky to the present*, pp. 235–43. Harwood.

Scholtissek, C., Schultz, U., Ludwig, S., and Fitch, W. M. (1993*c*). The role of swine in the origin of pandemic influenzas. In *Options in the control of influenza, II*, pp. 193–201. Excerpta Medica, Amsterdam.

Schwartz, R. M. and Dayhoff, M. O. (1978). Origins of prokaryotes, mitochondria and chloroplasts. *Science*, **199**, 395–403.

Slightom, J. L., Blechl, A. E., and Smithies, O. (1980). Human fetal ^{G}g- and ^{A}g-globin genes: complete nucleotide sequences suggest that DNA can be exchanged between these duplicated genes. *Cell*, **21**, 627–38.

Smith, T. F., Waterman, M. S., and Fitch, W. M. (1981). Comparative biosequence metrics. *J. Mol. Evol.*, **18**, 38–46.

Stewart, C.-B. and Wilson, A. C. (1987). Sequence convergence and functional adaptation of stomach lysozymes from foregut fermenters. *Cold Spring Harbor Symp. Quant. Biol.*, **52**, 891–9.

Zuckerkandl, E. and Pauling, L. (1962). Molecular disease, evolution, and genetic heterogeneity. In *Horizons in biochemistry*, pp. 189–225. Academic, New York.

9

Cross-species transmission and recombination of 'AIDS' viruses

Paul M. Sharp, David L. Robertson, and Beatrice H. Hahn

9.1 Introduction

Acquired immune deficiency syndrome (AIDS) is caused by two different human immunodeficiency viruses, HIV-1 and HIV-2, which are members of the lentivirus family of retroviruses. Related viruses are found in many species of non-human primates; these have been termed simian immunodeficiency viruses (SIV), with a subscript to denote their species of origin. (Here we use the term '"AIDS" viruses' to mean the entire group of primate lentiviruses; however, many of the SIVs are not known to cause immunodeficiency in what appear to be their natural hosts.) Lentiviruses have also been isolated from artiodactyls, perissodactyls, and carnivores. All retroviruses contain two copies of an RNA genome, approximately 10 kb in length, at each end of which lie long terminal repeats (LTR) encoding regulatory functions. Much of the sequence in between is taken up by the three major structural genes: *gag*, *pol*, and *env*. However, HIVs and SIVs also encode a number (five or six) of shorter accessory genes (see Hahn 1994).

Sequence comparisons reveal that primate lentiviruses form a distinct clade (Hirsch *et al.* 1993; Sharp *et al.* 1995). Within the primate lentiviruses (Fig. 9.1) five major, approximately equidistant, lineages are known (Sharp *et al.* 1994). Three of these are each represented by multiple isolates. One includes the many known strains of HIV-1, plus a virus isolated from a chimpanzee (SIV_{CPZ}). A second contains multiple strains of HIV-2, as well as viruses isolated from sooty mangabeys (SIV_{SM}) and captive macaques (SIV_{MAC}). The third comprises viruses from African green monkeys (SIV_{AGM}). The other two lineages are, as yet, represented by single isolates from a mandrill (SIV_{MND}) and a Sykes' monkey (SIV_{SYK}). Within each of the three lineages for which multiple strains have been characterized, further phylogenetic structure is apparent. Thus, HIV-1 strains fall into two distinct groups, M and O; within the M group, strains have been classified into at least eight different sequence subtypes (Myers *et al.* 1993; Louwagie *et al.* 1993; Sharp *et al.* 1994). Similarly, HIV-2 strains have been classified into (so far) five different subtypes (Sharp *et al.* 1994; Gao *et al.* 1994*b*). Within the SIV_{AGM} lineage, viruses isolated from the four different species of African green monkey cluster separately (Jin *et al.* 1994*a*).

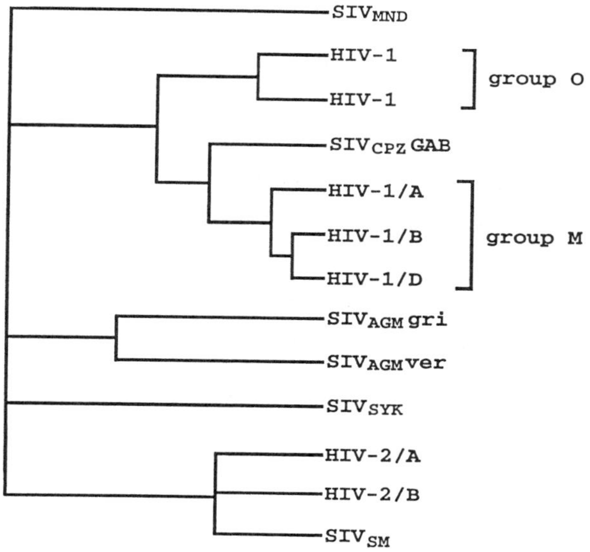

Fig. 9.1 Phylogeny of primate lentiviruses. Within the HIV-1 group M there are at least eight different sequence subtypes (A–H), of which just three are shown; within the HIV-2 group there are five known subtypes (A–E: see Fig. 9.3); within the SIV_{AGM} group there are four lineages (see Fig. 9.2). (Adapted from Jin *et al.* (1994*a*) and Sharp *et al.* (1994).)

Phylogenetic analyses of these viruses have been used for a number of purposes. For example, a phylogenetic approach has been used in attempts to assess the possibility that certain individuals have been the source of infection in transmission studies (Ou *et al.* 1992; Albert *et al.* 1993; Holmes *et al.* 1993; Hillis and Huelsenbeck 1994), and to track the global spread of HIV-1 and HIV-2 subtypes (Ou *et al.* 1993; Louwagie *et al.* 1993; Gao *et al.* 1994*a*). Here we concentrate on two other applications: the elucidation of cross-species transmission events, and the discovery of recombination between divergent viruses. The results we report have important implications for the biology and pathogenicity of AIDS viruses, and for the development of an AIDS vaccine.

9.2 Cross-species transmission of 'AIDS' viruses

For a general understanding of the evolution of the primate lentiviruses, and for specific insights into the origins and nature of AIDS in humans, we need to know how often cross-species transmissions have occurred, the species involved, and the direction of transmission. If these viruses had evolved in an entirely host-dependent fashion, one would expect their phylogeny to be the same as that of their species of origin. This is not the case. For example, African green monkeys (*Cercopithecus aethiops*) and Sykes' monkey (*Cercopithecus mitis*) are more closely related to each other than to other genera of Catarrhine monkeys, while the mandrill (*Mandrillus sphinx*) and sooty mangabey (*Cercocebus atys*) are more closely related to each other than to *Cercopithecus* (Adkins and Honeycutt 1994); however, the viruses isolated from these four species fall into distinct lineages that appear to have diverged at around the same time (Fig. 9.1). Furthermore, viruses infecting humans fall on two different major lineages. Thus, it is clear that cross-species transmission of primate lentiviruses has occurred.

It is generally believed that HIV-1 and HIV-2 represent zoonotic infections that have each arisen through relatively recent human acquisition of simian viruses. However, Ewald (1994) suggests that the common ancestor of the primate lentiviruses infected humans, and thus that viruses have been transmitted on numerous occasions from humans to other primates. He argues that this is the more parsimonious explanation for the current species distribution of the viruses: since the human viruses are found on two distinct lineages, one less cross-species transmission event is required if the ancestral virus was in humans. However, if cross-species transmissions have been relatively common during the evolution of primate lentiviruses, this parsimony argument is undermined.

Cross-species transmission in the SIV_{AGM} lineage

Within the lineage including SIV strains infecting African green monkeys (SIV_{AGM}) cross-species transmissions do indeed appear to have been rare. There are four varieties of African green monkey (vervet, grivet, sabaeus, and tantalus) which inhabit overlapping ranges across sub-Saharan Africa. These are described by most authors as species, though where they overlap they are believed to interbreed. Numerous SIV_{AGM} strains have been isolated from each of the four species. Phylogenetic analyses based on *env* protein sequences (Fig. 9.2) show that these viruses cluster in a species-specific manner (Muller *et al.* 1993; Jin *et al.* 1994*a*). Perhaps, this could be a biogeographic phenomenon? However, this does not appear to be the case. For example, vervet monkeys from East Africa have SIV strains much more closely related to those from vervet monkeys from South Africa than to SIV from grivet monkeys in East Africa (Soares *et al.*, in preparation). Thus, the various strains of SIV_{AGM} appear to have evolved in a host-dependent fashion, i.e. their common ancestor most likely infected the common ancestor of the four species of African green monkey. This in turn implies that this infection is quite ancient; this ancestral *C. aethiops* lived thousands and perhaps one or more million years ago. The antiquity of this host–virus relationship is also suggested by the fact that approximately one-half of wild adult African green monkeys are infected by SIV (Allan *et al.* 1991; Muller *et al.* 1993; Phillips-Conroy *et al.* 1994), and (less directly) by the observation that the virus is not pathogenic in these hosts (Phillips-Conroy *et al.* 1994).

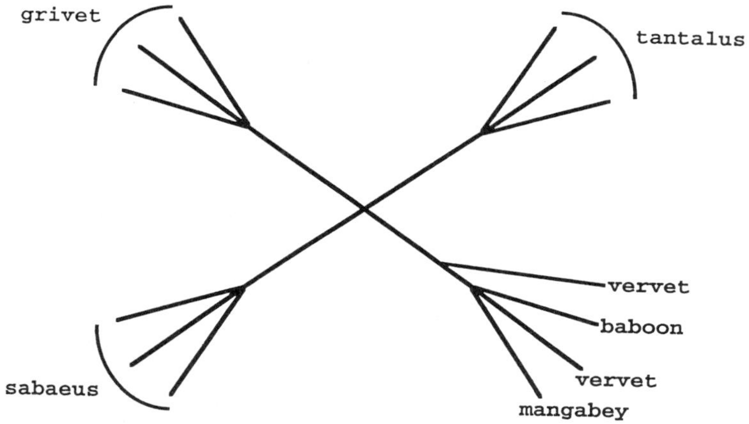

Fig. 9.2 Species-specific phylogenetic relationships among SIV_{AGM} strains from vervet, grivet, sabaeus, and tantalus monkeys, as derived from analysis of *env* protein sequences (adapted from Jin *et al.* (1994*a*)). In addition, the phylogenetic positions of single isolates from a yellow baboon (derived from Jin *et al.* (1994b)) and a white crowned mangabey (derived from Tomonaga *et al.* (1993), in which LTR sequences were analysed) are indicated. The tree is drawn unrooted; the most ancestral point is near the centre.

Nevertheless, instances of cross-species transmission involving SIV_{AGM} have recently been reported. First, a captive-born white-crowned mangabey (*Cercocebus torquatus lunulatus*) was found to be infected with an SIV_{AGM}-related virus (Tomonaga *et al.* 1993). Phylogenetic analysis of LTR sequences showed that this virus falls within the radiation of vervet viruses (Fig. 9.2). The authors suggested that the mangabey colony had probably been infected in the wild. However, white-crowned mangabeys are indigenous to West Africa, where the local variety of *C.aethiops* is the sabaeus monkey; the range of the vervet monkey is from East to South Africa. This strongly suggests that the cross-species transmission occurred in captivity. Secondly, analyses of the SIV_{AGM} strains carried by sabaeus monkeys have shown that these viruses have a mosaic genome that must have resulted from a recombination event (Jin *et al.* 1994*a*). The lineages involved were the ancestors of the viruses now found in other African green monkeys, and the ancestor of the SIV_{SM}/HIV-2 lineage. Moreover, the recombination appears to have occurred closer to the divergence of these lineages than to today (Jin *et al.* 1994*a*). In whatever host this recombination took place, there must have been cross-species transmission before or after the event. Finally, two of 279 wild-living yellow baboons (*Papio hamadryas cynocephalus*) that have been tested from Tanzania were reported to be seropositive for an SIV_{AGM}-like virus (Kodama *et al.* 1989). We have recently characterized a virus isolated from one of these animals (Jin *et al.* 1994*b*) which falls within the radiation of SIV_{AGM} strains from vervet monkeys (Fig. 9.2). The infected baboons lived in an area inhabited by vervets. The very low prevalence of SIV infection among these baboons, and the finding that none of 155 sera from Ethiopian olive and hamadryas baboons were seropositive (Kodama *et al.*

1989) suggests that baboons are not natural hosts for SIV. This virus represents the most direct evidence yet for a recent simian-to-simian cross-species transmission of SIV in the wild.

Cross-species transmission in the HIV-1 lineage

Known HIV-1 strains fall into two major groups: group M, containing diverse strains isolated from all continents, and group O consisting (as yet) of a small number of recently characterized isolates from Cameroon and Gabon. A single virus isolated from a chimpanzee in Gabon (SIV_{CPZ}GAB) also belongs to this lineage, and is more closely related to HIV-1 group M than to group O (Fig. 9.1). Ewald (1994) argues that these relationships can be most parsimoniously explained if the ancestral virus of this lineage infected humans, so that SIV_{CPZ}GAB represents a virus acquired by chimpanzees from humans, and only a single cross-species transmission event has occurred. However, most authors consider that HIV-1 is a relatively recent infection of humans, in which case the M and O groups would each have arisen through separate cross-species transmissions.

These alternative hypotheses as to whether humans are the natural host of HIV-1 lead to different conclusions concerning the biology of the host–virus interaction, and the antiquity of AIDS. For example, from phylogenetic and evolutionary rate studies, the common ancestry of the various HIV-1 strains in group M has been estimated to date back to about 1960 (Li *et al.* 1988). This agrees well with the earliest epidemiological evidence of AIDS. HIV-1 strains within group O exhibit approximately the same degree of genetic diversity as those within group M, and so if viruses from the two groups have been evolving at similar rates, the common ancestor of the O group would also have existed around 1960. The extent of divergence between the M and O groups of HIV-1 is two to three times greater than the diversity within each group. If the molecular clock has ticked at a constant rate throughout this divergence, the common ancestor of the M and O groups would date back to the latter half of the nineteenth century. However, this ancestor may well have been more ancient, since it is possible that the primate lentiviruses have evolved more slowly in the past (discussed below). If the M and O groups arose through two separate cross-species transmissions, those events could have occurred at any time up to the existence of the M and O common ancestors, i.e. as recently as the 1950s. Alternatively, under the single cross-species transmission hypothesis, that event must have occurred at least a century ago.

An important piece of evidence has now emerged. A second HIV-1-related SIV was isolated from a chimpanzee from Zaire (Peeters *et al.* 1992). This isolate has now been sequenced, and preliminary reports indicate that it falls on the HIV-1/SIV_{CPZ} lineage, but as an outgroup to all other strains (Myers and Korber 1994). Thus, whether the ancestral virus for this lineage infected humans or chimpanzees, a minimum of two cross-species transmission events are required to give the current distribution, and so the simple parsimony argument no longer weighs against the recent acquisition, on two occasions, of HIV-1 by humans. Other lines of evidence, such as the apparently low frequency of HIV-1 infection until recently, support the view that humans have been the recipients of these transmissions.

Cross-species transmission in the HIV-2 lineage

Viruses closely related to HIV-2 have been isolated from sooty mangabeys and several species of macaques. These viruses cause disease in macaques but not in sooty mangabeys. The native habitat of sooty mangabeys is West Africa, where it is estimated that around 30 per cent of these monkeys are infected (P. A. Marx, personal communication). The only macaques known to be infected with SIV come from primate research centres, and thus appear to have acquired the virus in captivity, presumably from sooty mangabeys. So far, HIV-2 is only endemic in West Africa, and except in urban sex workers the rate of human infection there is much lower than the rate of SIV infection in sooty mangabeys. Thus, it has been concluded that HIV-2 has arisen through human acquisition of SIV_{SM} (Hirsch *et al.* 1989; Gao *et al.* 1992). Humans in West Africa may often be exposed to monkey blood, since sooty mangabeys are hunted for food and kept as pets.

The majority of known HIV-2 strains fall into two distinct sequence subtypes (A and B), approximately equidistant from each other and from SIV_{SM}. This would not be significantly inconsistent with a single sooty mangabey-to-human transmission. However, we have recently described a number of new HIV-2 isolates (Gao *et al.* 1992, 1994*b*), including three viruses sufficiently divergent to each warrant separate subtype status (Fig. 9.3). Until very recently, the small number of SIV_{SM} isolates that had been characterized at the molecular level fell within this extended view of the HIV-2 radiation, all clustering with the one known example of subtype D. The various SIV strains from macaques also fall within this lineage. If the common ancestor of this group of viruses infected a sooty mangabey, we must invoke at least three, and perhaps as many as five (i.e. one per HIV-2 subtype), mangabey-to-human transmissions. Thus, the parsimony argument would seem to strongly favour a human origin for this entire group of viruses, with a single human-to-sooty mangabey transmission.

In this case, again, the parsimony argument may be undermined if, through poor sampling, our knowledge of the diversity of SIV_{SM} has been incomplete. This now appears to be the case. One of the SIV_{SM} isolates came from a pet animal from Liberia (Marx *et al.* 1991), while the others came from American primate research centres and may have also been of Liberian origin. (Interestingly, the HIV-2 isolate which is most closely related to the known SIV_{SM} strains, i.e. subtype D, also came from Liberia.) Now several new SIV_{SM} strains have been isolated from Cote d'Ivoire and Sierra Leone (Chen *et al.* 1994; Peeters *et al.* 1994): these do not cluster closely with the earlier strains, and some are clearly outside HIV-2 subtype D (Fig. 9.3). While the outgroup within the entire SIV_{SM}/HIV-2 radiation is still from humans (HIV-2 subtype E), it is clear that there have been sufficient cross-species transmissions within this lineage for the parsimony argument to no longer carry any weight. All other lines of evidence point to sooty mangabeys as the natural hosts of these viruses, and thus imply that there have been multiple occasions on which humans have acquired HIV-2. Detailed phylogenetic analyses also indicate that cross-species transmission to macaques (in captivity) must have occurred on a number of independent occasions (Gao *et al.* 1994*b*).

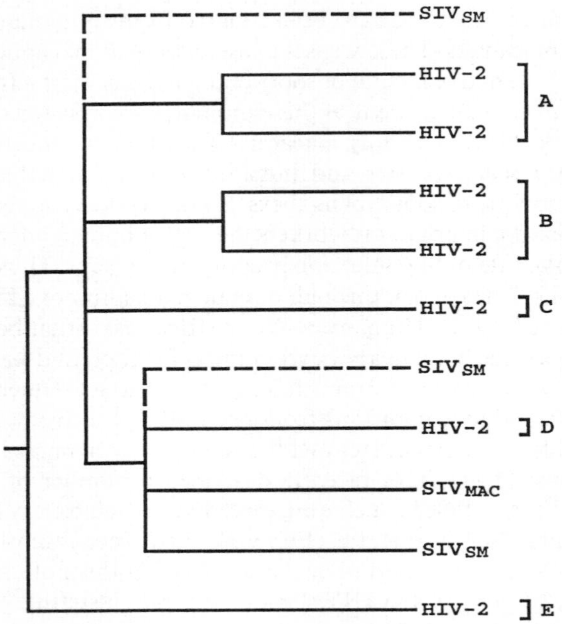

Fig. 9.3 Phylogenetic relationships among strains of HIV-2, SIV from sooty mangabeys (SIV_{SM}), and SIV from macaques (SIV_{MAC}). This tree is adapted from Gao *et al.* (1994*b*). The positions of very recently characterized SIV_{SM} isolates (see text), as inferred from Peeters *et al.* (1994) and Chen *et al.* (1994), are shown by broken lines.

Direction of cross-species transmissions

The phylogenetic analyses described above reveal that there have been a number of cross-species transmissions of primate lentiviruses, either in captivity or in the wild, within each of the three lineages for which multiple viruses have been characterized. However, these evolutionary trees alone do not indicate whether humans have been the donors or recipients of HIV-1 and HIV-2. Other evidence, such as epidemiological and biological data, point to humans as the recipients. Another line of molecular evolutionary analysis supports this.

It has been suggested on a number of occasions (for example by Ewald 1994) that one way to infer the direction of cross-species transmission is to suppose that when a virus enters a new host species its pattern of evolution will change.

Exposure to the new host is expected to produce selection for protein sequence changes, particularly in the *env* protein. Such selection would elevate the number of non-synonymous (amino acid replacing) nucleotide changes without necessarily affecting the number of synonymous (silent) nucleotide changes, and thus the ratio of non-synonymous to synonymous changes should increase. We have tested this hypothesis in the case of the clearest example of a recent natural cross-species transmission where the direction of transmission can be safely inferred, namely the vervet-to-baboon transmission discussed above (Jin *et al.* 1994*b*). Among multiple (partial) *env* sequences derived from the baboon the ratio of non-synonymous to synonymous differences is approximately 2, whereas comparable sequences from a naturally infected vervet yield a ratio of 0.5 (Soares *et al.*, in preparation). Thus, the prediction of an increased rate of non-synonymous change in the new host is borne out, and this lends some justification to the use of the same approach to infer the direction of cross-species transmission in other lineages. Ewald (1994) cites unpublished analyses suggesting that the ratios of non-synonymous to synonymous nucleotide changes are such as to indicate that chimpanzees and sooty mangabeys acquired HIV-1 and HIV-2, respectively, from humans. However, others have reported quite the opposite. High non-synonymous to synonymous ratios have been reported for HIV-1 (Balfe *et al.* 1990) and SIV in macaques (Burns and Desrosiers 1991). In particular, the ratio of non-synonymous to synonymous nucleotide changes in the region of *env* encoding the external envelope glycoprotein 120 has been found to be much higher in HIV-1 and SIV_{MAC} than in SIV_{AGM} and SIV_{SM} (Shpaer and Mullins 1993). Thus, the data of which we are aware point to humans (and macaques) being the recipients of SIV by cross-species transmission.

9.3 Recombination of 'AIDS' viruses

Retroviruses are highly recombinogenic (Hu and Temin 1990), and so recombination is expected to contribute to the generation of genetic diversity in HIV and SIV. However, recombination can only occur between viruses replicating within the same cell (and RNA genomes packaged into the same particle). Until very recently, there has been no direct evidence for co-infection (by which we mean either simultaneous co-infection or sequential superinfection) of single individuals with multiple divergent HIV-1 or HIV-2 strains. Thus, while recombination has been reported among the (relatively closely related) members of the quasispecies that evolves during the course of HIV infection (Vartanian *et al.* 1991), recombination among divergent viruses, for example those from different subtypes, has not been considered to be a significant source of new variation in primate lentiviruses.

If primate lentiviruses evolved in a strictly clonal fashion, evolutionary trees for different regions of the viral genome would be similar, at least within the power of resolution of the analysis. Hybrid genomes produced by recombination between divergent viruses would contain regions with different evolutionary histories. If the viruses involved were sufficiently divergent (and if appropriate sequences are available for analysis) past recombination events could be detected from significantly discordant phylogenetic relationships derived from different genes or parts of genes (Fig. 9.4). Taking this approach, evolutionary trees derived from different genomic regions have been found to be consistent with respect to most of the viruses investigated. Nevertheless, we have identified recombinant viruses in each of the three major lineages of primate lentiviruses for which multiple strains have been characterized.

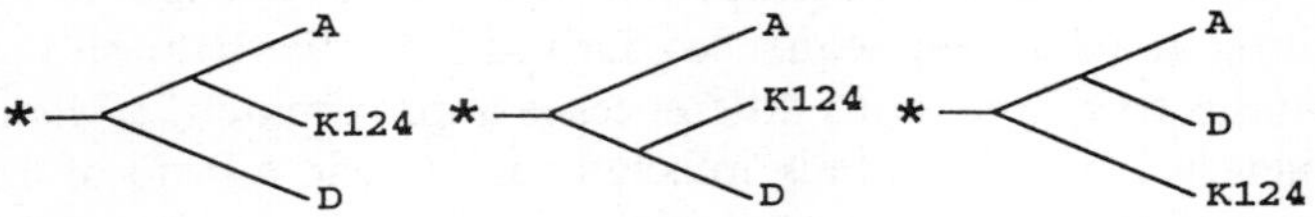

Region	Informative sites		
	A	D	O
1-1065	30	4	4
1096-2435	4	46	10
2480-2580	7	1	1

Fig. 9.4 Detecting recombination in HIV. $HIV-1_{K124}$ has been identified as a hybrid virus (Robertson *et al.* 1995*b*): in trees based on different regions of the *gag* and *env* genes, K124 clusters alternatively with subtype A and subtype D. To map the breakpoints between regions with different evolutionary histories, an alignment of four sequences was analysed: these were K124, consensus sequences of non-recombinant viruses from subtypes A and D, and an outgroup (SIV_{cpz}). Each phylogenetically informative site in the alignment could support one of the three trees shown (where * indicates the outgroup); if only subtype A and D viruses were involved in the recombination, sites supporting the third tree indicate the level of phylogenetic background noise. Breakpoints were localized to positions which maximized the heterogeneity (by chi-squared analysis) between adjacent regions with respect to the number of sites supporting clustering with sequence subtypes A or D. The results of this analysis for the *env* gene are shown; the chi-squared values (with one degree of freedom) for the two breakpoints are 54.1 and 28.4 (both $P \ll 0.001$). (See Robertson *et al.* (1995*a*, *b*) for full details.)

Examples of recombinant primate lentiviruses

Three particularly well-defined examples of recombinant primate lentiviruses exist, where in each case the complete viral genome has been sequenced and the positions of the crossovers between the recombining viruses have been localized.

A single example of HIV-1 (HIV_{MAL}) has long been suspected to be recombinant (Li *et al.* 1988). This strain was isolated from a boy with AIDS-related complex in Zaire, who was thought to have been infected by a blood transfusion (Alizon *et al.* 1986). We have recently described a detailed analysis of the

HIV-1$_{MAL}$ genome in which we confirmed that different regions of the genome have significantly different phylogenetic histories, and mapped the crossover points by examining the linear distribution of phylogenetically informative sites supporting alternative tree topologies (Robertson *et al.* 1995*a*). There are (at least) five such breakpoints, delimiting six genomic regions (Fig. 9.5). Three regions, at the 5′ end (the 5′ part of *gag*), in the middle (the 3′ part of *pol* plus the 5′ part of *vif*), and at the 3′ end (the extreme 3′ part of *env* plus the *nef* gene), all appear to be subtype A-like. Two regions, the 3′ part of *gag* and the 3′ part of *vif* through to the 3′ end of *env*, are subtype D-like. The remaining region, the 5′ part of *pol*, is not obviously closely related to subtypes A or D.

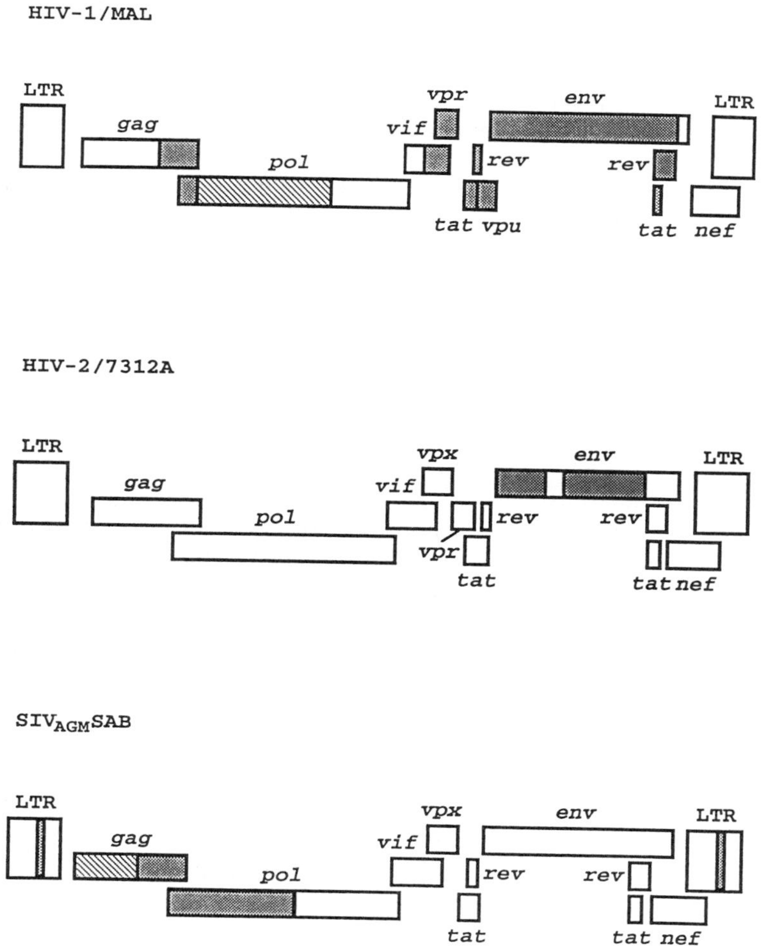

Fig. 9.5 Mosaic genomes of HIV-1$_{MAL}$, HIV-2$_{7312A}$, and SIV$_{AGM}$ sab. Boxes indicate genes, arranged according to their position along the viral genome. Different shading patterns indicate genomic regions inferred to have different evolutionary histories (see text for details). (Adapted from Jin *et al.* (1994*a*) and Robertson *et al.* (1995*a*).)

Similarly, a single example of HIV-2 (HIV-2_{7312A}, isolated from a male student from Cote d'Ivoire) was identified as a putative recombinant because phylogenetic analyses performed on partial *pol* and *env* sequences revealed quite discordant positions within the HIV-2 radiation (Gao *et al.* 1992). These results suggested that there was a crossover in the region between these two fragments: subsequent analysis (Robertson *et al.* 1995*a*) of a complete replication-competent provirus of HIV-2_{7312A} revealed that most of the genome is subtype B-like, but that most of the *env* gene is subtype A-like (Fig. 9.5). Again, detailed analysis of informative site configurations revealed multiple (four) putative recombination breakpoints, in this case all within the *env* gene (Fig. 9.5).

An example of a recombinant SIV_{AGM} has already been mentioned, in the context of cross-species transmission. Until recently, complete SIV_{AGM} genome sequences have been available only for viruses isolated from vervets and grivets. We cloned and sequenced a complete sabaeus provirus, and found discordant phylogenetic relationships from analyses of different regions of its genome (Jin *et al.* 1994*a*). A tree based on the complete *env* protein sequence confirmed the earlier finding that SIV_{AGM} sab represents a distinct lineage within the SIV_{AGM} radiation, approximately equidistant from SIV_{AGM} ver and SIV_{AGM} gri. In contrast, in both *gag*- and *pol*-derived trees SIV_{AGM} sab seemed no closer to other SIV_{AGM}s than to other primate lentiviruses. More detailed analyses revealed a putative crossover point near the middle of the *pol* gene (Fig. 9.5). In trees derived from sequences 3′ to this breakpoint, SIV_{AGM}sab clustered with other SIV_{AGM}s. However, in a tree derived from *pol* sequence 5′ to this breakpoint, and in a tree for the (partially overlapping) 3′ region of *gag*, SIV_{AGM} sab clustered with the SIV_{SM}/HIV-2 lineage. The 5′ region of *gag* from SIV_{AGM} sab did not appear to be more closely related to either SIV_{AGM} or SIV_{SM}, no matter how the region was subdivided for analysis. Interestingly, an important regulatory sequence within the LTR, the transactivation response (TAR) element, was found to be duplicated in SIV_{AGM} sab, with the potential to form a secondary structure far more like that found in HIV-2 and SIV_{SM} than in SIV_{AGM}s from vervets or grivets.

In each of the three cases discussed above, there appear to have been multiple crossovers during the recombination process. This is not surprising, since this is also seen in laboratory investigations of retroviral recombination (Hu and Temin 1990). In both HIV_{MAL} and SIV_{AGM} sab, there are regions of the genome which cannot be assigned to either of the 'parental' lineages that are inferred to have recombined. This could be because there have been further recombination events during the evolutionary history of these viruses, involving lineages whose descendants are not represented in the database. Alternatively, these regions could be very complex mosaics with several internal crossovers, which are very difficult to disentangle.

Recent evidence that recombinant HIVs may be widespread

The three examples discussed above may have seemed like unusual events, but as more sequence data for different strains of HIV-1 and HIV-2 accumulate there is increasing evidence for past recombination. For example, another possible example of a recombinant HIV-2 has recently been identified (Gao *et al.* 1994*b*). Strain FA was isolated from a female AIDS patient hospitalized with end-stage disease in Ghana. Phylogenetic analyses of partial *gag*, *pol*, and *env* sequences all showed FA to be a subtype A virus. However, in the *gag* and *pol* trees FA clustered with one particular subset of subtype A viruses, while in the *env* tree FA fell in a different group; in each case, the clustering was significant (as assessed by bootstraps). Sabino *et al.* (1994) have determined the *env* gene sequences of HIV-1 isolates from two epidemiologically linked individuals from Brazil, and shown that these genes appear to be the result of a recombination between viruses of sequence subtypes B and F. Diaz *et al.* (1995) have a recombinant *env* sequence (HIV_{DR106}) in an infant who was transfused with packed red blood cells from two different HIV-1 positive donors on the same day. In this case, both of the viruses involved were sequence subtype B, but despite this close relationship the recombination was discernible because both parental genomes (or at least sequences very closely related to them) were available for comparison.

Within the last year, the number of partial HIV-1 genomic sequences, in particular *gag* (for example Louwagie *et al.* 1993) and *env* (for example Louwagie *et al.* 1994; Gao *et al.* 1994*a*) gene sequences, published and deposited in the sequence database has grown considerably. We have now performed an extensive survey of these sequences which reveals a surprisingly large number of apparently recombinant viruses (Robertson *et al.* 1995*b*). Among 114 viruses for which near full-length *gag* or *env* gene sequences were available for analysis, 12 (including HIV-1_{MAL}) were found to be recombinants of sequences from different group M subtypes. As found above, in many cases the recombination event appears to have involved multiple crossovers. All eight subtypes of HIV-1 group M thus far described appear to have been involved in recombination events. No sequences have been found that were hybrids of group M and group 0 viruses. All of the putative recombinants originated from geographic regions where multiple sequence subtypes are known to co-circulate, including Central Africa, South America, and South-east Asia (Louwagie *et al.* 1993, 1995). It is important to realize that this survey has almost certainly underestimated the frequency of co-infections, since recombinants of divergent viruses from the same sequence subtype are much more difficult to detect.

In each of these cases where a complete viral genome has not been cloned prior to sequencing, it is possible that the individual was simultaneously co-infected with two different strains of HIV-1 (or HIV-2), which were differentially amplified in the polymerase chain reaction (PCR) procedure used to obtain DNA fragments for sequencing. Of course, this would not alter one of the most important implications of these results (discussed below), namely that co-infection occurs. Furthermore, in a co-infected individual recombination would be expected to occur, although not all mosaic genomes would necessarily be viable. Future analyses should aim to verify the presence of mosaic genomes *in vivo*, to exclude the possibility of tissue culture and/or PCR artefacts.

Implications of recombination

The discovery of extensive recombination during the history of the primate lentiviruses has immediate consequences for our understanding of HIV-1 (and HIV-2) pathogenesis and for vaccine development. Firstly, it clearly implies that co-infection with divergent HIV-1 strains is not as rare as previously thought. This raises the question of *when* during the course of HIV-1 infection superinfection can occur. Since co-infection was thought to be rare, it has been inferred that the host's immune response may play a role in protecting against infection by a second HIV-1 strain. Perhaps superinfection can only occur when exposure to a second virus occurs during the first weeks after initial infection, before an effective immune response is mounted. It is notable that two of the recombinant viruses (HIV-1_{MAL} and HIV-1_{DR106}) were obtained from individuals who had received blood transfusions which may have delivered divergent viruses simultaneously or within a very short time period. Alternatively, it is possible that the host's immune protection fails to extend to more divergent strains, for example members of different subtypes, an issue of obvious importance for vaccine development efforts.

The extensive genetic diversity of HIV-1 indicates that effective vaccines may need to include antigens that are representative of the particular locally circulating strains. Efforts are already under way to prepare subtype-specific vaccine formulations and to conduct large-scale efficacy trials in several different areas of the world. The unexpectedly high frequency of mosaic HIV-1 genomes among those viruses so far sequenced points to the need to devise measures to screen for recombinant viruses in these areas and to determine their possible impact on vaccine design and evaluation.

Finally, in other viruses recombination is known to generate new variants with significantly altered pathogenic properties. For example, co-infection with two avirulent herpes simplex viruses led to the production of virulent recombinants and a lethal disease course in experimentally infected mice (Javier *et al.* 1986), while reassortment of influenza A virus genes can generate pathogens with exceptional virulence and transmissibility (Kawaoka and Webster 1988). Recombination in conjunction with cross-species transmission may lead to even more dramatic effects. Thus, periodic exchanges of genetic material between influenza viruses infecting different mammals and birds has resulted in devastating pandemics in humans. Given this potential for recombination events to generate new pathogens, future studies should be aimed at determining whether genetic exchanges of the type reported above can influence clinically important properties of HIV, such as transmissibility, virulence, and replication potential.

9.4 Conclusions and perspectives

The detailed phylogenetic analyses of primate lentiviruses described above have revealed two particular complexities in the evolution of this group, i.e. that both cross-species transmissions and recombination events involving divergent viruses have been quite common. Some of the implications of these findings have already been discussed. One consequence of the discovery of numerous recombinant viruses that should be reiterated is that future characterization of primate lentiviruses should include careful phylogenetic investigation of possible genomic mosaicism.

The finding of multiple cross-species transmissions within the 'AIDS' virus family raises a number of questions concerning the source of these viruses: (1) from which species did humans acquire HIV-1; (2) which species did the common ancestor of the known primate lentiviruses infect; and (3) what has been the time-scale of the evolution of these viruses? To provide a framework in which to consider these questions, the conclusions we have reached above concerning which species have been the donors and recipients of cross-species transmissions are summarized in Table 9.1. In addition, we have indicated what is known about whether each virus causes disease in the host from which it was isolated. Note that in several cases there are insufficient data to make firm conclusions on one or both of these points. Nevertheless, it can be seen that the primate lentiviruses only seem to cause disease when transferred to a new host. Although it has been argued that lack of pathogenicity is not a necessary consequence of a long-term virus–host relationship (Ewald 1994), within this group of viruses the negative correlation between whether a virus is in its natural host, and whether it causes disease, is statistically perfect among those viruses for which good information is available (Table 9.1).

Table 9.1 Sources of all primate lentiviruses for which some nucleotide sequence data have been determined, and whether the viral infection is natural and pathogenic

Virus	Source species	Natural host?	Disease causing?
HIV-1	Human (*Homo sapiens*)	No	Yes
HIV-2	Human (*Homo sapiens*)	No	Yes
SIV_{CPZ}	Chimpanzee (*Pan troglodytes*)	Yes?	No?
SIV_{AGM}ver	Vervet monkey (*Cercopithecus aethiops*[a])	Yes	No
SIV_{AGM}gri	Grivet monkey (*Cercopithecus pygerythrus*[a])	Yes	No
SIV_{AGM}sab	Sabaeus monkey (*Cercopithecus sabaeus*[a])	Yes	No
SIV_{AGM}tan	Tantalus monkey (*Cercopithecus tantalus*[a])	Yes	No
SIV_{SYK}	Sykes' monkey (*Cercopithecus mitis*)	Yes	No
SIV_{SM}	Sooty mangabey (*Cercocebus atys*)	Yes	No
SIV_{WCM}	White crowned mangabey (*Cercocebus torquatus lunulatus*)	No	?
SIV_{MND}	Mandrill (*Mandrillus sphinx*)	Yes?	No
SIV_{BAB}	Yellow baboon (*Papio hamadryas cynocephalus*)	No	?
SIV_{MAC}	Macaques (various species[b])	No	Yes

[a] Members of the *Cercopithecus aethiops* super-species (African green monkeys).
[b] Captive rhesus (*Macaca mulatta*), pig-tailed (M. nemestrina), and stump-tailed (*M. arctoides*) macaques.

Considering the phylogenetic relationships of the primate lentiviruses (Fig. 9.1) the answer to the first of our questions may seem to be chimpanzees. However, the frequency of SIV infection in this species seems to be much lower than in African green monkeys or mangabeys (Peeters *et al.* 1992), and there is little information yet concerning whether SIV_{CPZ} infection in the wild is pathogenic. The alternative is that both chimpanzees and humans may have acquired this virus from a third, as yet unidentified, species. Since chimpanzees are known to hunt and eat small monkeys, presumably their opportunity for infection is similar to that of humans. Given the number of cross-species transmission events that can already be inferred to have occurred across the primate lentivirus phylogeny, the need to invoke two more events is not a great impediment to the hypothesis that chimpanzees are not the natural hosts of these viruses.

Again considering the evolutionary relationships of the primate lentiviruses (Fig. 9.1) the answer to the second of our questions may be guenons (the *Cercopithecus* monkeys). Given our conclusion that the radiation of the four types of SIV_{AGM} may have occurred with the divergence of their host species, the divergence of SIV_{AGM} from SIV_{SYK} just a little further back in the viral tree may also reflect host speciation. Guenons are the among the most numerous groups of monkeys in Africa, and several more *Cercopithecus* species have been shown to be seropositive for lentiviruses (Johnson *et al.* 1991). Under this scenario, each of the other three major lineages of SIV have arisen through cross-species transmissions from different species of *Cercopithecus*. Then, whether humans acquired HIV-1 via chimpanzees, or whether both humans and chimpanzees have independently become infected, remains in question. Given the empirical observations concerning pathogenicity of SIV in 'unnatural' hosts, more information about whether SIV_{CPZ} causes disease in the wild may provide some indication as to how long chimpanzees have been infected. In addition, determination of the nucleotide sequences of the as yet uncharacterized viruses present in other guenons should enable us to elucidate the entire history of the primate lentivirus group. If it emerges that both chimpanzees and humans have been relatively recently infected, and if it is found that this virus does not cause disease in wild chimpanzees, this would open an obvious line of inquiry into the factor(s) responsible for the difference between these two hosts.

Conclusions about the time-scale of the evolution of the primate lentiviruses (our third question) follow from the discussion of the initial hosts of this group. It is tempting to apply the molecular clock to place time estimates on the branching points in the primate lentivirus tree (Fig. 9.1). So far, the only nodes to which dates have been reliably placed are those within the radiation of HIV-1 group M. For example, the initial divergence (the split of subtype A from subtypes B and D in Fig. 9.1) has been placed at around 1960 (Li *et al.* 1988); this seems 'reliable' because several independent estimates of the rate of evolution are very similar, and because this time-scale fits well with epidemiological data. However, using this HIV-1-calibrated molecular clock would date the ancestry of the entire group of primate lentiviruses to only about 150 years ago (Sharp and Li 1988). If we are correct to conclude that the different species groups of SIV_{AGM} diverged at the same time as their hosts, which may have been thousands or millions of years ago, it is immediately apparent (from Fig. 9.1) that the rate of HIV/SIV evolution must have been very different at different points within the overall phylogeny. Ascertaining exactly which of the divergences in this phylogeny reflect host speciation events, and then dating the splits between these species from sequence comparisons, seems to be the only way to resolve this question.

Finally, the two main points discussed at length in this paper each have implications for the future. Recombination provides the opportunity for an 'evolutionary leap' in so far as the genetic consequence is far more drastic than the steady accumulation of individual mutations, and so a future hybrid virus may have significantly altered biological (and pathogenic) properties. It is also conceivable that SIV may cross into humans again, perhaps from a different simian source, and so yielding 'HIV-3'—we can only guess at the properties that virus might have.

Acknowledgements

We thank Francine McCutchan, Preston Marx, and Allen Mayer for making data available prior to publication. This work has been supported by a University of Nottingham studentship (to D.L.R.), by grants from the National Institutes of Health (RO1 AI 25291; RO1 AI 28273; NO1 AI 35170; to B.H.H.), and by shared facilities of the UAB Center for AIDS Research (P30 AI 27767). The content of this publication does not necessarily reflect the views or policies of the Department of Health and Human Services, nor does mention of trade names, commercial products, or organizations imply endorsement by the US Government.

References

Adkins, R. M. and Honeycutt, R. L. (1994). Evolution of the primate cytochrome *c* oxidase subunit II gene. *J. Mol. Evol.*, **38**, 215–31.

Albert, J., Wahlberg, J. and Uhlen, M. (1993). Forensic evidence by DNA sequencing. *Nature*, **361**, 595–6.

Alizon, M., Wain-Hobson, S., Montagnier, L., and Sonigo, P. (1986). Genetic variability of the AIDS virus: nucleotide sequence analysis of two isolates from African patients. *Cell*, **46**, 63–74.

Allan, J. S., Short, M., Taylor, M. E., Su, S., Hirsch, V. M., Johnson, P. R. *et al.* (1991). Species-specific diversity among simian immunodeficiency viruses from African green monkeys. *J. Virol.*, **65**, 2816–28.

Balfe, P., Simmons, P. Ludlam, C. A., Bishop, J. O., and Leigh Brown, A. J. (1990). Concurrent evolution of human immunodeficiency virus type 1 in patients infected from the same source: rate of sequence change and low frequency of inactivating mutations. *J. Virol.*, **64**, 6221–33.

Burns, D. P. and Desrosiers, R. C. (1991). Selection of genetic variants of simian immunodeficiency virus in persistently infected rhesus macaques. *J. Virol.*, **65**, 1843–54.

Chen, Z., Telfer, P., Reed, P., Zhang, L., Gettie, A., Ho, D. *et al.* (1994). First simian immunodeficiency virus from a free ranging sooty mangabey is equidistant from SIV and HIV-2 suggesting a new subtype. *12th annual symposium on non-human primate models for AIDS, Boston, MA*)

Diaz, R., Sabino, E. C., Mayer, A., Mosley, J. W., Busch, M. P., and the Transfusion Study Group (1995). Dual human immunodeficiency virus type-1 infection and recombination in a dually exposed transfusion recipient. *J. Virol.*, **69**, 3273–81.

Ewald, P. W. (1994). *Evolution of infectious disease*. Oxford University Press, New York.

Gao, F., Yue, L., White, A. T., Pappas, P. G., Barchue, J., Hanson, A. P. *et al.* (1992). Human infection by genetically diverse SIV_{SM}-related HIV-2 in West Africa. *Nature*, **358**, 495–9.

Gao, F., Yue, L., Craig, S., Thornton, C. L., Robertson, D. L., McCutchan, F. E. *et al.* (1994*a*). Genetic variation of HIV type 1 in four World Health Organization-sponsored vaccine evaluation sites: generation of functional envelope (glycoprotein 160) clones representative of sequence subtypes A, B, C and E. *AIDS Res. Human Retro.*, **10**, 1357–68.

Gao, F., Yue, L., Robertson, D. L., Hill, S. C., Hui, H., Biggar, R. J. *et al.* (1994*b*). Genetic diversity of Human Immunodeficiency Virus type 2: evidence for distinct sequence subtypes with differences in virus biology. *J. Virol.*, **68**, 7433–47.

Hahn, B. H. (1994). Viral genes and their products. In *Textbook on AIDS medicine*, (ed. S. Broder, T. Merigan, and D. Bolognesi), pp. 21–43. Williams and Williams, New York.

Hillis, D. M. and Huelsenbeck, J. P. (1994). Support for dental HIV transmission. *Nature*, **369**, 24–5.

Hirsch, V. M., Dapolito, G. A., Goldstein, S., McClure, H., Emau, P., Fultz, P. N. *et al.* (1993). A distinct African lentivirus from Sykes' monkeys. *J. Virol.*, **67**, 1517–28.

Hirsch, V. M., Olmsted, R. A., Murphy-Corb, M., Purcell, R. H., and Johnson, P. R. (1989). An African primate lentivirus SIV_{SM} closely related to HIV-2. *Nature*, **339**, 389–92.

Holmes, E. C., Zhang, C. Q., Simmonds, P., Rogers, A. S., and Leigh Brown, A. J. (1993). Molecular investigation of human immunodeficiency virus (HIV) infection in a patient of an HIV-infected surgeon. *J. Infect. Dis.*, **167**, 1411–14.

Hu, W.-S. and Temin, H. M. (1990). Retroviral recombination and reverse transcription. *Science*, **250**, 1227–33.

Javier, R. T., Sedarati, F., and Stevens, J. G. (1986). Two avirulent herpes simplex viruses generate lethal recombinants *in vivo*. *Science*, **234**, 746–8.

Jin, M. J., Hui, H., Robertson, D. L., Muller, M. C., Barre-Sinoussi, F., Hirsch, V. M. *et al.* (1994*a*). Mosaic genome structure of simian immunodeficiency virus from West African green monkeys. *EMBO J.*, **13**, 2935–47.

Jin, M. J., Rogers, J., Phillips-Conroy, J. E., Allan, J. S., Desrosiers, R. C., Shaw, G. M. *et al.* (1994*b*). Infection of a yellow baboon with SIV from African green monkeys: evidence for cross-species transmission in the wild. *J. Virol.*, **68**, 8454–60.

Johnson, P. R., Hirsch, V. M., and Myers, G. (1991). Genetic diversity and phylogeny of nonhuman primate lentiviruses. In *Annual review of AIDS research*, vol. 1, (ed. W. Koff), pp. 47–62. Dekker, New York.

Kawaoka, Y. and Webster, R. G. (1988). Molecular mechanism of acquisition of virulence in influenza in nature. *Microb. Pathogen.*, **5**, 311–18.

Kodama, T., Silva, D. P., Daniel, M. D., Phillips-Conroy, J. E., Jolly, C. J., Rogers, J.

et al. (1989). Prevalence of antibodies to SIV in baboons in their native habitat. *AIDS Res. Human Retro.*, **5**, 337–43.

Li, W.-H., Tanimura, M., and Sharp, P. M. (1988). Rates and dates of divergence between AIDS virus nucleotide sequences. *Mol. Biol. Evol.*, **5**, 313–30.

Louwagie, J., McCutchan, F. E., Peeters, M., Brennan, T. P., Sanders-Buell, E., Eddy, G. A. *et al.* (1993). Phylogenetic analysis of *gag* genes from 70 international HIV-1 isolates provides evidence for multiple genotypes. *AIDS*, **7**, 769–80.

Louwagie, J., Janssens, W., Mascola, J., Heyndrickx, L., Hegerich, P., van der Groen, G. *et al.* (1995). Genetic diversity of the envelope glycoproteins from human immunodeficiency virus type-1 (HIV-1) isolates of African origin. *J. Virol*, **69**, 263–71.

Marx, P. A., Li, Y., Lerche, N. W., Sutjipto, S., Gettie, A., Yee, J. A. *et al.* (1991). Isolation of a simian immunodeficiency virus related to human immunodeficiency virus type 2 from a pet sooty mangabey. *J. Virol.*, **65**, 4480–5.

Muller, M. C., Saksena, N. K., Nerrienet, E., Chappey, C., Herve, V. M. A., Durand, J.-P. *et al.* (1993). Simian immunodeficiency viruses from central and western Africa: evidence for a new species-specific lentivirus in tanatalus monkeys. *J. Virol.*, **67**, 1227–35.

Myers, G. and Korber, B. (1994). The future of human immunodeficiency virus. In *The evolutionary biology of viruses*, (ed. S. S. Morse), pp. 249–70. Raven, New York.

Myers, G., Korber, B., Wain-Hobson, S., Smith, R. F., and Pavlakis, G. N. (1993). *Human Retroviruses and AIDS.* Theoretical Biology and Biophysics, Los Alamos.

Ou, C.-Y., Ciesielski, C. A., Myers, G., Bandea, C. I., Luo, C.-C., Korber, B. T. M. *et al.* (1992). Molecular epidemiology of HIV transmission in a dental practice. *Science*, **256**, 1165–71.

Ou, C.-Y., Takebe, Y., Weniger, B. G., Luo, C.-C., Kalish, M. L., Auwanit, W. *et al.* (1993). Independent introduction of two major HIV-1 genotypes into distinct high-risk populations in Thailand. *Lancet*, **341**, 1171–74.

Peeters, M., Fransen, K., Delaporte, E., Vanden Haesevelde, M., Gershy-Damet, G.-M., Kestens, L. *et al.* (1992). Isolation and characterization of a new chimpanzee lentivirus: simian immunodeficiency virus isolate cpz-ant from a wild-captured chimpanzee. *AIDS*, **6**, 447–51.

Peeters, M., Jannsens, W., Fransen, K., Brandful, J., Heyndrickx, L., Koffi, K. *et al.* (1994). Isolation of simian immunodeficiency viruses from two sooty mangabeys in Cote d'Ivoire: virological characterization and genetic characterization and relationship to other HIV type 2 and SIVsm/mac strains. *AIDS Res. Human Retro.*, **10**, 1289–94.

Phillips-Conroy, J. E., Jolly, C. J., Petros, B., Allan, J. S., and Desrosiers, R. C. (1994). Sexual transmissions of SIV_{AGM} in wild grivet monkeys. *J. Med. Primatol.*, **23**, 1–7.

Robertson, D. L., Hahn, B. H., and Sharp, P. M. (1995*a*). Recombination in AIDS viruses. *J. Mol. Evol.*, **40**, 249–59.

Robertson, D. L., Sharp, P. M., McCutchan, F. E., and Hahn, B. H. (1995*b*). Recombination in HIV-1. *Nature*, **374**, 124–6.

Sabino, E. C., Shpaer, E. G., Morgado, M. G., Korber, B. T. M., Diaz, R., Bongertz, V. *et al.* (1994) Identification of HIV-1 envelope genes recombinant between subtypes B and F in two epidemiologically-linked individuals from Brazil. *J. Virol.*, **68**, 6340–6.

Sharp, P. M., Burgess, C. J., and Hahn, B. H. (1995). The molecular evolution of the Human Immunodeficiency Viruses. In *Molecular evolution of viruses*, (ed. A. J. Gibbs, C. H. Calisher, and F. Garcia-Arenal), pp. 438–54. Cambridge University Press.

Sharp, P. M. and Li, W.-H. (1988). Understanding the origins of AIDS viruses. *Nature*, **336**, 315.

Sharp, P. M., Robertson, D. L., Gao, F., and Hahn, B. H. (1994). Origins and diversity of human immunodeficiency viruses. *AIDS*, **8**, S27–S42.

Shpaer, E. G. and Mullins, J. I. (1993). Rates of amino acid change in the envelope protein correlate with pathogenicity of primate lentiviruses. *J. Mol. Evol.*, **37**, 57–65.

Tomonaga, K., Katahira, J., Fukasawa, M., Hassan, M. A., Kawamura, M., Akari, H. *et al.* (1993). Isolation and characterization of simian immunodeficiency virus from African white-crowned mangabey monkeys (*Cercocebus torquatus lunulatus*). *Arch. Virol.*, **129**, 77–92.

Vartanian, J.-P., Meyerhans, A., Asjo, B., and Wain-Hobson, S. 1991 Selection, recombination, and G→A hypermutation of human immunodeficiency virus type 1 genomes. *J. Virol.*, **65**, 1779–88.

10

Using interspecies phylogenies to test macroevolutionary hypotheses

Andy Purvis

10.1 Introduction

Have species differed in their chances of speciating and going extinct? If so, do any features, either of organisms or of environments, correlate with the differences? Have speciation and extinction rates changed significantly over the course of history? Such macroevolutionary questions have held an obvious fascination for centuries but, while the study of evolution within gene pools has moved on apace, attempts to study rigorously evolution at any higher level have generally been thwarted by a lack of information about the past. Recently, the cladistic and molecular revolutions have caused phylogenetic research to snowball. These new phylogenies, coupled with the development of suitable statistical null models, are enabling evolutionary biologists to revisit with profit fundamental questions about the evolution of diversity. Interspecies phylogenies differ from within-species genealogies in two respects that impact upon how they can be used: they can include all living members of a clade, but they are usually small. Within-species genealogies, which are considered in other chapters of this book, generally include only a very small fraction of extant lineages but this may none the less be a large number. Complete phylogenies allow more questions to be asked, but the small size of the datasets will limit the precision of the answers.

Phylogenies showing relationships among the extant species of a clade can yield insights into the tempo and mode of evolution because their shapes reflect the processes that generated them (Raup *et al.* 1973; Savage 1983; Dial and Marzluff 1989; Slowinski and Guyer 1989; Harvey *et al.* 1991; Hey 1992; Williams 1992; Nee *et al.* 1994*a*). Consider the two trees shown in Fig. 10.1. Both have eight tips, representing eight living species more closely related to each other than to any other living species, but they have very different shapes; Fig. 10.1a is extremely asymmetrical or unbalanced whereas Fig. 10.1b is totally symmetrical. Clearly, sister lineages have radiated at different rates in the first tree, but not in the second. Should we try to ascribe some macroevolutionary cause to the rate variation in the first tree? The answer is not immediately obvious. Although Fig. 10.1a is as unbalanced as it can be, we would not *expect* all trees to be totally symmetrical like that in Fig. 10.1b—such symmetry would suggest a programmed process like the division of a zygote, and we have no basis for expecting the radiation of clades to be regulated in any such way. So how asymmetrical, or indeed how symmetrical, must a tree be before we are sufficiently surprised to ask why? Clearly, to study macroevolution, we need more than phylogenies. We also need a null model of what those phylogenies should look like in the absence of any deterministic factors—a macroevolutionary analogue of a genetic drift model.

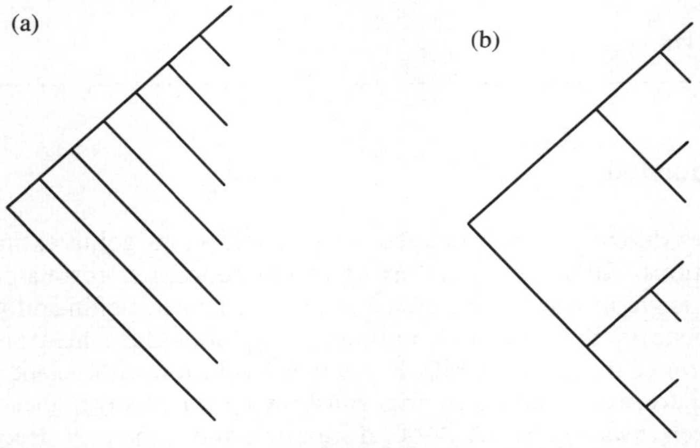

Fig. 10.1 Two interspecies phylogenies showing extremes of (a) imbalance or asymmetry, and (b) balance or symmetry. See text for discussion.

A suitable first null model is outlined in Section 10.2. It makes few assumptions and, as Sections 10.3–10.5 show, generates many testable predictions. What if the predictions are not in accord with what we see in our phylogenies? Section 10.6 considers the possibilities: the phylogenies may be wrong, the null model may not be a good yardstick against which to measure our observations, or some deterministic force may have been at work. The chapter ends with an assessment of the impact that new phylogenies are having on tests of macroevolutionary hypotheses, and an outline of research priorities.

10.2 The constant-rates birth–death process model

A suitable first null model is one where every lineage has the same probability λ of speciating, and the same probability μ of going extinct, in each time unit: this model is known variously as the Markov model (Simberloff *et al.* 1981) and the constant-rates birth–death process model (Purvis *et al.* 1995). For a clade to survive in the long term, it is necessary that λ be greater than μ. This null model has been used in phylogenetic studies of macroevolution for over 20 years (Raup *et al.* 1973) and is still generating new ways to test hypotheses (for example Kubo and Iwasa, 1995). Some of the tests that are described below make the further simplifying assumption that there has been no extinction: this pure-birth model has an even longer history (Yule 1924). Although the assumption of no extinction is probably unrealistic for interspecies phylogenies, the pure-birth model, which has a single parameter equivalent to $\lambda - \mu$, is often a close approximation to the birth–death process model. Additionally, some tests assume only that λ and μ have not varied among contemporaneous lineages; this assumption of lineage equivalence can be met even when λ and μ have changed over time.

In simulation trials, shapes and sizes of phylogenies can vary greatly from trial to trial, even when λ and μ are held constant (Raup *et al.* 1973). The variety reflects the stochastic nature of the birth–death process: the constancy of λ and μ does not mean that different clades will show the same *realized* per-lineage rates of speciation (b), extinction (d) or radiation ($r = b - d$): b, d, and r will be estimates of λ, μ and, $\lambda - \mu$ but the estimates may not be very accurate, especially with a small phylogeny. By chance, some lineages will speciate much sooner than others. Further, because every species has an equal chance of being the next to speciate, early chance differences in species number tend to become magnified greatly as clades grow exponentially. It follows that phylogenies are unlikely to be perfectly symmetrical. As a corollary, sister taxa must contain very different numbers of species before any process other than chance need be invoked (Slowinski and Guyer 1989; see also below).

10.3 Tests based on tree topology

Species richness of sister clades

Analytical results or simulations using the null model allow us to test whether a given phylogeny is consistent with the null hypothesis of lineage equivalence. Perhaps the simplest test (and certainly the one that has most often been derived (Farris 1976; Van Pelt and Verwer 1983; Slowinski and Guyer 1989; Nee *et al.* 1994*a*)) focuses on the numbers of species in two sister taxa. The null model predicts that all possible partitions of N species into two sister clades, for example (1, N-1), (2, N-2), and so on, are equiprobable. This result underlines that extreme imbalance is not unexpected: a (1, 7) or (7, 1) partition, as in Fig. 10.1a, is expected to occur two times out of seven—twice as likely as a perfectly balanced (4, 4) partition, and scarcely an event that demands explanation. Indeed, N must exceed 40 before even a totally unbalanced tree attains significance with this test (Kirkpatrick and Slatkin 1993). The test uses only very limited phylogenetic information, just the numbers of species in the two sister clades, which is both a strength and a weakness. Its strength is that it does not require a detailed phylogeny; for example Slowinski and Guyer (1989) were able to show that *Anolis* lizards are significantly more species rich than their sister group, *Chamaeleolis*, without needing to know the intrageneric phylogenies of either group. The weakness is that the test is not powerful because it cannot use more information when it is available.

Tree balance

If the pattern of sister taxon relationships is known in full, one can consider the balance or imbalance of the whole tree, rather than of just a single node. Kirkpatrick and Slatkin (1993) described six non-parametric tests for departure from the null model, each based on a different summary measure of tree shape. For a range of tree sizes ($N \leqslant 50$) they generated 10 000 null trees; computing the tree shape statistics for each one produced null distributions against which real trees can be compared. Because these tests use information from every branching point, they are very much more powerful than the sister clades test. For example, Fig. 10.1a is judged significantly ($p < 0.05$) asymmetric by all six tests. Additionally, these tests are two-tailed: excessive symmetry, as well as asymmetry, is inconsistent with the null model, and all six tests adjudge Fig. 10.1b to be too symmetric ($p < 0.05$). The tests will not always agree so well, because the measures on which they are based all capture slightly different aspects of tree shape. Kirkpatrick and Slatkin (1993) applied their tests to phylogenies of two groups of leaf beetles ($N = 14$ and 12 species), and concluded that both showed significant asymmetry.

Although these tests are more powerful than the first one considered, speciation rates must still be very different between sister taxa before the effect is likely to be noticed. Simulations indicate that, even with a fourfold rate difference, a tree with 20 species would be deemed consistent with the null model about 50 per cent of the time (Kirkpatrick and Slatkin 1993). Power increases with phylogeny size, but there are still very few fully resolved phylogenies of clades as large as 20 species. Single phylogenies, then, are unlikely to reject the null model, however badly it resembles the truth.

Although most single phylogenies are of limited use, collections of trees can allow powerful tests. If the frequencies with which different topologies occur depart significantly from the frequencies expected under the Markov model, we might conclude that speciation rates commonly vary among species. Savage (1983) compiled over 1000 phylogenies, mostly of insects or tetrapods, with between four and seven terminal taxa, and found the topology frequencies accorded well with null expectations. Heard (1992) considered 208 phylogenies with between 4 and 14 tips and found real trees of all sizes to be more unbalanced than null trees. However, both studies included many phylogenies that were incomplete; that is, they either did not have species as their tips or did not contain all the living species within the clade defined by the root node. Consequently, it is possible that the results, or the differences between them, were affected by some bias. Guyer and Slowinski (1991) analysed 120 nearly complete (no more than one species missing) five-species phylogenies, taken equally from tetrapods, insects, and angiosperms; each group showed significantly more asymmetric trees than expected under the null model. A collection of 39 complete phylogenies of at least eight species led to a similar conclusion (Mooers 1995).

10.4 Tests using a time-scale

The tests described so far consider only the topology of the phylogeny. The dates of lineage splits, whether absolute or relative, provide further information and are increasingly available. Sibley and Ahlquist's (1990) estimate of avian phylogeny is the most ambitious attempt so far to elucidate the relationships within a major group. Their dendrogram, based on DNA-DNA hybridization data from around 1700 of the nearly 10 000 bird species, provided the first testing ground for ways of using relative dates in tests of the null model (for example Harvey *et al.* 1991; Nee *et al.* 1992; Harvey and Nee 1993, 1994).

Do short branches give rise to short branches?

The most intuitive use of date information is to test whether speciation rate is heritable. Consider the sister lineages A and B in Fig. 10.2. Lineage A speciated first, giving rise to lineages A1 and A2; B later speciated to produce B1 and B2. If A's earlier speciation was due to some heritable trait not shared by B, then A1 and A2 should tend to persist for a shorter time than B1 and B2 before speciating further; in other words, short branches in the phylogeny should give rise to short branches more often than not. Under the null model, however, there should not be any consistent difference between the lengths of branches descended from A and those descended from B. Although Sibley and Ahlquist's (1990) phylogeny is far from complete, it includes 68 ancestral nodes that can contribute to a test comparing the shorter of A1 and A2 with the shorter of B1 and B2 (Harvey *et al.* 1991). Nodes below the family level were consistent with the null model, but significantly more than half of the nodes above the family level indicate heritability of branch lengths (Harvey *et al.* 1991). Very similar

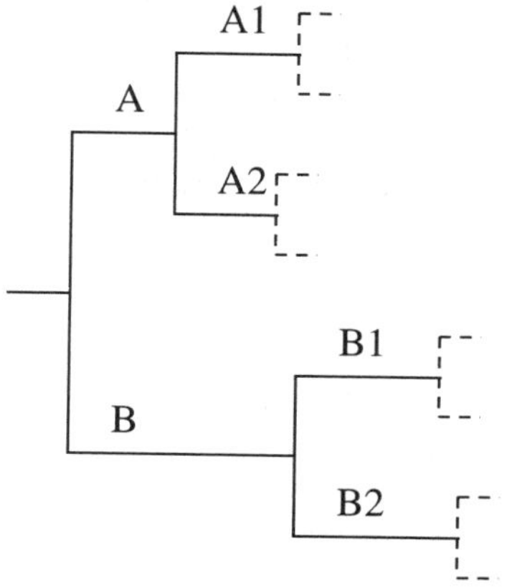

Fig. 10.2 Part of a phylogeny that can be used to test the heritability of speciation probability. Lineage A speciates before lineage B; if probabilities are heritable, branches A1 and A2 should tend to be shorter than B1 and B2. (After Harvey *et al.* 1991.)

logic underlies the Links test (Hey 1992). A 'link' is said to have occurred when the next lineage to speciate is one produced by the previous speciation event. When a clade contains N species ($N > 1$), the probability that the next speciation event will produce a link is simply $2/N$. Links should be rare unless N is small; Hey presents a quantitative test based on the likelihoods of links for each size of clade, which can be used to ask whether a phylogeny contains either too many links or too few (the latter implying that newly produced species had lower probabilities of further speciation). Hey tested the links patterns of six complete phylogenies ($N = 6$ to 26), and found all of them to be consistent with the null model.

Distribution of descendants among ancestors

Time-scales for phylogenies allow a powerful generalization of the sister clades comparison described in Section 10.3. Consider a set of species that formed a monophyletic group at some given time in the past, and their fate at some later time. Some may have left one or more descendant species in the later biota, whereas other clades may have gone extinct between then and now. Extinct lineages will generally not be represented in molecular phylogenies, so we can consider only those ancestors with living descendants. Under the assumption of lineage equivalence, the distribution of N descendants among k ancestors follows a broken-stick distribution (Nee *et al.* 1992, 1994*a*), easily simulated by repeatedly randomly breaking sticks of length N into k fragments whose lengths are positive integers. A dated phylogeny allows us to ask whether descendants are distributed among ancestors in accord with this expectation. Harvey *et al.* (1991) and Nee *et al.* (1992) asked this question for two successive time windows early on in the radiation of birds. At the start of the first, there were 24 bird lineages that have living descendants; this had grown to 56 by the end of the first time window. These 56 still-extant lineages grew to 137 during the second time window. The distribution of descendants among ancestors in the first window fitted the null model well, but two ancestors in the second—the Ciconiiformes (shorebirds) and the Passeri (songbirds)—produced far more descendants than expected. It seems that these lineages had particularly high probabilities of speciating, or low probabilities of extinction, or both (Nee *et al.* 1992).

The results of such time window analyses are sensitive to the particular window that is chosen, leading to the possibility that departures from the null model might be missed. If the phylogeny is not only dated but complete, the possibility can be overcome by using the END-EPI computer program (Rambaut *et al.*, submitted) which, in effect, analyses all possible sets of ancestors simultaneously. Use of this program with a composite phylogeny of the whole primate order (Purvis 1995) showed that the lineage that has given rise to the family Cercopithecidae (Old World monkeys) was a significant radiation (had a significantly higher rate of net cladogenesis than other lineages), and that there have apparently been further increases in the net rate of cladogenesis in four tribes within that family (Purvis *et al.* 1995).

Analysis of mean clade radiation rate

The mean radiation rate of a clade can be estimated from the clade's age (t) and current species richness (N), by treating clade growth as a pure birth process. The maximum likelihood estimate of $\lambda - \mu$ is simply $\ln(N)/t$. (Note that this, in common with the maximum likelihood estimates mentioned below, is biased (Sanderson and Donoghue 1996)). A 95 per cent confidence interval can be placed on the estimate (they are $-\ln(1-0.975^{1/N})/t$ and $-\ln(1-0.025^{1/N})/t$ (S. Nee, personal communication)) to give some indication of whether different clades have experienced similar growth parameters. If the relative or absolute times of all lineage appearances are known, they provide a more powerful means, also based on maximum likelihood, to test for significant differences among clades; the procedure is described by Purvis *et al.* (1995), who used it to show that Old World monkeys had radiated significantly more rapidly than strepsirhines, New World monkeys or apes.

Analysis of speciation and extinction rates

Perhaps surprisingly, complete phylogenies of only extant species allow inferences to be made about both speciation *and extinction* probabilities (Harvey *et al.* 1994): if both λ and μ have been constant, they can be estimated by maximum likelihood (Nee *et al.* 1994*a*,*b*) or other procedures (Kubo and Iwasa 1995). Figure 10.3 illustrates graphically how this is so. If $\lambda > \mu$, a clade will tend to grow exponentially at a rate that settles down to as $\lambda - \mu$ the clade becomes large. A semi logarithmic plot of the number of lineages extant at time t against t will therefore produce a straight line—with stochastic ups and downs—whose slope estimates $\lambda - \mu$; this is line 1 in Fig. 10.3. However,

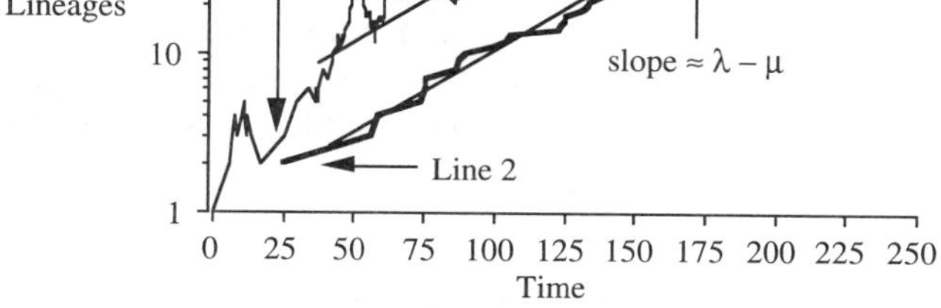

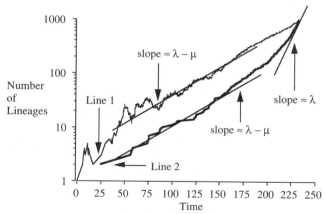

Fig. 10.3 The semilogarithmic lineages-through-time plot for a simulated phylogeny, illustrating how λ and μ can be estimated under the assumption that they have been constant. Line 1 shows the actual numbers of lineages extant at time t; the slope of this line estimates $\lambda - \mu$. Line 2 excludes lineages that have gone extinct. Over most of the time axis, the slope again estimates $\lambda - \mu$, but the curve steepens towards the present, where the slope of the tangent to the line estimates λ. (Modified after Harvey *et al.* 1994.)

lineages that have gone extinct will not generally be in our phylogeny. If only lineages with extant descendants (i.e. those in a complete molecular phylogeny) are considered, the plot looks somewhat different (line 2). Over most of the time axis the line is still straight with slope $\lambda-\mu$. It lies below line 1, because of lineages that have gone extinct. However, the line steepens toward the present because lineages that arose only recently have not yet had much chance of going extinct. At the present, the slope of the tangent to the line is λ (Harvey *et al.* 1994; Kubo and Iwasa 1995).

Both λ and μ can be estimated from the lineages-through-time plots by curve fitting (Kubo and Iwasa 1995) but significance testing is complicated by the problems associated with time-series data. The maximum likelihood procedure derived by Nee *et al.* (1994*a*) uses instead the intervals between the times of appearance of successive lineages with extant descendants to estimate two composite parameters $\lambda-\mu$ and μ/λ. A likelihood surface can be constructed with these axes (Nee *et al.* 1994*a*), and a confidence region (a two-dimensional confidence interval) put around the maximum likelihood values (Purvis *et al.* 1995). If different clades have shared the same λ and μ, the maximum likelihood estimates of these parameters from one clade should fall within the confidence region of another. This approach has been used to compare the estimates of λ and μ from major primate clades (Purvis *et al.* 1995), and produced very similar results to the comparison of clade net growth rates (see previous subsection).

Estimates of λ and μ have other uses too. Under the null model, knowing λ, μ, and N (the number of extant species) allows inferences about the number of extinct species, N_E, whose ancestry lay within the phylogeny being analysed (i.e. crown group, not stem group, species (Smith 1994)). N_E can be approximated by (Purvis *et al.*, submitted):

$$N_E = N.\ \frac{\mu/\lambda}{1-(\mu/\lambda)}.$$

The upper confidence limit for μ/λ translates directly into an upper confidence limit for N_E. This can be compared with the number of described fossil species within the crown group, providing an external 'reality check' on the model from data not used in its construction. If one were prepared to assume the model to be reasonable, one could even estimate the completeness of the clade's fossil record.

10.5 Testing the constancy of rates over time

Some of the tests outlined so far assume that λ and μ are constant over time (though others merely assume that they are constant across contemporaneous lineages (Guyer and Slowinski 1991; Nee *et al.* 1994*b*; Purvis *et al.* 1995)). This assumption can be tested if the phylogeny has a time-scale. Density-dependent cladogenesis (such as might be predicted under a niche-filling model) or smooth changes in rates over time leave characteristic 'footprints' in the lineages-through-time plots (Harvey *et al.* 1994; Kubo and Iwasa 1995). Statistical tests can use the expectation that, by hypothesis, the product of the number of lineages and the waiting time to the next speciation event is a constant: regression of this quantity against number of lineages or against time provides a simple test (Purvis *et al.* 1995). Tests such as this (others are described by Nee *et al.* (1995) and Kubo and Iwasa 1995) suggest that avian diversification has slowed down (Nee *et al.* 1992) but primate clade growth has not (Purvis *et al.* 1995). It is not possible to know whether diversification is due to increased λ or decreased μ; different combinations of the two can leave identical footprints (Kubo and Iwasa 1995). Episodes of explosive branching and, with more difficulty, mass extinction can also be identified from lineages-through-time plots for sufficiently large phylogenies (Harvey *et al.* 1994; Kubo and Iwasa 1995).

10.6 Explaining departures from the null model

What does it mean when a phylogeny suggests that contemporaneous lineages had different speciation and extinction probabilities? Leaving aside the possibility of a Type I error, some assumption that has gone into the null model must be incorrect. The most exciting possibility is that some lineages have experienced factors that affect their chances of speciation and/or extinction; this possibility, a necessary condition for clade selection (Williams 1992), is discussed shortly. First, though, I consider other, more prosaic, explanations—the phylogeny may be wrong, or the constant-rates birth–death process may not be a suitable null model in the first place.

Errors in the phylogeny?

All of the tests described above assume that the phylogeny is known. In practice it is only estimated and, phylogeny estimation being a difficult business, sometimes estimated poorly. If errors are unbiased with respect to tree shape, they will merely add extra uncertainty to parameter estimates in some tests. However, it has been suggested that errors in phylogenies are likely to produce trees that are more unbalanced than they should be (Shao and Sokal 1990). Two lines of evidence support this idea. Firstly, if parsimony is applied to a data matrix that contains random data, the resulting tree will effectively be a random choice from the set of possible trees; the average choice is more unbalanced than the average tree from the null model (Guyer and Slowinski 1991). It seems that some misleading characters (homoplasy) in a matrix might be enough to bias tree shape: cladograms that are only poorly supported by the data used to construct them (because of homoplasy) tend to be more unbalanced than well-supported trees (Mooers *et al.* 1995). Interestingly, the data matrices used to construct some published trees contain significantly less hierarchical signal than would be expected in random data (A. Ø. Mooers, personal communication)! Guyer and Slowinski (1991) found that their small phylogenies (mentioned in Section 10.3) showed a collective degree of imbalance that, while being too great for the null model, was consistent with the trees having been chosen randomly.

A second cause for concern is that simulations of five common tree-building procedures—maximum likelihood, neighbour-joining, parsimony, UPGMA (unweighted pair group method using arithmetic averages), and distance metrics—show all of them to be biased, even when their assumptions are met (Huelsenbeck and Kirkpatrick, in press). The first three methods seem particularly biased, all tending to unbalance trees, especially when rates of evolution are high. Although UPGMA seems to fare well when the molecular clock assumption is correct, it is very liable to produce asymmetrical trees otherwise (Mooers *et al.* 1994). At present, it is not clear either quite where the biases reported by Mooers *et al.* (1995) and Huelsenbeck and Kirkpatrick (in press) come from, or whether they are sufficient to cause the discrepancies between observed topology frequencies and null expectations; more work is needed urgently.

The wrong null model?

Another possible reason for observations departing from expectations is that the expectations are derived from a poor null model. Ideally, the null model should fail only when case-specific factors are at play; factors that are intrinsic to the processes of speciation and extinction, in any taxon, ought to be represented in the null model. Perhaps the constant-rates birth–death process model is often a poor fit to data because some such factor is missing from it? Losos and Adler (1995) point out that the null model precludes any refractory or lag period in which a newly formed lineage is less likely to undergo further lineage splitting. Such a refractory period could arise if, for instance, species formed as peripheral isolates took time to attain the average geographic range; equally, one might expect newly formed and hence geographically localized lineages to have atypically high extinction rates. Refractory periods can markedly alter expectations of tree shapes if they are not small compared to the mean length of a branch on the tree (Losos and Adler 1995). Little is known about how long real refractory periods might be, but it seems likely that they should be considered in at least rapid radiations such as cichlid fishes (Losos and Adler 1995). Including a refractory period (which could be estimated from the phylogeny simultaneously with other parameters) in the null model would complicate testing, but might produce a benefit by more accurately identifying patterns that require explanation. However, it is worth noting that the inclusion of a refractory period makes expected trees more balanced: trees that are judged significantly asymmetrical under the constant-rates model would still be so under the more general null model that Losos and Adler suggest.

Case-specific factors?

The remaining explanation for failure of the null model is that some case-specific factor has been at work in the clade under study. Obviously, we would like to know what that factor is, be it an evolutionary novelty (a key innovation (Liem 1973); see also Erwin (1992) and Heard and Hauser (1995)) or an environmental change. As well as throwing light on the evolution of particular groups, the accumulation of case studies might show us how common clade selection (*sensu* Williams 1992) might be, which deviations from the null model predominate, and whether particular categories of key innovations (for example feeding adaptations, changes in life history) or environmental changes (for example dispersal events, vicariance events) keep cropping up regularly.

How can hypotheses about such factors be tested? Comparisons between sister taxa, one possessing the hypothesized factor and the other not, provide a simple and valid way to proceed (Cracraft 1981, 1984). Mitter *et al.* (1988) used this approach to test whether phytophagy has promoted insect diversification. They were able to make 13 sister taxon comparisons; in 11 of these, the phytophagous clade contained more species than its presumed sister clade. A sign test comfortably rejects the null hypothesis that phytophagy and species richness are independent (other, more powerful, tests are possible which consider the magnitude (Barraclough *et al.* 1995) or improbability under the null model (Slowinski and Guyer 1993; but see Nee *et al.* 1996) of the observed differences in species richness). Similar analyses have suggested that the evolution of carnivorous parasitism by insects does not have such a marked effect on diversification (Wiegmann *et al.* 1993), that plant lineages that produce latex and resin canals are more diverse than their sister taxa (Farrell *et al.* 1991), and that sexual dichromatism is associated with species richness in passerine birds (Barraclough *et al.* 1995).

Sister clade comparisons are powerful only if the proposed cause has appeared in a reasonable number of different lineages. If, instead, the proposed cause was unique, convincing statistical testing is out of the question (Williams 1992; Nee *et al.* 1996), beyond testing whether the diverse clade is indeed too diverse for the null model. Logical tests are still possible, however, which can reject hypotheses as being implausible (Doyle and Donoghue 1993; Sanderson and Donoghue 1994; Heard and Hauser 1995). To be a plausible cause of increased diversification for a clade, a characteristic should be shared by all basal lineages within the significant radiation, but absent from sister groups (i.e. it should be a synapomorphy for the radiation). Additionally, there should be a mechanism whereby the proposed cause affects speciation and/or extinction probabilities—the plausibility of this mechanism might also be testable. Few hypotheses, however, have been tested as rigorously as they might be (Heard and Hauser 1995). A notable exception is provided by Sanderson and Donoghue's (1994) rejection of some popular key innovation hypotheses for the diversity of flowering plants.

The studies mentioned so far in this section have all tested the effects of traits that are either present or absent. Clade success is also likely to be affected by continuous variables: body size, for instance, has often been singled out (for example Stanley 1973; Dial and Marzluff 1988; Martin 1992). Tests for such variables are currently less well developed, and studies using them with phylogenies are so far thin on the ground. In two studies of high-level avian phylogeny, diversity was not significantly related to either body size (Nee *et al.* 1992) or the proportion of species that breed colonially (Mooers and Møller, in press).

10.7 Discussion

This chapter has outlined several ways in which interspecies phylogenies of living species, such as molecular data are now producing, can be used to investigate macroevolutionary patterns and processes. Workers can choose from a battery of tests according to the nature and quantity of available phylogenetic data (Sanderson and Donoghue 1995). But there is a long history of attempts to tackle these issues using taxonomic information before suitable phylogenies became available (for example Stanley 1973; Van Valen 1975; Dial and Marzluff 1989). Why is it so vital to use a phylogeny? The answer is that taxonomies may not reflect evolutionary history accurately—they may not even *try* to do so—so it is difficult to know whether a non-random pattern in a taxonomy is due to bias in evolution or bias in taxonomic practice (Doyle and Donoghue 1993). Two mammalian examples serve to make the point. Across mammal genera, there is a significant negative correlation between the body size and species richness (Van Valen 1973; Martin 1992). The relationship might be taken to indicate that speciation is more rapid among smaller-bodied mammals, but it might merely indicate a tendency of taxonomists to split groups of large organisms more readily than groups of small ones. As a second example, 1803 of the 4052 species of placental mammal listed by Corbet and Hill (1991) are in a single order, Rodentia. The time window analysis of Section 10.4, with 19 ancestors (one for each order) and 4052 descendants, rejects strongly the null model of lineage equivalence. One might naturally go on to ask why there are so many rodent species, perhaps concluding that it is due to their small size (Kochmer and Wagner 1988). However, consideration of recent molecular phylogenies suggests that the result should not be trusted for three reasons. First, the sister clade of the Rodentia comprises most of the other orders and contains around 2000 species (Li *et al.* 1990; Bulmer *et al.* 1991)—clearly it is misguided to ask why there are so *many* rodents. Secondly, the order may not even be monophyletic (Graur *et al.* 1991), in which case no comparison between orders is possible. Lastly, not all of the early branches within the Rodentia are particularly species rich (Sarich 1986). If one clade has become unusually diverse, it is probably a lineage within the Myomorpha: attempts to explain the diversity should therefore focus on differences between myomorphs and their sister group, rather than the differences between rodents and other placental mammals. Phylogenies, in summary, help us to ask the right questions, and to ask them of the right taxa. They also help us to ask them in the right way: as in comparative studies of adaptation (Harvey and Pagel 1991), tests of association based on phylogenies can give very different results from those which ignore the hierarchy (Nee *et al.* 1992; Mooers and Møller, in press).

We must, then, base our tests on the best phylogenies that are available. This imperative suggests some research priorities. We need more, bigger and better phylogenies: more trees—especially more complete trees—let us study more groups, increasing our chances of seeing any general patterns that there may be. Bigger trees make many tests more powerful and make others feasible (phylogenies of living species will need to be very large if we are to demonstrate a mass extinction, for instance (Kubo & Iwasa 1995)). Trees can continue to get better: the apparent bias in phylogeny estimation procedures merits close scrutiny, and more attention could usefully be paid to estimating dates of splits.

We also need to develop further the statistical tests. The most obvious case in point is the assumed perfection of phylogenies: we need tests that allow this assumption to be relaxed, or at least we need to investigate how badly it might mislead us. Also, there is currently no powerful way to relate continuous character variation to diversity. Additionally, although the phylogenies will hopefully get ever closer to the ideals required by theory, the theory underlying the tests must also move towards where the data are now. Currently, for instance, dates can be used only if they are available for every node, and trees have to be completely bifurcating for many tests: workers have to choose between less powerful tests or extra assumptions (see Purvis *et al.* 1995).

The next few years will surely see an explosion of phylogenetic tests of macroevolutionary hypotheses. The literature already contains countless hypotheses (see, for example Heard and Hauser 1995), many of them supported by pre-phylogenetic tests: it will be interesting to see which survive re-examination. As studies accumulate, we shall begin to see whether any generalities emerge. Have different radiations had similar causes, or have they all been due to unique combinations that defy generalization? We may be about to find out.

Acknowledgement

This work was funded by the NERC (GR3/8515).

References

Barraclough, T. G., Harvey, P. H., and Nee, S. (1995). Sexual selection and taxonomic diversity in passerine birds. *Proc. R. Soc.*, **B259**, 211–15.

Bulmer, M., Wolfe, K. H., and Sharp, P. M. (1991). Synonymous nucleotide substitution rates in mammalian genes: implications for the molecular clock and the relationships of mammalian orders. *Proc. Natl. Acad. Sci. USA*, **88**, 5974–78.

Corbet, G. B. and Hill, J. E. (1991). *A world list of mammalian species*. Natural History Museum, London.

Cracraft, J. (1981). Pattern and process in paleobiology: the role of cladistic analysis in systematic paleontology. *Paleobiology*, **7**, 456–8.

Cracraft, J. (1984). Conceptual and methodological aspects of the study of evolutionary rates, with some comments on bradytely in birds. In *Living fossils*, (ed. N. Eldredge and S. M. Stanley), pp. 95–104. Springer, New York.

Dial, K. P. and Marzluff, J. M. (1988). Are the smallest organisms the most diverse? *Ecology*, **69**, 1620–4.

Dial, K. P. and Marzluff, J. M. (1989). Nonrandom diversification within taxonomic assemblages. *Syst. Zool.*, **38**, 26–37.

Doyle, J. A. and Donoghue, M. J. (1993). Phylogenies and angiosperm diversification. *Paleobiology*, **19**, 141–67.

Erwin, D. H. (1992). A preliminary classification of evolutionary radiations. *Hist. Biol.*, **6**, 133–47.

Farrell, B. D., Dussourd, D. E., and Mitter, C. (1991). Escalation of plant defense: do latex and resin canals spur plant diversification? *Am. Nat.*, **138**, 881–900.

Farris, J. S. (1976). Expected asymmetry of evolutionary rates. *Syst. Zool.*, **25**, 196–8.

Graur, D., Hide, W. A. and Li, W.-H. (1991). Is the guinea pig a rodent? *Nature*, **351**, 649–52.

Guyer, C. and Slowinski, J. B. (1991). Comparison of observed phylogenetic topologies with null expectations among three monophyletic lineages. *Evolution*, **45**, 340–50.

Harvey, P. H. and Nee, S. (1993). New uses for new phylogenies. *Eur. Rev.*, **1**, 11–19.

Harvey, P. H. and Nee, S. (1994). Comparing real with expected patterns from molecular phylogenies. In *Phylogenetics and ecology*, (ed. P. Eggleton and R. Vane-Wright), pp. 219–31. The Linnean Society, London.

Harvey, P. H. and Pagel, M. D. (1991). *The comparative method in evolutionary biology*. Oxford University Press.

Harvey, P. H., Nee, S., Mooers, A. Ø., and Partridge, L. (1991). These hierarchical views of life: phylogenies and metapopulations. In *Genes in ecology: the 33rd symposium of the British Ecological Society*, (ed. R. J. Berry, T. J. Crawford, and G. M. Hewitt), pp. 123–37. Blackwell Scientific, Oxford.

Harvey, P. H., May, R. M., and Nee, S. (1994). Phylogenies without fossils. *Evolution*, **48**, 523–9.

Heard, S. B. (1992). Patterns in tree balance among cladistic, phenetic, and randomly generated phylogenetic trees. *Evolution*, **46**, 1818–26.

Heard, S. B. and Hauser, D. L. (1995). Key evolutionary innovations and their ecological mechanisms. *Hist. Biol.*, **10**, 151–73.

Hey, J. (1992). Using phylogenetic trees to study speciation and extinction. *Evolution*, **46**, 627–40.

Huelsenbeck, J. P. and Kirkpatrick, M. Do phylogenetic methods produce trees with biased shapes? *Evolution*. (In press.)

Kirkpatrick, M. and Slatkin, M. (1993). Searching for evolutionary patterns in the shape of a phylogenetic tree. *Evolution*, **47**, 1171–81.

Kochmer, J. P. and Wagner, R. H. (1988). Why are there so many kinds of passerine birds? Because they are small. A reply to Raikow. *Syst. Zool*; **37**, 68–9.

Kubo, T. and Iwasa, Y. (1995). Inferring the rates of branching and extinction from molecular phylogenies. *Evolution*, **49**, 694–704.

Li, W. -H., Gouy, M., Sharp, P. M., O'hUigin, C., and Yau-Wen, Y. (1990). Molecular phylogeny of Rodentia, Lagomorpha, Primates, Artiodactyla and Carnivora and molecular clocks. *Proc. Natl. Acad. Sci. USA*, **87**, 6703–7.

Liem, K. F. (1973). Evolutionary strategies and morphological innovations: cichlid pharyngeal jaws. *Syst. Zool.* **22**, 425–41.

Losos, J. B. and Adler, F. R. (1995). Stumped by trees? a generalized null model for patterns of organismal diversity. *Am. Nat.*, **145**, 329–42.

Martin, R. A. (1992). Generic species richness and body size in North American mammals: support for the inverse relationship of body size and speciation rate. *Hist. Biol.*, **6**, 73–90.

Mitter, C., Farrell, B., and Wiegmann, B. (1988). The phylogenetic study of adaptive zones: has phytophagy promoted insect diversification? *Am. Nat.*, **132**, 107–28.

Mooers, A. Ø. (1995). Tree balance and tree completeness. *Evolution*, **49**, 379–84.

Mooers, A. Ø. and Møller, A. P. Colonial breeding and speciation in birds. *Evol. Ecol.*, (in press).

Mooers, A. Ø., Nee, S., and Harvey, P. H. (1994). Biological and algorithmic correlates of phenetic tree pattern. In *Phylogenetics and ecology*, (ed. P. Eggleton and D. Vane-Wright), pp. 233–51. The Linnean Society, London.

Mooers, A. Ø., Page, R. D. M., Purvis, A., and Harvey, P. H. (1995). Character congruence and the balance of cladistic trees. *Syst. Biol.*, **44**, 332–42.

Nee, S., Mooers, A. Ø., and Harvey, P. H. (1992). The tempo and mode of evolution revealed from molecular phylogenies. *Proc. Natl. Acad. Sci. USA*, **89**, 8322–6.

Nee, S., May, R. M., and Harvey, P. H. (1994*a*). The reconstructed evolutionary process. *Phil. Trans. R. Soc.*, **B344**, 305–11.

Nee, S., Holmes, E. C., May, R. M., and Harvey, P. H. (1994*b*). Extinction rates can be estimated from molecular phylogenies. *Phil. Trans. R. Soc.*, **B344**, 77–82.

Nee, S., Holmes, E. C., Rambaut, A., and Harvey, P. H. (1995). Inferring population history from molecular phylogenies. *Phil. Trans. R. Soc.*, **B349**, 25–31.

Nee, S., Barraclough, T. G. and Harvey, P. H. (1996). Temporal changes in biodiversity: detecting patterns and identifying causes. In *Biodiversity: a biology of numbers and difference*, (ed. K. J. Gaston), pp. 230–52. Oxford University Press.

Purvis, A. (1995). A composite estimate of primate phylogeny. *Phil. Trans. R. Soc.*, **B348**, 405–21.

Purvis, A., Nee, S., and Harvey, P. H. (1995). Macroevolutionary inferences from primate phylogeny. *Proc. R. Soc.*, **B260**, 329–33.

Purvis, A., Nee, S., and Harvey, P. H. Using fossils to test models of cladogenesis. *Paleobiology*. (Submitted.)

Rambaut, A., Harvey, P. H., and Nee, S. End-Epi: an application for reconstructing population dynamic histories from phylogenies. *Comp. Appl. Biosci.* (Submitted.)

Raup, D. M., Gould, S. J., Schopf, T. J. M., and Simberloff, D. S. (1973). Stochastic models of phylogeny and the evolution of diversity. *J. Geol.*, **81**, 525–42.

Sanderson, M. J. and Donoghue, M. J. (1994). Shifts in diversification rate with the origin of angiosperms. *Science*, **264**, 1590–13.

Sanderson, M. J. and Donoghue, M. J. (1996). Reconstructing shifts in diversification on phylogenetic trees. *Trends Ecol. Evol.*, **11**, 15–20.

Sarich, V. M. (1986). Rodent macromolecular systematics. In *Evolutionary relationships among rodents; a multidisciplinary approach, NATO ASI series A92*, (ed. W. P. Luckett and J. L. Hartenberger), pp. 423–52.

Savage, H. M. (1983). The shape of evolution: systematic tree topology. *Biol. J. Linn. Soc.*, **20**, 225–44. Plenum, New York.

Shao, K. -T. and Sokal, R. R. (1990). Tree balance. *Syst. Zool.*, **39**, 266–76.

Sibley, C. J. and Ahlquist, J. E. 1990 *Phylogeny and classification of birds: a case study in molecular evolution*. Yale University Press, New Haven, CT.

Simberloff, D., Hecht, K. L., McCoy, E. D., and Connor, E. F. (1981). There have been no statistical tests of cladistic biogeographical hypotheses. In *Vicariance biogeography: a critique*, (ed. G. Nelson and D. E. Rosen), pp. 40–63. Columbia University Press, New York.

Slowinski, J. B. and Guyer, C. (1989). Testing the stochasticity of patterns of organismal diversity: an improved null model. *Am. Nat.*, **134**, 907–21.

Slowinski, J. B. and Guyer, C. (1993). Testing whether certain traits have caused amplified diversification: an improved method based on a model of random speciation and extinction. *Am. Nat.*, **142**, 1019–24.

Smith, A. B. (1994). *Systematics and the fossil record: documenting evolutionary patterns*. Blackwell Scientific, Oxford.

Stanley, S. M. (1973). An explanation for Cope's Rule. *Evolution*, **27**, 1–26.

Van Pelt, J. and Verwer, R. W. H. (1983). The exact probabilities of branching patterns under terminal and segmental growth hypotheses. *Bull. Math. Biol.*, **45**, 269–5.

Van Valen, L. (1973). Body size and numbers of plants and animals. *Evolution*, **27**, 27–35.

Van Valen, L. (1975). Group selection, sex and fossils. *Evolution*, **29**, 87–94.

Wiegmann, B. M., Mitter, C., and Farrell, B. (1993). Diversification of carnivorous parasitic insects: extraordinary radiation or specialized dead-end? *Am. Nat.*, **142**, 737–54.

Williams, G. C. (1992). *Natural selection: domains, levels, challenges*. Oxford University Press.

Yule, G. U. (1924). A mathematical theory of evolution, based on the conclusions of Dr J. C. Willis F.R.S. *Phil. Trans. R. Soc.*, **A213**, 21–87.

11

Using phylogenetic trees to reconstruct the history of infectious disease epidemics

Eddie C. Holmes, Paul L. Bollyky, Sean Nee, Andrew Rambaut, Geoff P. Garnett, and Paul H. Harvey

11.1 Introduction

Phylogenetic analysis of molecular sequence data gives insights into the spread of infectious diseases. During an epidemic, mutations accumulate in the genomes of pathogens and particularly in viruses, because many lack fidelity during replication. Some of these mutations confer phenotypic differences, such as allowing the virus to infect different cell types, to evade host immune responses, or to be transmitted by different routes, while others can be used to document the history of transmission events. Molecular epidemiology aims to recover this information in order to reconstruct the history of a pathogen's spread through host populations and to make predictions about its future progress.

Epidemiological information can be retrieved through the reconstruction of phylogenetic trees of sequences taken from infected individuals. These trees provide information about the extent of genetic diversity within populations, where different viral strains might have originated, and whether new strains are entering populations as an epidemic continues (Holmes and Garnett 1994). On a finer level, the trees can be used to ask whether different risk groups possess characteristic strains and to determine the contact networks between infected individuals, such as that involving the patients of an HIV infected dentist in Florida (Ou *et al.* 1992). Patterns of sequence divergence recovered from phylogenetic trees can also form the basis for studies of more direct clinical importance, such as examining whether certain strains are transmitted more by some routes than others, and whether these strain differences may lead to different clinical outcomes.

In all these cases the most important element is the determination of phylogenetic pattern, on which are mapped other aspects of viral diversity such as geographic distribution and virulence. Here, using the methods described in Chapter 5, we illustrate a new use for phylogenetic trees in molecular epidemiology—the reconstruction of the dynamics of viral transmission from an analysis of branching structure. Our basic premise is that the branching pattern of a phylogeny—the distribution of lineage-splitting events across the tree—provides information about the frequency of transmission events through time. If we assume that each lineage on a tree represents a single infected individual and that a lineage splits every time a transmission event occurs, then transmission events which have occurred in the recent past are expected to be represented by recently diverged sequences, whereas those which occurred longer ago will be depicted as older lineage divisions (Harvey *et al.* 1994*a*: Nee *et al.* 1994). We consider the case studies of three viruses which are imposing serious health costs on human populations: human immunodeficiency virus type 1 (HIV-1), hepatitis C virus (HCV), and hepatitis B virus (HBV). Nucleotide sequence data for these viruses were taken from infected individuals world-wide; each sequence analysed came from a different infected individual.

11.2 The sequence data

1. *World-wide variation of HIV-1*. The data from Louwagie *et al.* (1993), consisting of 72 sequences from the *gag* gene of HIV-1 taken from infected individuals throughout the world, represent seven of the eight known genetic subtypes. Although sequence data were available from the complete *gag* coding region, approximately 1500 base pairs in length, difficulties in alignment between some of the more divergent isolates meant that an unambiguous region of 1254 base pairs was used for phylogenetic analysis.

2. *World-wide variation of HCV: E1 Gene*. 64 sequences of 576 base pairs taken from the E1 (envelope) gene of HCV from individuals infected throughout the world, including representatives of the six major genotypes of HCV (data mainly from Bukh *et al.* (1993)).

3. *World-wide variation of HCV: NS-5 gene*. 76 sequences of 222 base pairs belonging to the NS-5 gene of HCV taken from individuals infected throughout the world; includes sequences assigned to all six major genotypes of HCV (data from Simmonds *et al.* (1993a)).

4. *World-wide variation of HBV*. 78 sequences of 681 base pairs taken from the S (surface antigen – HBsAg) gene from infected individuals throughout the world. These sequences were all taken from GenBank (a full list of the sequences used is available from the authors on request) and cover the six known genotypes of the virus (Norder *et al.* 1993).

11.3 Analysis of phylogenetic tree structure

Where necessary nucleotide sequences were aligned using the CLUSTALV package (Higgins *et al.* 1992), prior to checking by eye. Phylogenetic trees were then reconstructed using two methods taken from the PHYLIP package (version 3.55c; Felsenstein 1993): the neighbour-joining clustering method (program

NEIGHBOR) and the Fitch–Margoliash clustering method with contemporaneous tips (program KITSCH). Input distances for the programs were estimated from a model of base substitution which allows different frequencies of the four nucleotides and different rates of transition and transversion (program DNADIST). The KITSCH program, which assumes a constant rate of nucleotide substitution, is needed in the analysis of phylogenetic tree structure because the nodes of the tree must be aligned relative to each other in time.

Different transmission dynamics result in different branching patterns, which can be compared by plotting the logarithm of the number of lineages in the phylogeny against the time at which they appear (see Nee *et al.*, Chapter 5 this volume). Such a comparison is useful even if only a small sample of lineages is taken, as is likely to be the case in the analysis of viral genomes, even though the resolution achieved will not be as great as with a complete sample (Nee *et al.* 1994). When only a small sample of lineages is taken and the infected population is growing exponentially there is a distinct shallowing off in the number of lineages towards the present (the 'epidemic' transmission model). If, on the other hand, the infected population remains at a constant size there will be upturn in the number of lineages curve (the 'endemic' transmission model). By performing other transformations of the number-of-lineages axis or the time axis (formulations given in Chapter 5) it is possible to further distinguish between infections which are growing exponentially at a constant rate, an accelerating rate, or a decelerating rate, as well as between those that have maintained a constant size, have increased in a linear fashion, or have declined in size.

Two computer packages were used in the analysis of lineage-through-time plots. The BIRTH–DEATH simulation package illustrates what these plots are expected to look like given different rates of lineage birth and death, whereas the ENDEMIC–EPIDEMIC package takes empirical trees (such as those produced by the PHYLIP package), estimates the numbers of lineages-through-time, and then performs the various transformations. The packages are available by sending a blank Macintosh disc to Andrew Rambaut, Department of Zoology, University of Oxford, Oxford OX1 3PS, UK or through the Internet at http://evolve.zps.ox.ac.uk/.

11.4 Results

World-wide variation of HIV-1

HIV-1 is characterized by extensive genetic variability within single patients, within infected communities, and on a global scale where the virus has been classified into a number of genetic subtypes. Broadly, HIV-1 can be divided into two genetic subgroups, 'O' and 'M', which are so phylogenetically divergent as to be separated from each other by homologous viral sequences found in chimpanzees. The 'M' subgroup, containing the vast majority of viruses sequenced to date, has been divided into at least eight genetic subtypes or clades (denoted A to H) (Myers *et al.* 1993; Louwagie *et al.* 1993) which vary in their geographic distribution. For example, subtype A is most commonly found in Central and West Africa while subtype B predominates in the developed world. The subgroup 'O' strains are found in infected individuals from the Cameroon and Gabon. Despite the great efforts to classify this virus it is unclear whether any of these subtypes are characterized by different phenotypic properties. However, whether or not important biological differences exist between the subtypes, this variation is likely to hinder immunization programmes (see Chapter 9).

A neighbour-joining tree reconstructed using the 72 *gag* sequences sampled from around the world is shown in Fig. 11.1a, with the equivalent KITSCH tree presented in Fig. 11.1b. Visual inspection indicates that lineage splitting has occurred along the entire length of the tree. The lineages-through-time plot of this tree is shown in Fig. 11.1c. The shallowing off of the line toward the present indicates that HIV-1 may be spreading exponentially at a roughly constant rate around the world. However, when the epidemic transformation is applied the resulting line has a slight upward curvature (Fig. 11.1d).

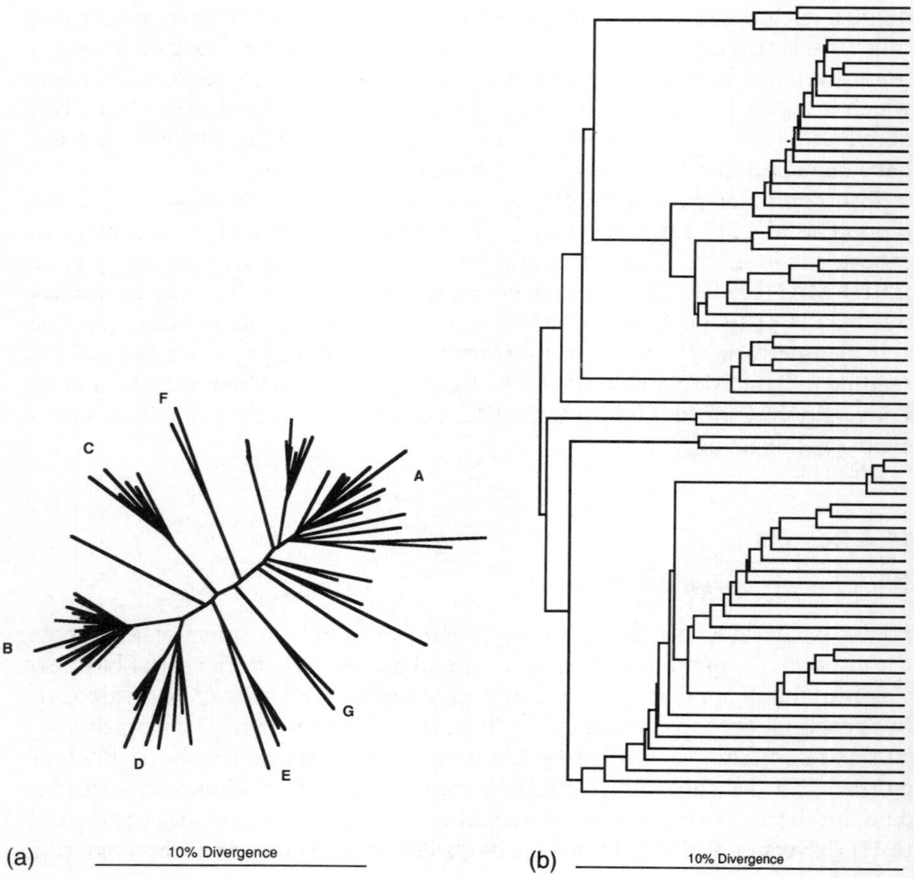

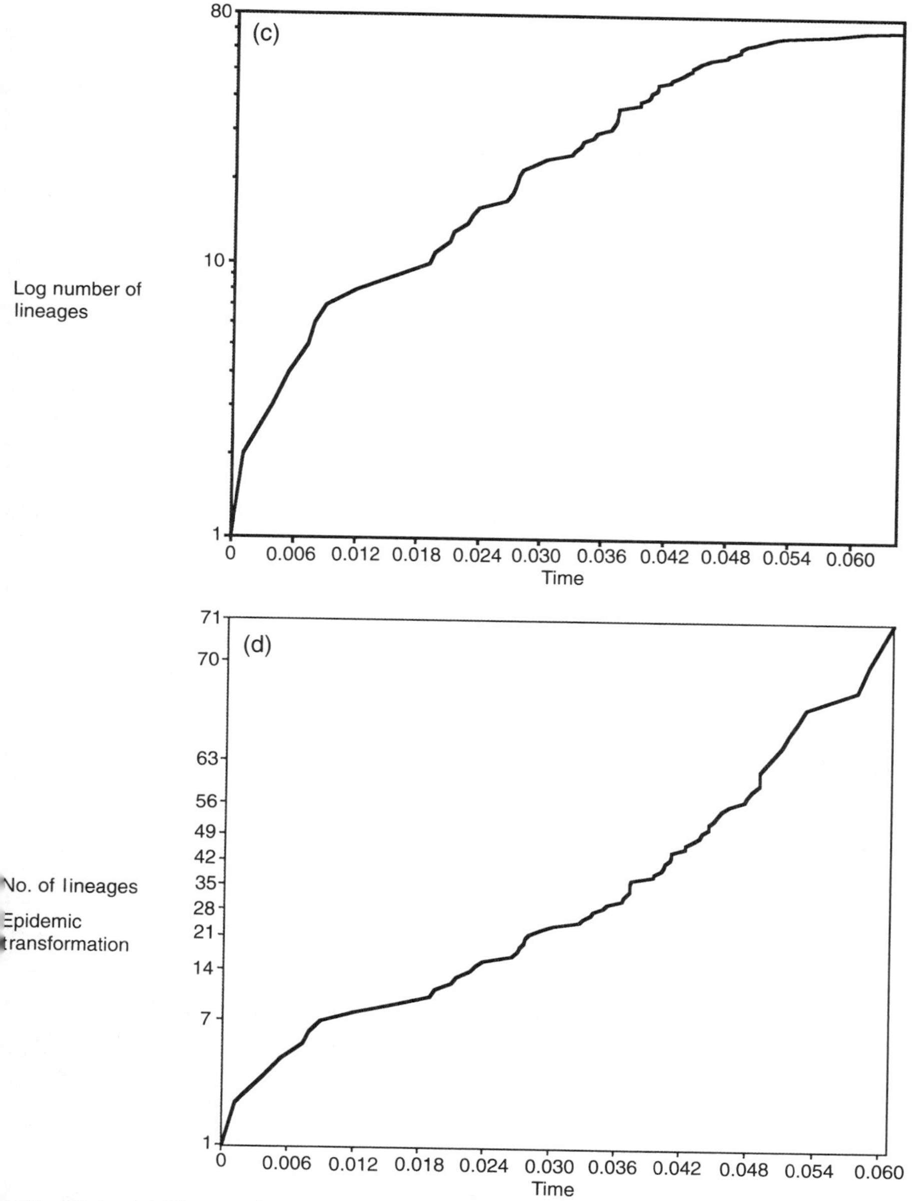

Fig. 11.1 (a) Unrooted neighbour-joining tree of 72 *gag* gene sequences of HIV-1 taken from around the world. Clusters corresponding to the subtypes are indicated. (b) KITSCH tree of the same data. Because the KITSCH method is time dependent the root is automatically assigned to the longest branch. (c) Semi-logarithmic lineages-through-time plot of the KITSCH tree. Time is measured as percentage base-pair substitution from the root of the tree to the present. (d) Epidemic transformation of the number-of-lineages axis.

It is also of interest to determine whether the different subtypes of HIV-1 are characterized by different rates of spread, as this may indicate differences in transmission potential and possibly virulence. Unfortunately, with the available *gag* sequence data, only two subtypes possess sufficient sequences for such an exercise: the African subtype A (27 sequences) and the western subtype B (17 sequences). The lineage-through-time plots and epidemic transformations for both these subtypes suggest a constant rate of population growth (analyses not shown), although many more sequences are required before the question of subtype differences can be answered with confidence.

World-wide variation of HCV

Hepatitis C virus, first isolated in 1989 (Choo *et al.* 1989), is a positive-sense, single-stranded, enveloped RNA virus of approximately 9.5 kb in length whose genome is organized as a single continuous open reading frame. The gene products at the 5′ end of the genome encode structural proteins (capsid, envelope) while the 3′ proteins are non-structural (Choo *et al.* 1991). HCV is phylogenetically related to flaviviruses and pestiviruses, and has a mutation rate of about 10^{-3} substitutions per site per year, similar to that observed in HIV-1 (Miller and Purcell 1990; Ogata *et al.* 1991).

HCV is predominantly transmitted through the transfer of blood or blood products, such as the clotting factors used by haemophiliacs or needle-sharing by injecting drug users. Unlike HIV and HBV, however, sexual and vertical transmission seem to constitute minor pathways of infection (Bresters *et al.* 1993; Ohto *et al.* 1994). A feature shared between all three viruses is that they induce long-term persistent infections so that disease symptoms may take many years to develop. Approximately 50 per cent of HCV infected individuals develop a chronic infection and of these about 20 per cent of these experience serious liver abnormalities and most importantly liver cancer (hepatocellular carcinoma) (Brown and Dusheiko 1993).

Recently there has been a concerted effort to document the extent of world-wide genetic variation in HCV as an aid to vaccine development. Particular attention has been paid to sequencing parts of the E1 (envelope) and NS-5 genes from infected individuals from diverse localities (Bukh *et al.* 1993; Simmonds et al. 1993*a*). A neighbour-joining tree of 64 E1 sequences taken world-wide is shown in Fig. 11.2a and an equivalent NS-5 tree from 76 infected individuals is shown in Fig. 11.3a. At the tips of both trees are distinct clusters of sequences. To date, six major clusters of sequences, called types, can be recognized which, at the sequence level, differ by an average of 50 per cent. Within these major types there are a number of more closely related clusters of sequences, referred to as subtypes. Both types and subtypes are also observed in other regions of the genome (Chan *et al.* 1992; Simmonds *et al.* 1993*b*, 1994).

A relative time axis can be placed on the E1 and NS-5 trees using the ultrametric KITSCH algorithm (Figs. 11.2b and 11.3b respectively). Visual inspection reveals that most of the branching activity (transmission) has occurred near the present, rather than across the entire length of the tree.

On the basis of these constant-rate trees it is possible to plot the number of lineages-through-time (Figs. 11.2c and 11.3c respectively). These plots have a distinctive structure which is different from that seen in HIV-1 world-wide: they start off with a very gentle increase and suddenly steepen toward the present. Similar patterns are observed under the epidemic transformation (Figs. 11.2d and 11.3d). This distinctive structure is not an artefact of only looking at certain regions of the genome, because the same pattern is seen in the 5′ non-coding region as well as the core, NS-3 and NS-4 genes (Holmes *et al.*, in preparation). Nor is it an artefact due to the biased sampling of a single population, as the same tree structures and lineage-through-time plots are seen when specific populations are analysed separately (Holmes *et al.*, in preparation). Simulation studies reveal that these lineages-through-time plots are what might be expected when an endemically transmitted virus suddenly becomes an epidemically transmitted virus (Harvey *et al.* 1994*b*). In other words, for much of its evolutionary history HCV may have infected a roughly constant number of humans such that we can say it is in an endemic state. Then, at some point in the relatively recent past, there was a dramatic increase in the viral population size and the virus became transmitted in an epidemic (exponentially growing) state (Holmes *et al.*, in preparation). Indeed, when the epidemic transformation is applied only to the steep part (the recent past) of the NS-5 lineages-through-time plot a straight line is observed (Fig. 11.3e).

World-wide variation of HBV

Hepatitis B virus (HBV) is a major public health problem in many areas of the world and especially the Far East and tropical Africa where between 10 and 15 per cent of the population are chronic HBV carriers (Sherlock 1993). Unlike the other viruses which cause hepatitis, HBV has a DNA genome of approximately 3.2 kb. Furthermore, this genome is organized in a complex manner with double-stranded and single-stranded regions as well as translation in more than one reading frame. The virus encodes four major open reading frames; the S (HBsAg—surface antigen) gene, the P (polymerase) gene, the C (HBcAg—core) gene, and the X gene which makes a protein not found in the virus particle (Gerlich 1993). The course of a clinical infection is similar to that in HCV although the factors determining the final outcome of disease are complex. In endemic areas approximately 95 per cent of infected infants will become chronic carriers with up to 40 per cent of HBV infections resulting in hepatocellular carcinoma (Buendia *et al.* 1993). Finally, HBV also evolves a great deal more slowly than HIV and HCV, with the average synonymous substitution rate for the S gene estimated to be 5.75×10^{-5} substitutions per site per year (Gojobori *et al.* 1994).

Early serological studies of the HBV surface antigen revealed nine antigenic subtypes world-wide, while more recent sequencing studies have shown that

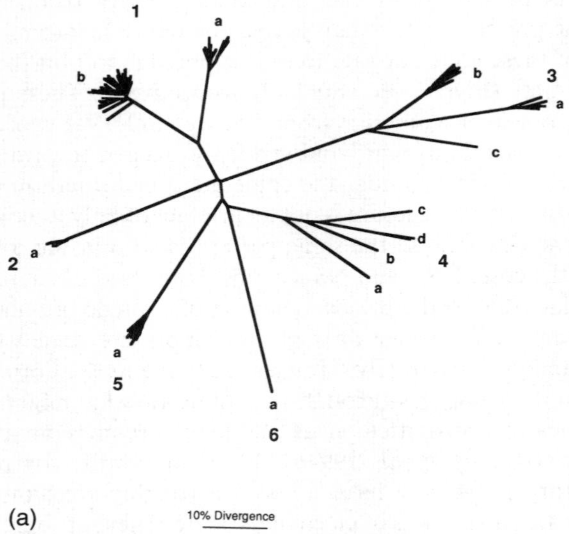

(a)

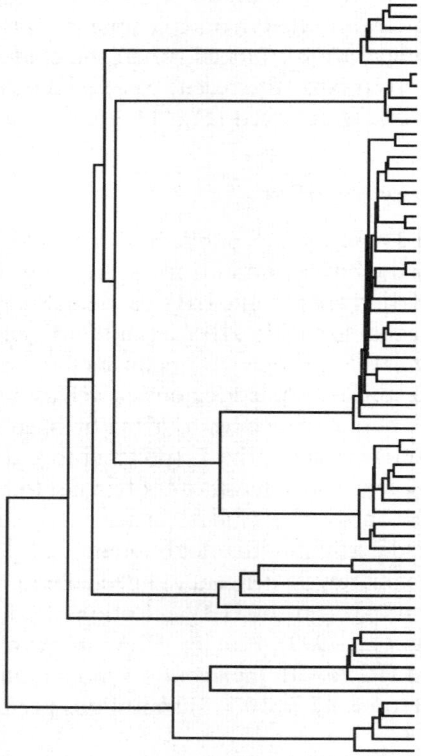

(b)

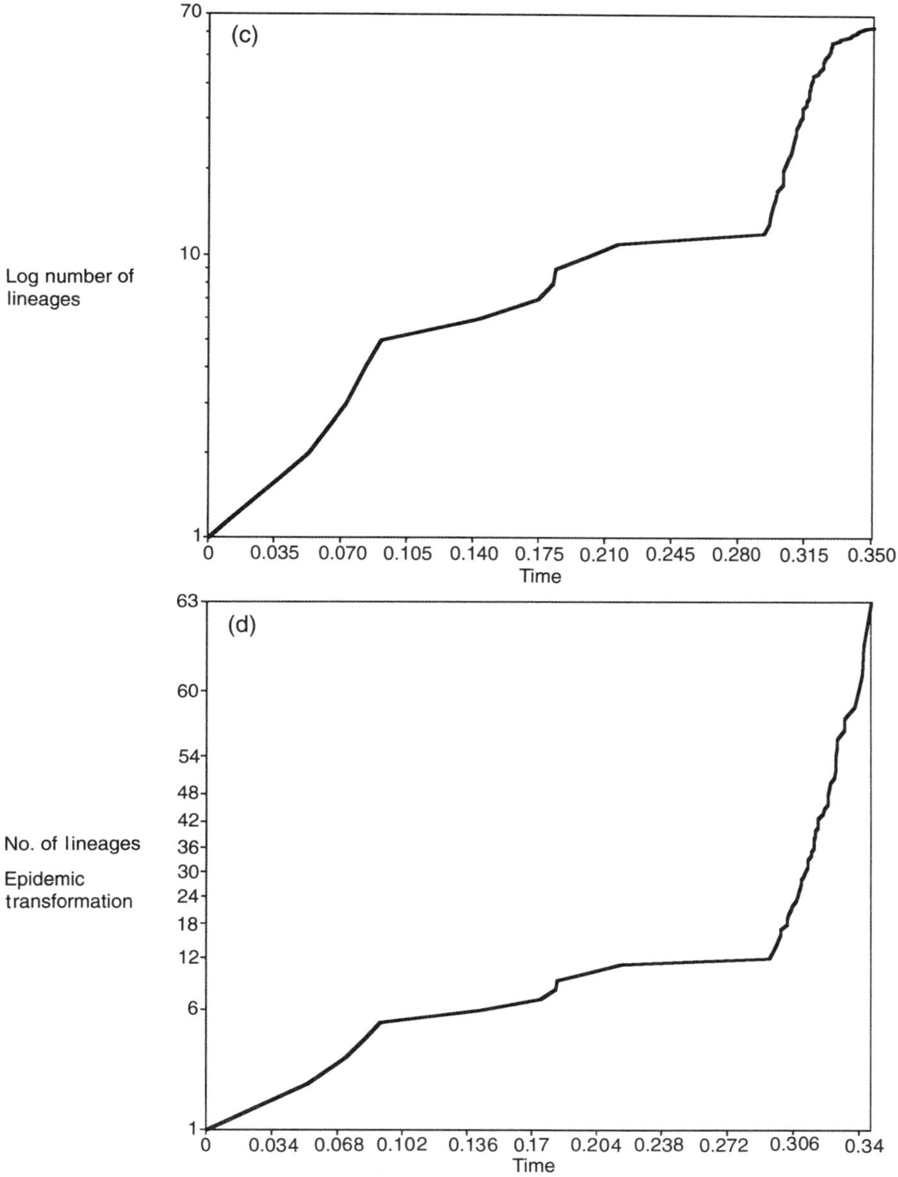

Fig. 11.2 (a) Unrooted neighbour-joining tree of 64 isolates of HCV E1 gene taken from around the world. Clusters corresponding to the types and subtypes are indicated. (b) KITSCH tree of the same data. (c) Lineages-through-time plot of the KITSCH tree. The gently rising slope which characterizes the early part of the plot is indicative of a virus which is spreading in an endemic manner (infecting a constant fraction of the population) while the sudden increase in the recent past indicates that the virus changed to one spreading in an epidemic manner (growing exponentially at a constant rate). Axes as in Fig. 11.1 (d) Epidemic transformation of the E1 sequence data.

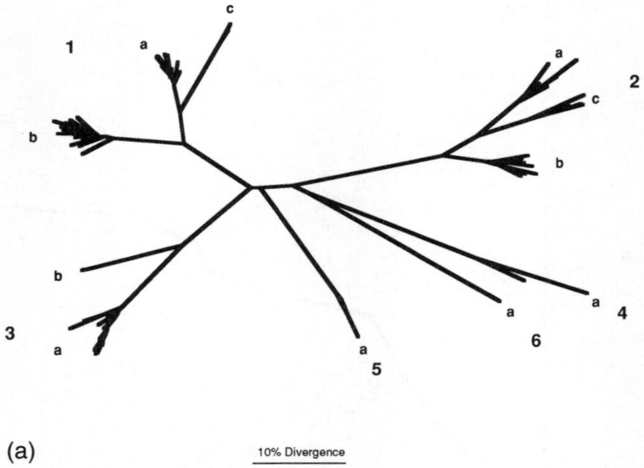

(a)

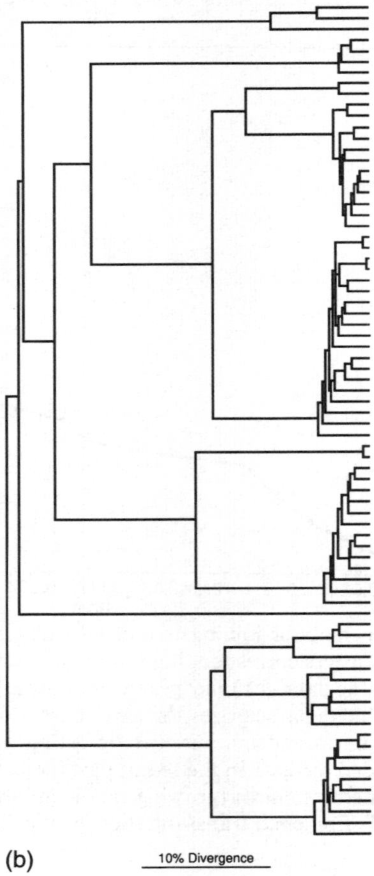

(b)

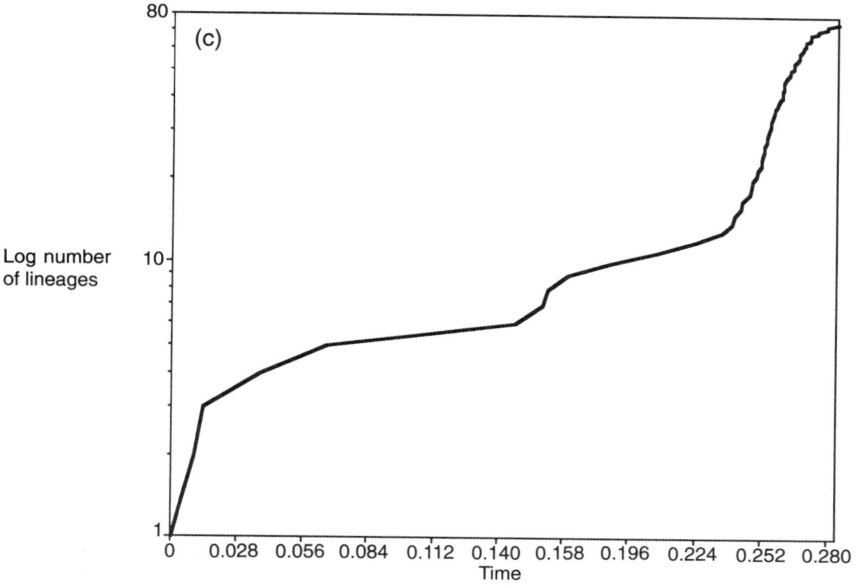

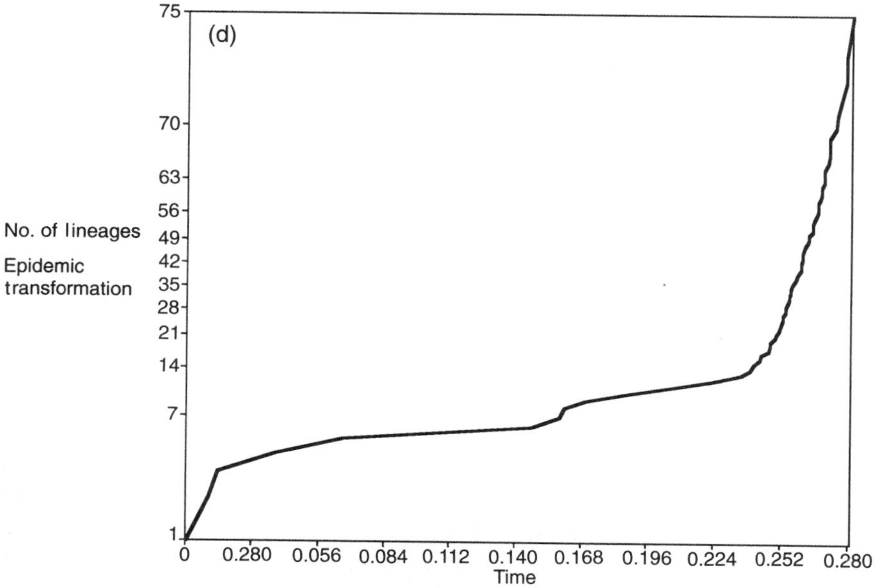

Fig. 11.3 (a) Unrooted neighbour-joining tree of 76 isolates of HCV NS-5 gene taken from around the world. Clusters corresponding to the types and subtypes are indicated. (b) KITSCH tree of the same data. (c) Lineages-through-time plot of the KITSCH tree. Axes as in Fig. 11.1 (d) Epidemic transformation of the NS-5 sequence data. (e) Epidemic transformation of the steep part of the lineages-through-time plot. The straight line indicates that HCV has been spreading exponentially at a constant rate in the recent past.

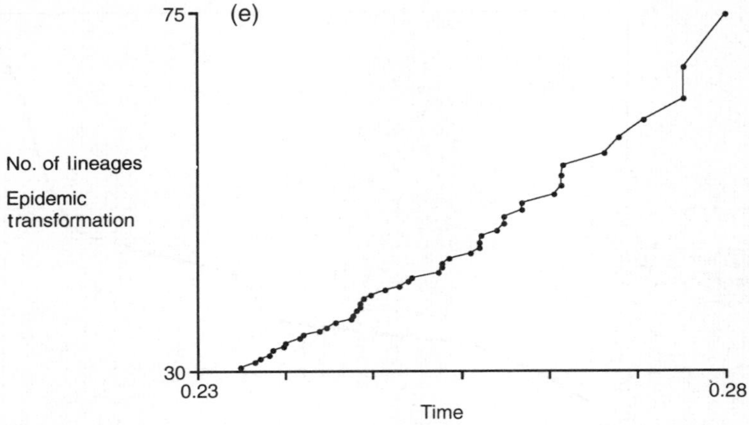

these serotypes correspond to six genotypes (numbered A to F in Fig. 11.4a) which differ by up to 10 per cent in nucleotide sequence (Norder *et al*. 1993). As with HCV, these genotypes vary in frequency around the world, although there is no evidence of a correlation between genotype and virulence at present. Genotypes A and D have the widest geographic distribution, being found on most continents, genotypes B and C, however, are restricted to South-east Asia, the Far East, and the Pacific, while genotype E is found in the western part of sub-Saharan Africa, and genotype F appears to be indigenous to New World populations (Norder *et al*. 1993).

The neighbour-joining tree of 78 S gene sequences is presented in Fig. 11.4a and the equivalent KITSCH tree is given in Fig. 11.4b. The lineages-through-time plot derived from this KITSCH tree is presented in Fig. 11.4c and the epidemic transformation in Fig. 11.4d. The gradual curvature observed under the epidemic transformation suggests that the rate of spread of HBV around the world has been exponential but accelerating.

11.5 Discussion

The molecular epidemiology of HIV-1, HCV, and HBV

Different dynamics of viral transmission (rates of population growth) leave different signatures in the branching structure of phylogenetic trees. The differences in the tree structures of world-wide isolates of HIV-1, HCV, and HBV, and highlighted in the logarithmic and epidemic transformations, suggest that these viruses have very different epidemiological histories. What processes can explain this difference? One way to attempt to answer this question is to relate the spread of these viruses to real time. There have been a number of attempts to date the spread of HIV-1 in human populations using both epidemiological and molecular information. Although there are no firm conclusions, most estimates put the initial divergence of the major subtypes of

HIV-1 (subgroup 'M') to be after World War II, and perhaps in the late 1950s or early 1960s (Li *et al.* 1988; Myers and Pavlakis 1992). Performing a similar exercise for HCV reveals that this virus has been present in human populations for hundreds (perhaps thousands) of years and that the explosion of HCV transmission occurred up to 50 years ago, but with a mean of about 25 to 30 years ago. Therefore HIV-1 and HCV began spreading at about the same time so that the epidemic take-off in HCV corresponds to the start of the HIV-1 epidemic in human populations. It seems reasonable to think that such a correlation is due to shared aspects of the epidemiology of HIV and HCV, and in particular that new and susceptible host populations were encountered, perhaps because of the increased use of commercial blood products, the increased use of needle-sharing by injecting drug users, the increased frequency of unprotected sex (especially for HIV transmission), and modern transportation methods which facilitate the movement of infected populations. Consequently, the apparent difference in the epidemiological histories of HIV-1 and HCV, as reflected in their differing lineages-through-time plots, may result from HCV having an older association with human populations characterized by an initial period where population growth was low.

Although HBV is known to have had a long association with human populations—a date of 3000 years ago has been proposed from one molecular study (Orito *et al.* 1989)—the rate of transmission of this virus appears to have accelerated. In the developed world this increased rate of spread will be due to the same factors which facilitated the spread of HIV and HCV, although the high rate of spread appears to have started earlier in the case of HBV, perhaps because of the use of unsterilized needles and syringes in health care during the late nineteenth and early twentieth centuries (Purcell 1994). In the developing world the virus is predominantly transmitted vertically from mother to child and horizontally among very small children. In these circumstances the increasing rate of HBV transmission will reflect the increasing size of the human population.

Future directions

To date we have only been able to make general statements about the transmission dynamics of viruses in populations—that is, we can distinguish between viruses that are increasing in frequency and those that have maintained a constant population size. It is desirable to make more precise statements about transmission *rates*, such as those provided by longitudinal serological studies. An important consideration in this respect is that the basic reproductive rate of a pathogen, R_0 (the number of new hosts infected by a particular host over the lifetime of the host's infection; Anderson and May (1991)) may be estimated if the lineage birth rate and the lineage death rate are known. We also hope to use these techniques to estimate the time between infections, allowing us to distinguish between mechanisms limiting transmission, such as the latent period, and the time taken to transmit infection or to make contact with

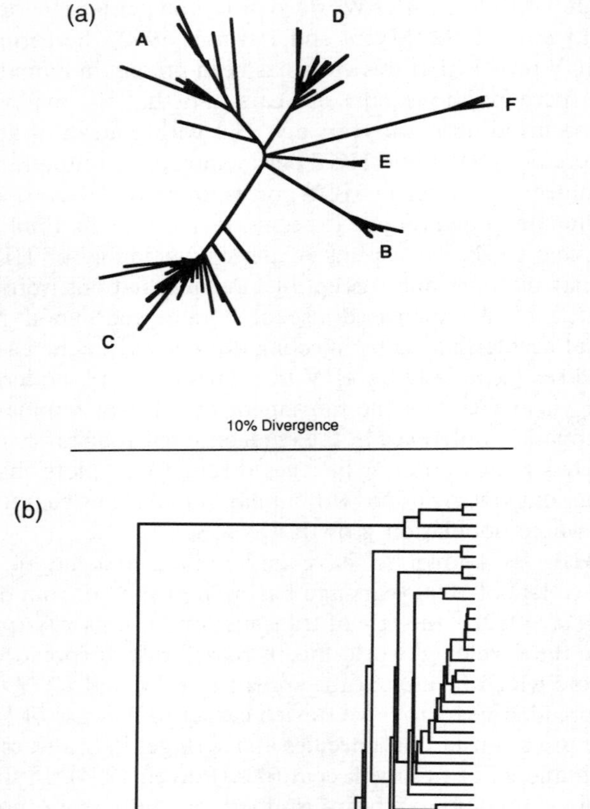
(a)
A
D
F
E
B
C
10% Divergence
(b)
10% Divergence

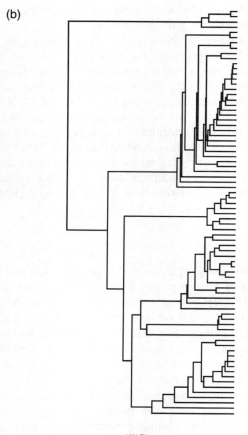

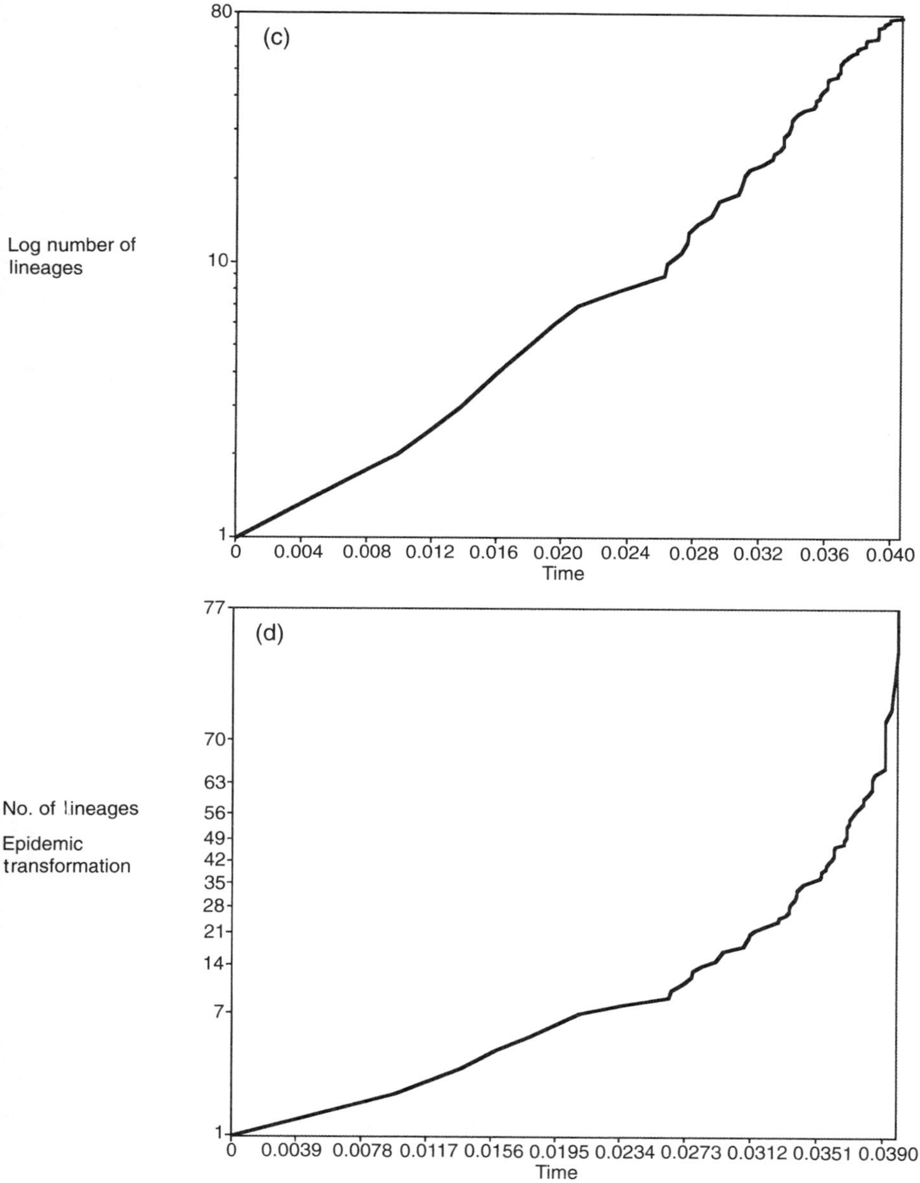

Fig. 11.4 (a) Unrooted neighbour-joining tree of 78 isolates of HBV S gene taken from around the world. Clusters corresponding to the genotypes are indicated. (b) KITSCH tree of the same data. (c) Lineages-through-time plot of the KITSCH tree. Axes as in Fig. 11.1. (d) Epidemic transformation of the S gene sequence data.

susceptibles. For example, when there is a standing network of transmission, such as a group of injecting drug users readily sharing needles, we might expect the waiting times between new infection events to be short and the reconstructed trees to be 'bushy', with many lineages diverging within a short time period (near the root) and then levelling off as the number of susceptible hosts declines. Such a situation has been reported in the HIV epidemic seen in injecting drug-users in Edinburgh (Holmes *et al.* 1995). In contrast, it might be expected that individuals who were infected via sexual contact to be connected by longer branches as the waiting times between transmission events are generally greater.

Finally, although the initial studies of viral sequence diversity world-wide do provide a useful sketch of the sequence variation that might exist in infected populations, they are limited with respect to sampling. The geographic distribution of sequences is often highly skewed: some localities, particularly where acquisition of samples is relatively easy as in many developed countries, may be over-represented. This will mean that the tree may contain too many closely related sequences. On the other hand, sequences may be absent from other areas completely. Another limitation is that the individuals from which viruses have been taken are more likely to be those with symptomatic infections, and thus seeking treatment, than those who are asymptomatic and therefore have no requirement to seek medical attention. Because of the long incubation period of these infections this may mean that strains which arose earlier in the epidemic are more likely to form the basis of the samples taken. It is clear therefore that a more representative sampling of viral strains be undertaken so that a real molecular epidemiology of viruses can be established.

Acknowledgements

We thank Dr Peter Simmonds of the Department of Medical Microbiology, University of Edinburgh, for providing the HCV sequence data. This work was supported by grants from the Wellcome Foundation, The Royal Society, and the BBSRC.

References

Anderson, R. M. and May, R. M. (1991). *Infectious diseases of humans.* Oxford University Press.

Bresters, D., Mauser–Bunschoten, E. P., Reesink, H. W. *et al.* (1993). Sexual transmission of hepatitis C virus. *Lancet*, **342**, 210–11.

Brown, D. and Dusheiko, G. (1993). Hepatitis C virus: diagnosis. In *Viral hepatitis*, (ed. A. J. Zuckerman and H. C. Thomas), pp. 283–301. Churchill Livingstone, Edinburgh.

Buendia, M. A., Paterlini, P., Tiollais, P. and Bréchot, C. (1993). Hepatitis B virus: liver cancer In *Viral hepatitis*, (ed. A. J. Zuckerman and H. C. Thomas), pp. 137–64. Churchill Livingstone, Edinburgh.

Bukh, J., Purcell, R. H. and Miller, R. H. (1993). At least 12 genotypes of hepatitis C virus predicted by sequence analysis of the putative E1 gene of isolates collected worldwide. *Proc. Natl. Acad. Sci. USA*, **90**, 8234–8.

Chan, S.-W., McOmish, F., Holmes, E. C. *et al.* (1992). Analysis of a new hepatitis C virus type and its phylogenetic relationship to existing variants. *J. Gen. Virol.*, **73**, 1131–41.

Choo, Q.-L., Kuo, G., Weiner, A. J., Overby, L. R., Bradley, D. W., and Houghton, M. (1989). Isolation of a cDNA clone derived from a blood-borne Non-A, Non-B viral hepatitis genome. *Science*, **244**, 359–62.

Choo, Q.-L., Richman, K. H., Han, J. H. *et al.* (1991). Genetic organisation and diversity of the hepatitis C virus. *Proc. Natl. Acad. Sci. USA*, **88**, 2451–5.

Felsenstein, J. (1993). PHYLIP (Phylogeny Inference Package) Version 3.5c. Distributed by the author. Department of Genetics, University of Washington, Seattle

Gerlich, W. (1993). Hepatitis B virus: structure and molecular virology. In *Viral hepatitis*, (ed. A. J. Zuckerman and H. C. Thomas), pp. 83–113. Churchill Livingston, Edinburgh.

Gojobori, T., Yamaguchi, Y., Ikeo, K., and Mizokami, M. (1994). Evolution of pathogenic viruses with special attention to rates of synonymous and nonsynonymous substitutions. *Japan. J. Genet.*, **69**, 481–8.

Harvey, P. H., Holmes, E. C., Mooers, A.Ø., and Nee, S. (1994*a*). Inferring evolutionary processes from molecular phylogenies. In *Models in phylogeny reconstruction*, Systematics Association Special Volume Series 52, (ed. R. W. Scotland, D. J. Siebert, and D. M. Williams), pp. 310–33. Systematics Association, London.

Harvey, P. H., Holmes, E. C., and Nee, S. (1994*b*). Model phylogenies to explain the real world. *BioEssays*, **16**, 767–70.

Higgins, D. G., Bleasby, A. J., and Fuchs, R. (1992). CLUSTALV: improved software for multiple sequence alignment. *CABIOS*, **8**, 189–91.

Holmes, E. C. and Garnett, G. P. (1994). Genes, trees and infections: molecular evidence in epidemiology. *Trends Ecol. Evol.* **9**, 256–60.

Holmes, E. C., Zhang, L. Q., Robertson, P. *et al.* (1995). The molecular epidemiology of HIV-1 in Edinburgh, Scotland. *J. Infect. Dis.*, **171**, 45–53.

Li, W.-H., Tanimura, M., and Sharp, P. M. (1988). Rates and dates of divergence between AIDS virus nucleotide sequences. *Mol. Biol. Evol.*, **5**, 313–30.

Louwagie, J., McCuthan, F. E., Peeters, M. *et al.* (1993). Phylogenetic analysis of *gag* genes from 70 international HIV-1 isolates provides evidence for multiple genotypes. *AIDS*, **7**, 769–80.

Miller, R. H. and Purcell, R. H. (1990). Hepatitis C virus shares amino acid sequence similarity with pestiviruses and flaviviruses as well as members of two plant virus supergroups. *Proc. Natl. Acad. Sci. USA*, **87**, 2057–61.

Myers, G. and Pavlakis, G. N. (1992). Evolutionary potential of complex retroviruses. In *Viruses: the Retroviridae*, Vol. 1, (ed. R. R. Wagner, and H. Fraenkel-Conrat), pp. 51–105. Plenum, New York.

Myers, G., Korber, B. T. M., Wain-Hobson, S., Smith, R. F., and Pavlakis, G. N. (1993). *Human retroviruses and AIDS*. Los Alamos National Laboratory.

Nee, S., Holmes, E. C., May, R. M., and Harvey, P. H. (1994). Extinction rates can be estimated from molecular phylogenies. *Phil. Trans. R. Soc.*, **B344**, 77–82.

Norder, H., Hammas, B., Lee, S.-D. *et al.* (1993). Genetic relatedness of hepatitis B viral strains of diverse geographical origin and natural variations in the primary structure of the surface antigen. *J. Gen. Virol.*, **74**, 1341–8.

Ogata, N., Alter, H. J., Miller, R. H., and Purcell, R. H. (1991). Nucleotide sequence and mutation rate of the H strain of hepatitis C virus. *Proc. Natl. Acad. Sci. USA*, **88**, 3392–6.

Ohto, H., Terazawa, S., Sasaki, N. *et al.* (1994). Transmission of hepatitis C virus from mothers to infants. *New Engl. J. Med.*, **330**, 744–50

Okuda, K. (1993). Hepatitis C virus: liver cancer. In *Viral hepatitis*, (ed. A. J. Zuckerman and H. C. Thomas), pp. 269–81. Churchill Livingstone, Edinburgh.

Orito, E., Mizokami, M., Ina, Y. *et al.* (1989). Host-independent evolution and a genetic classification of the hepadnavirus family based on nucleotide sequences. *Proc. Natl. Acad. Sci. USA*, **86**, 7059–62.

Ou, C.-Y., Ciesielski, C. A., Myers, G. *et al.* (1992). Molecular epidemiology of HIV transmission in a dental practice. *Science*, **256**, 1165–71.

Purcell, R. H. (1994). Hepatitis viruses: changing patterns of human disease. *Proc. Natl. Acad. Sci. USA*, **91**, 2401–6.

Sherlock S. (1993). Clinical features of hepatitis. In *Viral hepatitis*, (ed. A. J. Zuckerman and H. C. Thomas), pp. 1–17. Churchill Livingstone, Edinburgh.

Simmonds, P., Holmes, E. C., Cha, T. A. *et al.* (1993*a*). Classification of hepatitis C virus into 6 major genotypes and a series of subtypes by phylogenetic analysis of the NS-5 region. *J. Gen. Virol.*, **74**, 2391–9.

Simmonds, P., McOmish, F., Yap, P. L. *et al.* (1993*b*). Sequence variability in the 5′ non-coding region of hepatitis C: identification of a new virus type and restrictions on sequence diversity. *J. Gen. Virol.*, **74**, 661–8.

Simmonds, P., Smith, D. B., McOmish, F. *et al.* (1994). Identification of genotypes of hepatitis C virus by sequence comparisons in the core, E1 and NS-5 regions. *J. Gen. Virol.*, **75**, 1053–61.

12

Relating geographic patterns to phylogenetic processes

A. Malhotra, R. S. Thorpe, H. Black, J. C. Daltry, and W. Wüster

12.1 Introduction

Studies of geographic variation within species have provided many insights into evolutionary processes. These include the importance of natural selection for locally varying conditions and the role of allopatric differentiation in the divergence of lineages. Until recently, the majority of studies on geographic variation were concerned with morphology, but this may be influenced by both processes and their relative importance is difficult to assess. Various quantitative tests have been evaluated for this purpose, for example patterns of congruence between character sets (Thorpe 1991), the pattern of anagenesis in phylogenetic trees (Thorpe 1991), and tests of association between observed morphological patterns and those expected from various hypothesized causes such as vicariance or ecological adaptation (Dow *et al.* 1987; Brown and Thorpe 1991*a,b*; Brown *et al.* 1991; Malhotra and Thorpe 1991*a*; Sokal *et al.* 1991; Thorpe 1991; Thorpe and Brown 1991; Thorpe and Baez 1993). These methods have been discussed in detail previously (Thorpe *et al.* 1991, 1994*a*) and will not be described here. Although these methods can be useful, it was not until the development and increasing accessibility of molecular techniques (for example PCR-based sequencing of the fast-evolving mitochondrial genome) that intraspecific molecular phylogenies could be readily constructed and it became possible to assess the causes of geographic variation against an independently derived hypothesis of phylogeny.

Recent years have seen the publication of numerous phylogeographic studies on diverse organisms, for example swallowtail butterflies (Sperling and Harrison 1994), sparrows (Zink and Dittmann 1994), white-tailed deer (Ellsworth *et al.* 1994), grashopper mice (Riddle *et al.* 1993), salamanders (Moritz *et al.* 1992), and African elephants (Georgiadis *et al.* 1994), with further examples given in a recent review by Avise (1994). A common feature of these studies is that the molecular phylogenies obtained (from restriction fragment length polymorphisms or, more recently, from sequence information) are interpreted in terms of various biogeographic and historical scenarios. Many studies go no further than this. However, a molecular phylogeny can also be used in quantitative tests, giving a new perspective on the causes of geographic differentiation between populations. This is illustrated with two examples: morphological evolution in the Canary island lacertid *Gallotia galloti* and venom evolution in the Malayan pit viper *Calloselasma rhodostoma*. We then use fine-scale geographic patterns in *G. galloti* within the geologically complex island of Tenerife to choose between alternative historical and geological hypotheses concerning its origin. Similar work on the anole *Anolis oculatus* from Dominica, however, highlights several current concerns regarding the use of mitochondrial DNA in phylogenetic studies and suggests several avenues for future development of this approach.

12.2 Canary Island lizards: colonization sequence and the correlates of evolution

The Canary Islands have four extant species of diurnal, generally herbivorous, lacertid of the endemic genus *Gallotia* (Baez 1987) recognized on morphological criteria. The small *G. atlantica* is found on the eastern islands and the large *G. stehlini* on Gran Canaria. In the west the medium-sized lacertid *G. galloti* is found on several islands while the large *G. simonyi* is found only on Hierro. Prior to investigating the geographic variation in *G. galloti* it was necessary to validate these 'species' and to establish that the group is monophyletic. Molecular phylogenies based on cytochrome b (Fig. 12.1a), cytochrome oxidase I sequence (Fig. 12.1b), 12s rRNA sequence (Fig. 12.1c), combined sequence from all three of these mitochondrial genes (Fig. 12.1d), 6-cut RFLPs of the mitochondrial genome (Fig. 12.1e), and RAPD analysis of the nuclear genome (Fig. 12.1f), all show the western species *G. galloti* to be monophyletic (Thorpe *et al.* 1993*a,b*, 1994*a,b*). *G. atlantica* is indicated as a sister species to *G. galloti* in all except the cytochrome oxidase tree (Fig. 12.1b)

The western Canarian lacertid, *G. galloti*, can therefore be considered to be composed of a monophyletic series of populations found on the ecologically heterogenous islands of Tenerife, La Palma, El Hierro, and La Gomera. Many morphological characters from the scalation, colour pattern, and body dimensions vary substantially among islands as well as within Tenerife. For example, the mature males on El Hierro tend to be small and blackish with blue spots on the legs and relatively robust heads while in north Tenerife they tend to be large with bright yellow dorsal cross bars, blue cheeks, and relatively small heads.

The western Canary Islands are known to have erupted from the sea independently, are separated by deep channels and are not thought to have ever been connected to each other or with the African mainland (see references in Thorpe *et al.* 1994*b*). Thus the lizard's present distribution can be interpreted in terms of over-water colonization, but the colonization sequence cannot be recovered from a phylogenetic tree based on morphology as this may be compromised by ecogenesis. A Fitch–Margoliash tree based on combined mtDNA sequence information available (over 1005 base pairs (bp) from the

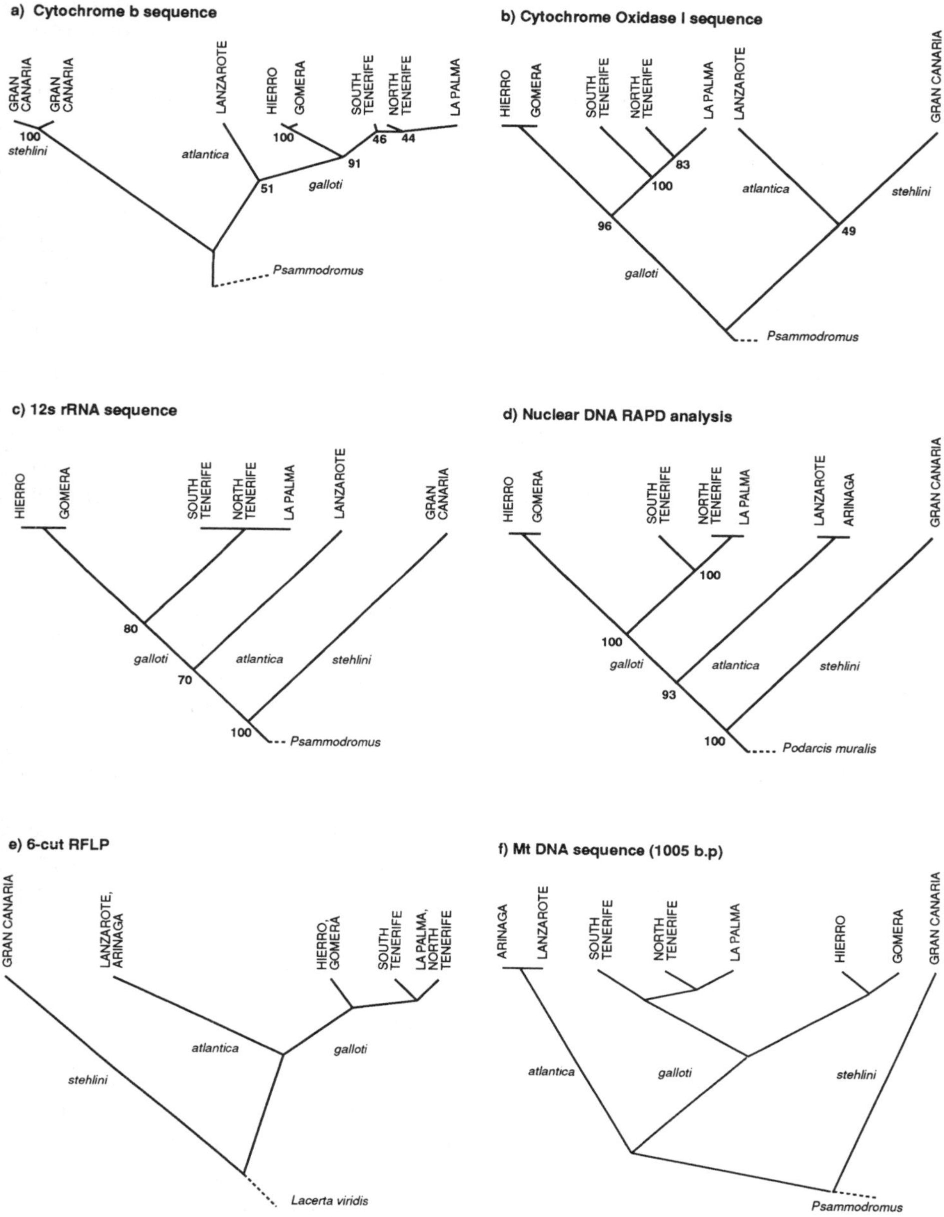

Fig. 12.1 Trees (a) to (d) are Wagner trees based respectively on (a) cytochrome b sequence (406 bp), (b) cytochrome oxidase sequence (Thorpe *et al.* 1994*b*), (c) 12s rRNA sequence (Thorpe *et al.* 1994*b*), and (d) RAPD analysis of the nuclear genome (Thorpe *et al.* 1994*b*). Trees (e) and (f) are Fitch–Margoliash trees based on 6-cut RFLPs (Thorpe *et al.* 1993*b*, 1994*a*) and 1005 bp from the above three mtDNA genes (Thorpe 1994*a*,*b*).

cytochrome b, cytochrome oxidase, and 12s rRNA genes) indicates that there is a distinct intraspecific phylogeny with northern (comprising populations from Tenerife and La Palma) and southern (populations from Gomera and Hierro) lineages (Fig. 12.1f). The colonization sequence can then be deduced from this tree following the method of Thorpe *et al.* (1994*a*) using tree topology plus branch length or topology plus geography. This gives us an origin for the species in south Tenerife, dispersal to north Tenerife, then a westward colonization to La Palma. In the south, an independent colonization event resulted in the colonization of Gomera and subsequently of Hierro. Not only is the sequence compatible with the geological history (Tenerife being the oldest island and the islands to the west progressively younger) but if one uses a molecular clock assumption (Brown *et al.* 1982) then so is the timing (Thorpe *et al.* 1993*b*, 1994*a*,*b*) with the geological origin of each island preceding the time of arrival of the lizards on it.

Does the morphological variation observed in this species reflect this dispersal history, or is it more influenced by marked differences in environmental conditions among the islands? Since the phylogenetic tree was constructed from a dataset independent of morphology, it can be used without circularity to test these alternative hypotheses. A partial Mantel matrix correlation test was performed (Thorpe 1991; Thorpe *et al.* 1994*a* and references therein). This method uses a randomization procedure to estimate probabilities for the correlation between matrices, standard significance tables being invalidated by the non-independence of matrix elements. Here we use an extension of the simple two-matrix Mantel test, a partial Mantel test which is based on a multiple regression (B.F.J. Manly, personal communication) and allows up to eight independent matrices to be compared with the dependent matrix simultaneously, thus largely removing the intercorrelation problems associated with multiple hypothesis tests. Each morphological character (represented by a similarity matrix between individuals) was tested simultaneously against the phylogeny (represented by a patristic distance matrix derived from the above Fitch–Margoliash tree) and two natural selection hypotheses represented by matrices of environmental variation (derived from data on environmental richness or biodiversity and climate). While many characters reflect both the phylogeny and environment, there are a few characters that are predominantly influenced by a single factor. For example, blue spots on legs can be shown to be closely associated with phylogeny, irrespective of ecology, while body size appears to be primarily related to environmental richness, with lizards of both lineages being smaller on the depauperate western islands. Other characters are associated mainly with wet, lush climates, for example lizards from north Tenerife and La Palma have gracile heads, blue cheeks, and yellow dorsal bars (Fig. 12.2).

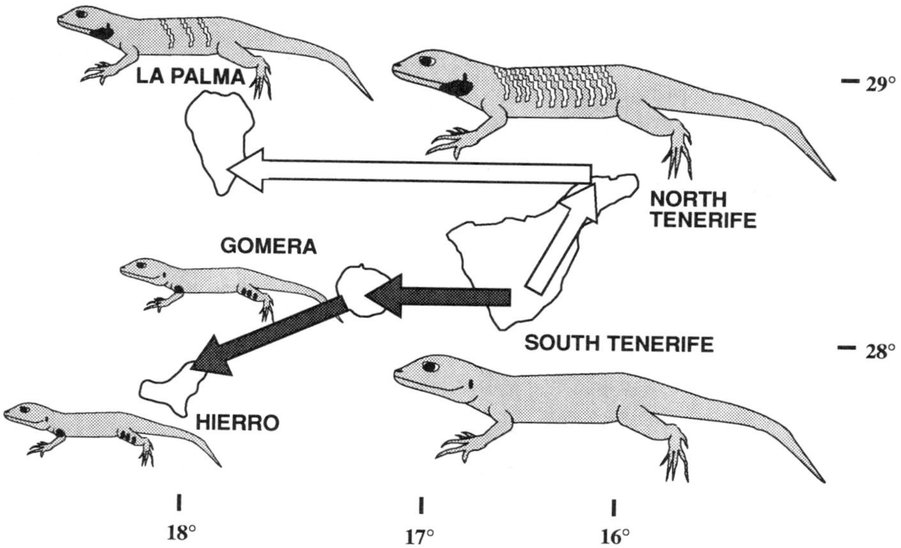

Fig. 12.2 Schematic representation of evolution in *G. galloti*. The arrows indicate the hypothesized colonization sequence of the two lineages (northern lineage, empty arrows; southern lineage, shaded arrows). Blue spots on the legs, thought to be for conspecific signalling, are present in the southern lineage only (phylogenesis) while sexually mature males are large in biodiverse islands (Tenerife), smaller in more depauperate areas (La Palma), and smallest in the most depauperate islands (Gomera, Hierro), irrespective of phylogenetic lineage. Similarly, lizards from wet, lush areas have disruptive yellow bars, blue cheeks, and gracile heads (ecogenesis) although phylogenetic lineage has an effect in the number of bars present.

12.3 The causes of venom variation in the Malayan pit-viper

A similar approach can be taken for other types of geographic variation; for example, the study of venom variation in venomous snakes. It has been well documented (Warrell 1989) that clinical symptoms produced by a snake bite can vary quite markedly between different populations of the same species, yet the causes of venom variation are poorly understood. Do they reflect neutral molecular change, or adaptation to local conditions such as available prey types? This study examined venom variation in the Malayan pit viper *Calloselasma rhodostoma*, which occurs in lowland forests and plantations in Java, northern West Malaysia, Thailand, Laos, Cambodia, and southern Vietnam (Fig. 12.3) and is the leading cause of venomous bites in many areas (Viravan *et al.* 1992). Venom and tissue biopsies were obtained from 131 Malayan pit vipers caught in the field, from 36 localities in Java, Malaysia, Thailand, and Vietnam. Faecal material was also collected to build up a picture of the species' prey, supplemented by the identification of some 200 prey items in the gastrointestinal tracts of museum and road-killed specimens. The diet shows distinct ontogenetic and geographic variation (Fig. 12.3), and minor sexual differences. A 767 base pair section of the mitochondrial cytochrome b gene was amplified by PCR and cut using seven 5- and 4-cut restriction enzymes. The patristic distances between localities on a Fitch–Margoliash tree constructed from these data represent their phylogenetic relationships. Individual venom

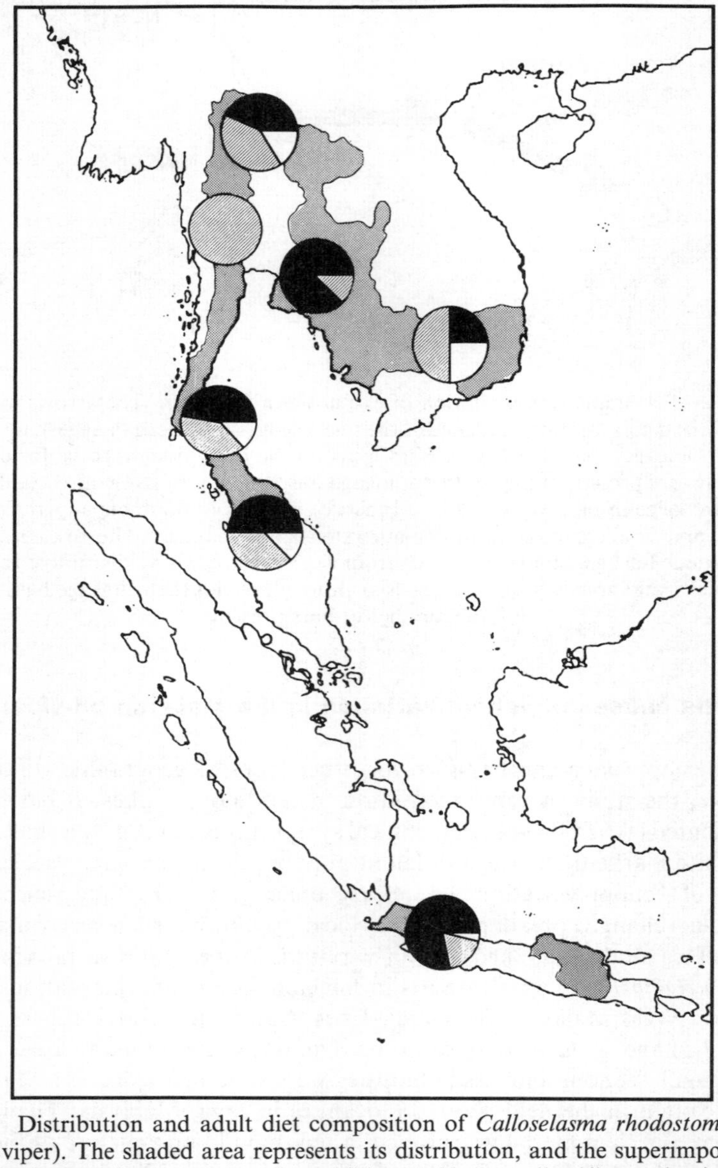

Fig. 12.3 Distribution and adult diet composition of *Calloselasma rhodostoma* (Malayan pit viper). The shaded area represents its distribution, and the superimposed pie charts show the relative proportion of amphibians (white), reptiles (diagonal stripes) and endotherms (black) in the diet of adult snakes from different areas.

samples were isoelectrically focused (IEF) across a high-resolution polyacrylamide gel (pH 3.5–9.5) and the protein banding patterns were compared. Ontogenetic differences were found, with the most marked changes occurring at sexual maturity, and variation was also obvious among venom samples collected from different geographic areas. IEF profiles from adult snakes (snout–vent length (SVL) exceeding 400 mm) were then used to construct distance matrices representing similarity/dissimilarity of venom composition among localities. Thirteen bands were found to vary between groups (the criteria being that variable bands are those which are present in at least two groups and absent from no less than two). In addition, two bands were found to be only present in females, and were excluded from further analysis. Matrices representing the overall venom composition (combining all bands) were constructed, in addition to matrices representing variation in individual bands. Similarity matrices were also produced to represent diet (the mean proportion of amphibians, reptiles, and mammals consumed by adults of each population), phylogenetic relationships (patristic distances), and geographic proximity. A partial Mantel test with the overall venom composition of adults as the dependent variable and diet, geographic distance, and patristic distance as independent variables, found diet alone to be significantly partially correlated ($P<0.001$ after Bonferroni correction). When individual variable bands are tested they are seen to vary in their association with different factors, for example some bands are associated with geographic distance, probably reflecting the opportunity for genetic exchange between spatially close localities. Almost half of the bands are significantly associated with average adult diet at each locality. Only one band is solely associated with phylogenetic relationships, which calls into question the use of venom as a taxonomic tool (Tu and Adams 1968; Chen *et al.* 1984). Although the variable proteins have not yet been identified it has been shown that there are also geographic differences in the necrotic and toxic effects of the venom (unpublished data).

Different prey taxa vary in their susceptibility to venom (Minton and Minton 1969) and it appears that natural selection has caused snake populations to produce venom locally adapted to their particular diet. Studies of captive-bred *C. rhodostoma* indicate that this venom/prey association is inherited rather than environmentally induced (unpublished data). Intraspecific variation in venom has important implications for snake-bite therapy (Anderson *et al.* 1993; Wüster *et al.* 1992), and while it would be helpful to be able to predict the distribution of clinically significant proteins on the basis of phylogenetic relationships or geographic distribution, this study suggests that painstaking biochemical analysis may be the only solution in some cases.

12.4 Within-island geographic variation in lizards

So far the cases discussed have been on a relatively large geographic scale, where we might expect to be able to find enough genetic variation between populations to be amenable to analysis. However, as long as genetic variation is detectable, this approach can also be used to study fine-level processes, such as those occurring within islands. Two examples are given.

Gallotia galloti *within Tenerife*

The island of Tenerife has a geologically complex, but well-documented history. It has only existed in its present form for a comparatively short period of time, although the oldest rocks found date as far back as 15.7 million years ago (Ancochea *et al.* 1990). On the basis of the presence of these ancient basaltic rocks, three possible precursor islands have been identified (Fig. 12.4a). These are Anaga, Teno, and the Roque del Conde (also known as Adeje). Later extensive volcanic activity, which continued until around 200 000 years ago, created the Las Cañadas edifice between these precursors, eventually joining them. It is possible that Teno and Adeje might originally have been a single island, with intervening rocks having been overlain by later volcanic deposits. Fossils found on Tenerife date the arrival of lacertids on the island to at least five million years ago, well before the eruption of Teide, hence they must have been present on one or more of the precursor islands.

However, morphological studies do not help us to decide whether more than one lineage is present on the island. Tenerife today is an environmentally heterogenous island, with the huge 3718 m cone of Teide creating a semipermanent layer of thick cloud at mid-altitude to the north. This results in a dramatic ecotone between lush forests on the north-facing slopes and barren semidesert on the southern slopes. Previous studies of within-island morphological variation in Tenerife *G. galloti* have revealed a strong association with these environmental patterns. In particular, it was proposed that the north–south variation in colour pattern was due to an evolutionary trade-off between sexual selection for bright colours and natural selection for crypsis, mediated by both visual characteristics and thermal considerations imposed by the different habitat types (Thorpe and Brown 1989, 1991). An independent assessment of the phylogeny is required in order to test the various scenarios for the origin and spread of the species on Tenerife.

A detailed investigation of microgeographic genetic variation across the island of Tenerife was undertaken, with tissue samples being collected from 63 localities across the island, from all three precursor islands, and intervening localities. Of the three mitochondrial genes sequenced for the earlier between-island study, cytochrome b had exhibited the finest level of resolution, so a 406 bp PCR fragment of the cytochrome b gene was subjected to a combination of denaturing gradient gel electrophoresis (DGGE) and direct sequencing. A total of 19 haplotypes were found, of which 16 were rare haplotypes found at only one or two localities. The three most common haplotypes showed a clear geographic distribution across the island (Fig. 12.4a). A pairwise distance matrix was generated, incorporating within-locality variation, and used to produce a Fitch–Margoliash tree. The tree shows that there are two main

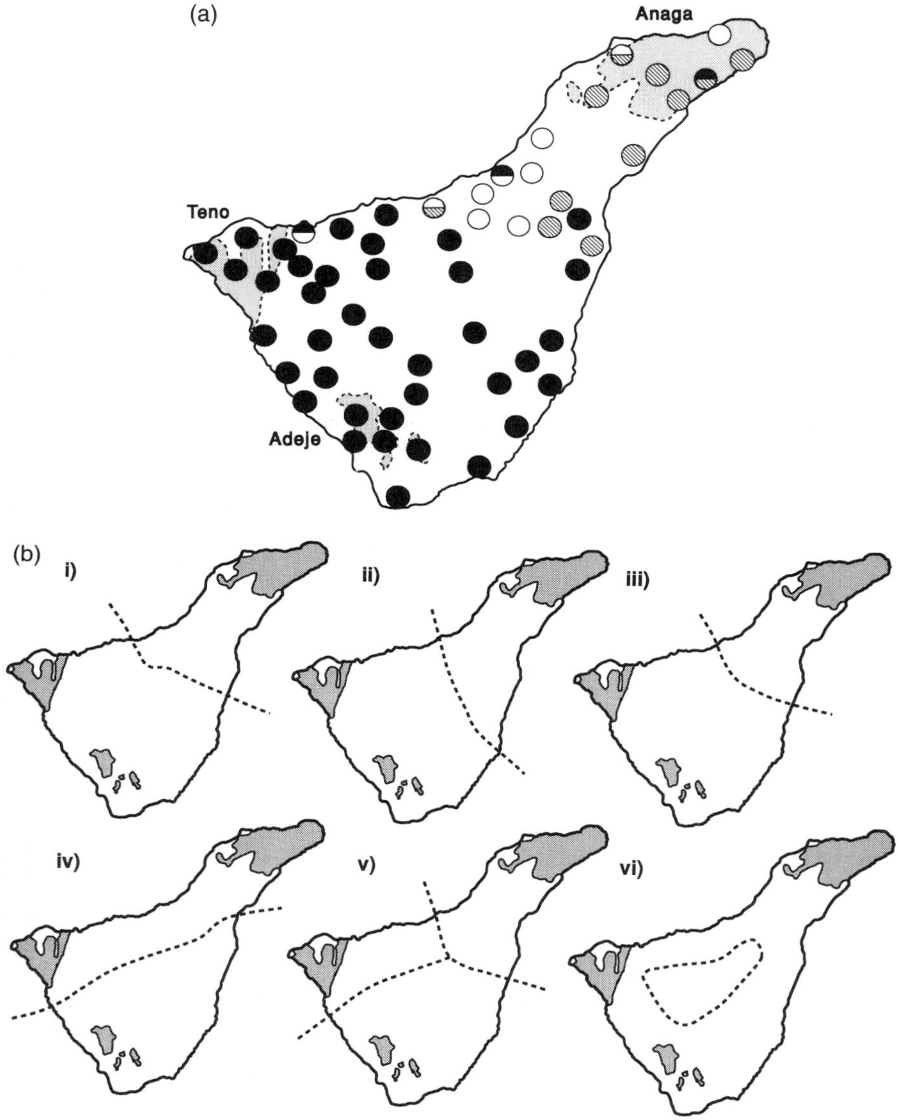

Fig. 12.4 (a) The island of Tenerife, showing haplotype distribution across the island. Haplotype A is indicated by full circles, haplotype B by empty circles, and haplotype B′ by hatched circles. Where more than one haplotype is present at any locality, the predominant haplotype is indicated. If haplotypes are present in equal proportions, they are indicated as appropriate. The position of older rocks corresponding to the precursor islands are indicated by broken lines. (b) The expected patterns produced by the various historical hypotheses. Broken lines separate localities into categories. (i) to (v) are phylogenetic hypotheses based on dispersion and secondary contact from precursor islands, described in the text. Localities were assigned to categories on the basis of which ancient basaltic rock they were closest to. Hypotheses (i), (ii), and (iii) produce very similar patterns, with only one or two centrally positioned localities distinguishing them. (vi) Vicariance due to the mid-altitude cloud layer.

lineages, the first comprising haplotype A and the second haplotypes B and B′ and rare variants. The patristic distances were then input into a Mantel test as the dependent variable, with several independent variables derived from historical hypotheses describing different patterns of secondary contact between possible precursor populations from (i) Anaga and Adeje, (ii) Anaga and Teno, (iii) Anaga and joint Teno–Adeje, (iv) Teno and Adeje, and (v) all three precursor islands separately (Fig. 12.4b). Also considered were (vi) a vicariance hypothesis (Pasteur and Salvidio 1985) which postulates that a ring of cloud around the mountain has resulted in two allopatric populations as population densities would be expected to be extremely low at the affected altitudes and (vii) geographic proximity, the opportunity for gene flow being greater between geographically closer populations. The results showed none of the three Anaga versus western island hypotheses (i–iii), or the hypothesis of geographic proximity, could be rejected. To eliminate the effects of intercorrelation of hypotheses, a partial Mantel test was then performed on these hypotheses, with the result that only the Anaga versus Teno and Adeje hypothesis (hypothesis (iii)) could not be rejected. The timing of the split between the two lineages (just less than one million years ago) is consistent with the geological evidence which indicates that Anaga may not have been linked with the rest of the island until later in the eruptive phase, possibly as late as 100 000 years ago.

We can now use the data on the distribution of the phylogenetic lineages within the island to extract the phylogenetic component from morphological data, to see if lineage has an effect on the association found previously. This can be done in two ways. First, phylogeny (represented as a matrix of patristic distances among localities) is included as an independent variable together with matrices representing habitat type and geographic proximity. The colour pattern matrix (dependent variable) is then compared with these three independent matrices using a partial regression in a simultaneous Mantel test. Second, tests for association of morphology and ecology can be carried out within each lineage separately (localities allocated according to which of the two main lineages predominates). When this is done using six colour pattern characters of mature males (Thorpe and Brown 1989), we find that there is a lineage effect, since lineage is significantly correlated with colour pattern in the former test. However, regardless of the method of extracting lineage, the hypothesis that colour pattern variation in *G. galloti* reflects natural selection for habitat type cannot be rejected.

Anolis oculatus *in Dominica (West Indies)*

Like Tenerife, Dominica is a volcanic island, with the combination of a high central mountain barrier and a constant wind direction resulting in the presence of a wide variety of habitats, from dry scrub to rainforest (Malhotra and Thorpe 1991*a*). The lizard *Anolis oculatus*, a small semiarboreal insectivore, is endemic to the island and displays a quite remarkable degree of geographic variation in colour pattern, body shape and size, and scalation (Malhotra and Thorpe

1991*a*; Malhotra 1992), although these are not necessarily congruent. Hypothesis testing has shown this to be highly associated with ecological variation (Malhotra and Thorpe 1991*a*), and further evidence for natural selection was provided by manipulative field experiments involving the translocation of populations between habitats (Malhotra and Thorpe 1991*b*).

At the same time, we looked at patterns of variation in a closely related species, *Anolis marmoratus*, from the island immediately to the north of Dominica, Basse Terre, which is part of the Guadeloupean archipelago. This is an equally variable species, and we showed that many morphological characters showed parallel variation in relation to the same ecological factors as in *Anolis oculatus* (Malhotra and Thorpe 1994). These parallels can be interpreted in terms of parallel selection regimes since the islands have been independent since their origin and hence a common historical explanation is extremely unlikely.

However, there is a marked and relatively congruent cline in morphological characters of *A. oculatus* (body proportions, scalation, and colour pattern) along the Caribbean coast of Dominica which does not fit this pattern, appearing out of proportion to the relatively subtle ecological clines along this cline compared to the rest of the island. To investigate the possibility of secondary contact zones (although no evident explanation for it exists), we sequenced 267 bp of the cytochrome b gene of mtDNA along a 22 km transect covering the area of the cline. This revealed the presence of a high level of genetic divergence (overall 12 per cent, with a maximum corrected pairwise difference of 9 per cent) between northern and southern populations (Malhotra and Thorpe 1994). The geographic pattern of sequence divergence (Fig. 12.5) is congruent with morphological and environmental (moisture) gradients.

Since similar clines in morphology had also been observed in *A. marmoratus* along the Caribbean coast of Basse Terre, we sequenced cytochrome b for a similar transect. Unexpectedly, a similar pattern of cytochrome b divergence was found, which was also correlated with at least some morphological clines and with moisture gradients. It is extremely unlikely that similar phylogenetic events (for example multiple colonizations, vicariance resulting from volcanic activity) have taken place on both islands, with their positions bearing a similar relationship to environmental gradients (Brown *et al.* 1991). Another unusual feature is the high level of differentiation found. Among the 24 sequences from 11 populations of *oculatus*, there are 21 different haplotypes, with only two being shared between pairs of neighbouring populations. Similarly in *marmoratus* there are 12 different haplotypes from six populations, with only three being shared between neighbours. If we use a commonly quoted rate for cytochrome b evolution (2.5 per cent per million years, which we used in the Canary island study without contradiction from other available evidence), then the time of divergence of the southern and northern lineages in *oculatus* would be of the order of four million years. The volcanic activity associated with the origin of Dominica is thought to have begun in the late Pliocene, with major, explosive eruptions associated with the appearance of most of the main volcanic cones in existence today continuing to as late as 100 000 years ago (Martin-Kaye 1969). The maintenance of this degree of genetic isolation between populations separated by a distance of only a few kilometres without any major barriers to dispersal is also difficult to explain.

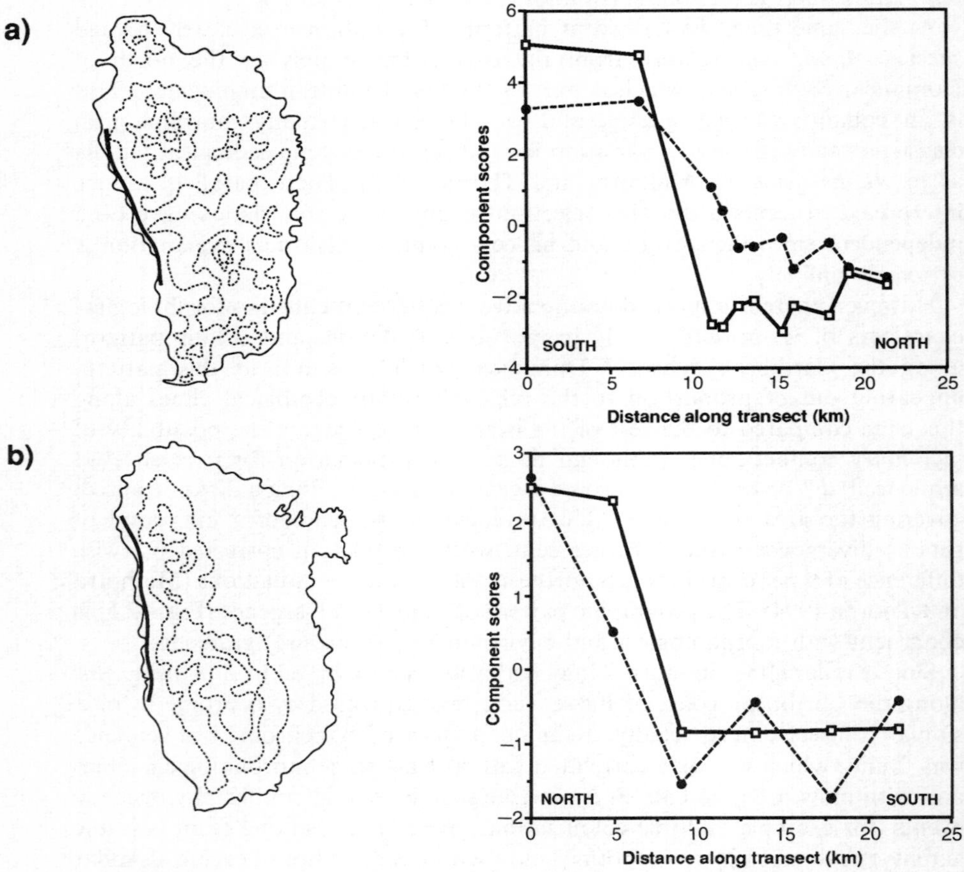

Fig. 12.5 The Lesser Antillean islands of (a) Dominica and (b) Guadeloupe, with contours at 300 m intervals indicated by the dotted lines. The positions of the transect are marked by a heavy line. The adjacent graphs show the variation along the transect in the cytochrome b sequence (full line) and scalation characters (broken line) in *Anolis oculatus* and *Anolis marmoratus* respectively. Note that the transects are plotted in opposite orientations since the ecological clines are in opposite directions in each island.

More information on divergence in other parts of the mitochondrial genome as well as that of nuclear markers may help to find an explanation for these patterns, and is the subject of ongoing research. However, it is possible that some of the assumptions involved in the phylogenetic approach used so far may be violated. The possibility of the non-neutral evolution of the mitochondrial genome (for example selection) cannot be rejected at this stage, especially in the light of recent papers which have questioned the neutrality of cytochrome b evolution (Ballard and Kreitman 1994).

12.5 Discussion and conclusions

Molecular phylogenies can be useful in helping us to understand the evolutionary processes underlying observed geographic patterns of population differentiation. This is particularly well illustrated by the *Gallotia* example where the molecular phylogeny is used at a variety of hierarchical levels (species across islands, island populations of the same species, and populations within islands) both to validate the taxonomic units being studied and to indicate the biogeographic processes underlying their geographic distribution. Combining molecular phylogenies with quantitative hypothesis testing methods throws light on the significance of adaptation to spatially varying environmental conditions in the *Gallotia* and *Calloselasma* examples and highlights the lack of recoverable phylogenetic information from morphology in the presence of strong selection for current environmental conditions.

The existence of extensive intraspecific variation at the scale of the studies reported here sounds a warning for the interpretation of higher-level mitochondrial phylogenies, as the problem of inadequate sampling of intraspecific variation (Melnick *et al.* 1993) leading to misinterpretations of species relationships becomes correspondingly larger. Intraspecific molecular studies are also increasingly challenging accepted assumptions about mitochondrial sequence evolution (such as neutrality and rates of evolution) which have relevance for phylogenetic reconstruction methods; the development of representations other than dichotomous trees (Hoelzer and Melnick 1994; Fitch, Chapter 8 this volume) and associated alternative methods of analysis (Excoffier *et al.* 1992; Excoffier and Smouse 1994) may provide a particularly valuable impetus to this area of research.

Acknowledgements

We thank B. F. J. Manly for providing his partial Mantel program RT-MANT, T. Griffiths for providing some of the sequences, and M. Baez and D. Lobidel for providing some tissue samples. This work was funded by the NERC (GR3/6943 and GR9/665, research studentship to HB), SERC (GR/F/19968, research studentships to AM, JCD), the Leverhulme Trust, the Royal Society, the EC, the Carnegie Trust, and the Bonhote Trust.

References

Ancochea, E., Fuster, J. M., Ibarrola, E., Cendrero, A., Coello, J., Hernan, F. *et al.* (1990). Volcanic eruption of the island of Tenerife (Canary Islands) in the light of new K-Ar data. *J. Volcan. Geothermal Res.*, **44**, 231–49.

Anderson, S. G., Gutierrez, J. M., and Ownby, C. L. (1993). Comparison of the immunogenicity and antigenic composition of ten central american snake venoms. *Toxicon*, **31**, 1051–9.

Avise, J. A. (1994). *Molecular markers, natural history and evolution*. Chapman and Hall, New York.

Baez, M. (1987). Les reptiles des Iles Canaries. *Bull. Soc. Zool. France*, **112**, 153–64.

Ballard, J. W. O. and Kreitman, M. (1994). Unravelling selection in the mitochondrial genome of *Drosophila*. *Genetics*, **138**, 757–72.

Brown, R. P. and Thorpe, R. S. (1991*a*). Within-island microgeographic variation in the colour pattern of the skink, *Chalcides sexlineatus*: pattern and cause. *J. Evol. Biol.*, **4**, 557–74.

Brown, R. P. and Thorpe, R. S. (1991*b*). Description of within-island microgeographic variation in the body dimensions and scalation of the Gran Canarian skink, *Chalcides sexlineatus*, with testing of causal hypotheses. *Biol. J. Linn. Soc.*, **44**, 47–64.

Brown, R. P., Thorpe, R. S., and Baez, M. (1991). Lizards on neighbouring islands show parallel within-island micro-evolution. *Nature*, **352**, 60–2.

Brown, W. M., Prager, E. M. Wang, A., and Wilson, A. C. (1982). Mitochondrial DNA sequences of primates: Tempo and mode of evolution. *J. Mol. Evol*, **18**, 225–39.

Chen, Y., Wu, X., and Zhai, E. (1984). Classification of *Agkistrodon* species in China. *Toxicon*, **22**, 53–61.

Dow, M. M., Cheverud, J. M., and Friedlaender, J. (1987). Partial correlation of distance matrices in studies of population structure. *Am. J. Phys. Anthropol.*, **72**, 343–52.

Ellsworth, D. L., Honeycutt, R. L., Silvy, N. J., Bickham, J. W., and Klimstra, W. D. (1994). Historical biogeography and contemporary patterns of mitochondrial DNA variation in white-tailed deer from the southeastern United States. *Evolution*, **48**, 122–36.

Excoffier, L. and Smouse, P. E. (1994). Using allele frequencies and geographic subdivision to reconstruct gene trees within a species—molecular variance parsimony. *Genetics*, **136**, 343–59.

Excoffier, L., Smouse, P.E., and Quattro, J. M. (1992). Analysis of molecular variance inferred from metric distances among DNA haplotypes: application to human mitochondrial DNA restriction data. *Genetics*, **131**, 479–91.

Georgiadis, N., Bischof, L., Templeton, A., Patton, J., Karesh, W., and Western, D. (1994). Structure and history of African elephant populations. 1. Eastern and southern Africa. *J. Hered.*, **85**, 100–4.

Hoelzer, G. and Melnick, D. J. (1994). Patterns of speciation and limits to phylogenetic resolution. *Trends Ecol. Evol.*, **9**, 104–7.

Malhotra, A. (1992). What causes geographic variation: a case study of *Anolis oculatus*. PhD thesis, University of Aberdeen.

Malhotra, A. and Thorpe, R. S. (1991*a*). Microgeographic variation in *Anolis oculatus* on the island of Dominica, West Indies. *J. Evol. Biol.*, **4**, 321–35.

Malhotra, A. and Thorpe, R. S. (1991*b*). Experimental detection of rapid evolutionary response in natural lizard populations. *Nature*, **353**, 347–8.

Malhotra, A. and Thorpe, R. S. (1994). Parallels between island lizards suggests selection on mitochondrial DNA and morphology. *Proc. R. Soc.*, **B257**, 37–42.

Martin-Kaye, P. H. A. (1969). A summary of the geology of the Lesser Antilles. *Overseas Geol. Min. Resour.*, **10**, 172–206.

Melnick, D. J., Hoelzer, G. A., Absher, R., and Ashley, M. V. (1993). MtDNA diversity in Rhesus monkeys reveals overestimates of divergence time and paraphyly with neighbouring species. *Mol. Biol. Evol.*, **10**, 282–95.

Minton, S. A. and Minton, S. R. (1969). *Venomous reptiles*. Scribners, New York.

Moritz, C., Schneider, C. J., and Wake, D. B. (1992). Evolutionary relationships within the *Ensatina eschscholtzii* complex confirm the ring species interpretation. *Syst. Biol.*, **41**, 273–91.

Pasteur, G. and Salvidio, S. (1985). Notes on ecological genetics of *Gallotia galloti* populations from Tenerife. *Bonn. Zool. Beit.*, **36**, 553–6.

Riddle, B. R., Honeycutt, R. L., and Lee, P. L. (1993). Mitochondrial-DNA phylogeography in northern grasshopper mice (*Onychomys leucogaster*)—the influence of quaternary climatic oscillations on population dispersion and divergence. *Mol. Ecol.*, **2**, 183–93.

Sokal, R. R., Oden, N. L., and Wilson, C. (1991). Genetic evidence for the spread of agriculture in Europe by demic diffusion. *Nature*, **351**, 143–5.

Sperling, F. A. H. and Harrison, R. G. (1994). Mitochondrial DNA variation within and between species of the *Papilio machaon* group of swallowtail butterflies. *Evolution*, **48**, 408–22.

Thorpe, R. S. (1994). Clines and cause: microgeographic variation in the Tenerife gecko *Tarentola delalandii*. *Syst. Zool.*, **40**, 172–87.

Thorpe, R. S. and Baez, M. (1993). Geographic variation in scalation of the lizard *Gallotia stehlini* within the island of Gran Canaria. *Biol. J. Linn. Soc.*, **48**, 75–87.

Thorpe, R. S. and Brown, R. P. (1989). Microgeographic variation in the colour pattern of the lizard *Gallotia galloti* within the island of Tenerife: distribution, pattern and hypothesis testing. *Biol. J. Linn. Soc.*, **38**, 303–22.

Thorpe, R. S. and Brown, R. P. (1991). Microgeographic clines in the size of mature male *Gallotia galloti* (Squamata: Lacertidae) on Tenerife: causal hypotheses. *Herpetologica*, **47**, 28–37.

Thorpe, R. S., Brown, R. P., Malhotra, A., and Wüster, W. (1991). Geographic variation and population systematics: distinguishing between ecogenetics and phylogenetics. *Boll. Zool.*, **58**, 329–35.

Thorpe, R. S., McGregor, D., and Cumming, A. M. (1993*a*). Population evolution of Canary Island lizards (*Gallotia galloti*): 4-base endonuclease restriction fragment length polymorphisms of mitochondrial DNA. *Biol. J. Linn. Soc.*, **49**, 219–27.

Thorpe, R. S., McGregor, D., and Cumming, A. M. (1993*b*). Molecular phylogeny of the Canary Island lacertids (*Gallotia*): mitochondrial DNA restriction site divergence in relation to sequence divergence and geological time. *J. Evol. Biol.*, **6**, 725–35.

Thorpe, R. S., Brown, R. P., Day, M. L., MacGregor, D. M., Malhotra, A., and Wüster, W. (1994*a*). Testing ecological and phylogenetic hypotheses in microevolutionary studies: an overview. In *Phylogenetics and ecology*, (ed. P. Eggleton and R. Vane-Wright), pp. 189–206. Academic, New York.

Thorpe, R. S., McGregor, D. P., Cumming, A. M. and Jordan, W. C. (1994*b*). DNA evolution and colonization sequence of island lizards in relation to geological history: mtDNA RFLP, cytochrome b, cytochrome oxidase, 12s rRNA sequence, and nuclear RAPD analysis. *Evolution*, **48**, 230–40.

Tu, A. T. and Adams, B. L. (1968). Phylogenetic relationships among venomous snakes of the genus *Agkistrodon* from Asia and North American continent. *Nature*, **217**, 760–2.

Viravan, C., Looareesuwan, S., Kosakarn, W., Wuthiekanun, V., McCarthy, C. J., Stimson, A. F. *et al.* (1992). A national hospital-based survey of snakes responsible for bites in Thailand. *Trans. R. Soc. Trop. Med. Hyg.*, **86**, 100–6.

Warrell, D. A. (1989). Snake venoms in science and clinical medicine I. Russell's viper: biology, venom and treatment of bites. *Trans. R. Soc. Trop. Med. Hyg.*, **83**, 732–40.

Wilson, A. C., Cann, R. L., Carr, S. M., George, M., Gyllensten, U. B., Helm-Bychowski, K. M. *et al.* (1985). Mitochondrial DNA and two perspectives on evolutionary genetics. *Biol. J. Linn. Soc.*, **26**, 375–400.

Wüster, W., Otsuka, S., Malhotra, A., and Thorpe, R. S. (1992). Population systematics of Russell's viper: a multivariate study. *Biol. J. Linn. Soc.*, **47**, 97–113.

Zink, R. M. and Dittmann, D. L. (1994). Gene flow, refugia, and evolution of geographic variation in the song sparrow (*Melospiza melodia*). *Evolution*, **47**, 717–29.

13

Uses of molecular phylogenies for conservation

Craig Moritz

13.1 Introduction

Phylogenies of species are producing important insights into taxonomy, the evolution of characters, and the long-term patterns and dynamics of adaptation and divergence (Brooks and McLellan 1991; Harvey and Pagel 1991). Molecular characters provide a rich source of information for analysis of species phylogeny and, within limits, can indicate the approximate timing as well as the sequence of events (Hillis and Moritz 1990). Molecular systematics can also reveal the phylogeny of alleles within species, revealing previously inaccessible details of historical biogeography and population processes (Avise 1989, 1994; Hudson 1990). Whether these new tools and concepts will prove of practical and general use in conservation remains to be demonstrated. In this chapter I consider the use of molecular phylogenies in two areas relevant to wildlife conservation; describing biodiversity and inferring population processes.

13.2 Describing biodiversity

Identification of 'evolutionarily significant units'

The definition of conservation units within species is fundamental to prioritize and conduct management (for example Daugherty *et al.* 1990). Moritz (1994*a*, *b*) suggested a distinction between two types of conservation unit (Fig. 13.1): *management units* (MUs), representing sets of populations that are currently demographically independent, and *evolutionarily significant units* (ESUs) which represent historically isolated sets of populations that together encompass the evolutionary diversity of a taxon. Both types of unit are significant for conservation, the former primarily to short-term management and the latter more to strategic, long-term issues. For example, MUs are the logical unit for monitoring responses of populations to impacts and management. Their value is as important functional components of the (usually) larger evolutionary entity. ESUs constitute the larger entities that conservation actions seek to preserve and can be seen as a complement to species as defined under broader criteria (Moritz 1994*b*) or as essentially equivalent to species under a phylogenetic species concept (Volger and DeSalle 1994).

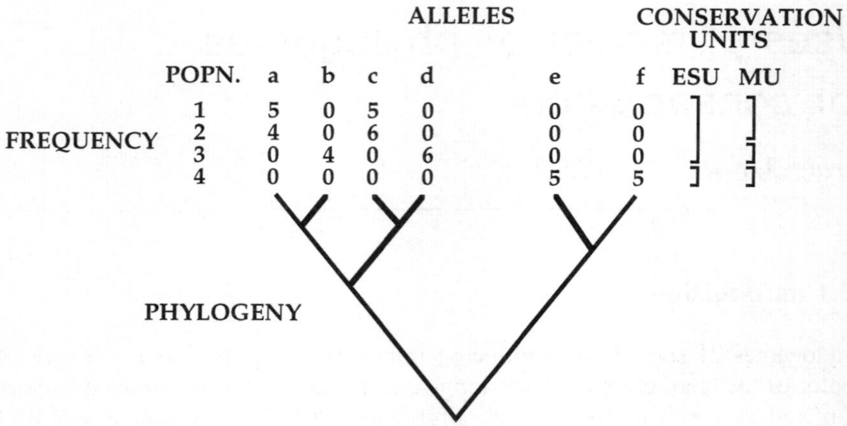

Fig. 13.1 Different types of conservation unit defined by allele phylogeny and frequency. Management units can be recognized as populations with distinct allele frequencies, for example populations 1 + 2 versus 3 versus 4; evolutionarily significant units are defined by having reciprocally monophyletic mtDNA alleles, for example populations 1 + 2 + 3 versus 4 as well as divergence in allele frequencies at nuclear loci. (Modified from Moritz (1994*a*).)

Whereas MUs often are best defined by differences in allele frequency (Hudson *et al.* 1992), the suggested definition of ESUs incorporates information on allele phylogeny, specifically the phylogeny of mtDNA alleles in relation to their distribution (see also Dizon *et al.* 1992; Volger and DeSalle 1994). The definition proposed (Moritz 1994*b*) is that *ESUs should be reciprocally monophyletic for mtDNA alleles and also differ significantly for the frequency of alleles at nuclear loci* (Fig. 13.1). Reciprocal monophyly of mtDNA was selected not because this is evolutionarily significant in itself, but because theory and simulations suggest that isolated sets of populations reach this condition after a specific amount of time, of the order of 4*N* generations (Neigel and Avise 1986), although allele coalescence may be more rapid in a declining population (Avise *et al.* 1984). This criterion is stringent, has the advantage of being qualitative rather than quantitative, and has a sound basis in population genetics theory. Whether nuclear genes should show concordant phylogenetic structuring (Avise and Ball 1990) is open to debate. This may be overly restrictive given that nuclear genes are expected to retain ancestral polymorphisms for longer than mtDNA, and clearly demarked species that are reciprocally monophyletic for mtDNA often retain ancestral polymorphisms at nuclear loci (for example Slade *et al.* 1994).

A clear example of a species with multiple ESUs, as defined above, is provided by the ghost bat, *Macroderma gigas*. This species has undergone a marked contraction of its range and is now restricted to a series of disjunct breeding populations around the humid north coast of Australia. Each regional population has monophyletic alleles for mtDNA (Worthington Wilmer *et al.* 1994) and significant divergence of allele frequencies at microsatellite loci (J. Worthington Wilmer, unpublished data). Thus, each regional population is a separate ESU and warrants conservation effort.

This approach has proved useful for documenting the evolutionary diversity within species, but, in general, resources or political will are inadequate to conserve all ESUs. A significant and more practical extension of this approach is to identify geographic areas between which many species show phylogeographic structure (Avise 1992, 1996; Brooks *et al.* 1992), i.e. to identify evolutionarily significant areas. These could be targeted separately for establishing protected areas or development of off-park conservation measures. A dramatic example is provided by the wet tropical rainforests of north-east Australia. Comparison of mtDNA sequences from several species of rainforest-restricted vertebrates revealed geographically congruent phylogeographic structuring about a historical barrier predicted by previous paleoclimatological modelling (Joseph *et al.* 1995). The rainforests to the north and south of this site are already included in a World Heritage Area and protected accordingly, but the molecular data have enhanced appreciation of the evolutionary significance of the smaller northern block of rainforest.

Representation and conservation value

The above use of molecular information in defining ESUs was confined to consideration of phylogenetic *pattern*; it was not the intention to ascribe conservation value on the basis of the magnitude of sequence divergence between populations. By contrast, several authors (for example Faith 1992; Crozier 1992) have derived algorithms for assessing the conservation priority of taxa, and thus areas, based on branch lengths of molecular phylogenies (for example Crozier and Kusmierski 1994). This builds on the suggestion that the priority accorded to taxa should take account of their phylogenetic distinctiveness (Vane-Wright *et al.* 1991).

The usual goal is to measure biodiversity in order to ensure that reserve systems are representative. The 'taxic diversity' approach of Vane-Wright *et al.*, Faith, and others emphasizes phylogenetically divergent lineages or a set of such lineages that best represents the breadth of the evolutionary diversity. Molecular phylogenies can contribute significantly to the estimation of species phylogenies for this purpose, regardless of branch lengths, but the inclusion of molecular distance does, at least crudely, add a time dimension. However, it should be borne in mind that single-gene trees are subject to variation in rates of evolution (Gillespie 1986) and also may depart from the true phylogeny of the species (Pamilo and Nei 1988; Slade *et al.* 1994).

A very different, and potentially conflicting, approach accords higher priority to *currently diversifying* taxa, as these may provide the basis for 'faunal reconstruction' following the current mass extinction (Erwin 1991; Brooks *et al.* 1992). Molecular phylogenies can be used for this purpose to distinguish between geographic areas that contain rapidly speciating lineages ('evolutionary fronts'; Erwin 1991) versus those with predominantly old lineages. For example, using the DNA:DNA hybridization data of Sibley and Ahlquist (1990), Fjeldså (1994) identified areas of Amazonian rainforest with concentrations of young species and suggested that these could provide a better basis for protection and reservation than the more dispersed areas with old species which would receive high priority under the taxic diversity approach.

To a large extent, the conflict between conserving old versus young lineages stems from differences in conservation goals and philosophy. Taxic diversity explicitly focuses on pattern, attempting to identify areas that will include the greatest phylogenetic breadth of species without making assumptions about future processes. By contrast, Fjeldså (1994) (see also Brooks *et al.* 1992) argues that an understanding of process, and the use of this information, is necessary to take differences in vulnerability and sustainability into account. Towards this end, Nee *et al.* (1994) have developed methods for measuring lineage speciation and extinction rates from molecular phylogenies and suggest that these could be used to identify clades of species that are under threat.

The contradictions between the 'taxic diversity' and 'evolutionary front' approaches suggests that to use phylogenetic criteria as a primary means to attach conservation value to species is fraught with unresolved ethical and conceptual problems and, thus, may be premature. The issue of conserving pattern versus process is considered further below.

13.3 Inferring population processes

Species management typically requires information on population size and connectivity, particularly for harvested or declining species. These parameters are central to the dynamics and viability of metapopulations, but connectivity in particular is difficult to measure using traditional ecological methods. Traditional population genetics provides measures of migration (gene flow) among populations (Slatkin 1987) and, at least potentially, changes in population size (Nei *et al.* 1975; cf. Leberg 1992). More recently, there has been intense interest in methods for estimating gene flow and trends in population size from molecular phylogenies (for example Slatkin and Maddison 1989; Ball *et al.* 1990; Slatkin and Hudson 1991; Nee *et al.* 1995; see also Chapter 5), assuming neutrality (cf. Rand *et al.* 1994).

In applying these measures for conservation, one needs to bear in mind that (i) it is difficult to discriminate between current and historical processes and (ii) most species in need of conservation have undergone dramatic changes in population size, structure, and connectivity within the recent past. For example, 10 species of Australian marsupial have gone extinct in the past 200 years and a further 27 species have undergone dramatic (>50 per cent) reductions in range, many in the past 100 years (Kennedy 1992).

Estimation of gene flow

Slatkin and Maddison (1989) introduced an approach for measuring gene flow among populations from the geographic distribution of alleles in relation to their phylogeny. The basic concept is to estimate the minimum number of migration events consistent with the phylogeography and, from this, to estimate the value of *Nm*, the average number of migrants per generation at equilibrium under an island model. Subsequently this approach was developed to test for different types of population structure, for example isolation-by-distance (Slatkin and Maddison 1990) and panmixia (Maddison and Slatkin 1990). In general, these methods are unable to distinguish between historical and current gene flow (but see Slatkin 1993) and are best regarded as providing a long-term perspective on connectivity, that may or may not correspond with present-day processes.

In a different approach, Neigel *et al.* (1991) developed a method for estimating single-generation dispersal distance from the variance in geographic ranges of specific mtDNA lineages and this was subsequently extended to distinguish between non-equilibrium and equilibrium situations (Neigel and Avise 1993). These methods do not assume an equilibrium between gene flow and drift, but do require a constant rate of base substitution. The behaviour of these models for species undergoing a rapid contraction in size or range is not clear, although it is notable that the variance in geographic distribution of lineages did not appear to be affected by cyclic range contractions simulating Pleiostocene events (Neigel and Avise 1993).

The distinction between long- and short-term processes of migration is well illustrated by an analysis of mtDNA variation in red kangaroos, *Macropus rufus* Clegg *et al.*, in preparation). There is no phylogeographic structure of mtDNA from the west coast of Australia to central Queensland, covering most of the species range and a distance of >3000 km. Yet, in larger sample sizes screened for RFLPs, there was evidence for significant heterogeneity of allele frequencies over distances as small as 60 km. This apparent contradiction can be resolved by proposing that kangaroos exhibit considerable long-distance gene flow in the long term, creating a phylogenetically random geographic distribution of alleles, but may have transiently isolated populations, resulting in locally heterogeneous allele frequencies. For the purpose of management, this distinction is critical, with the short-term population structure being more significant for identifying and managing stocks subject to control and harvesting.

Estimation of population trajectories

Another development stems from the realization that population expansion affects a number of parameters for DNA sequences (Fig. 13.2). Because of increased retention of gene lineages, expanding populations are expected to contain DNA sequences with a star-like phylogeny and a Poisson distribution of pairwise differences between alleles, unlike stable populations which are more likely to have strongly structured allele phylogenies with non-Poisson (for example geometric or multimodal) distributions of pairwise differences (Slatkin and Hudson 1991; Rogers and Harpending 1992; Felsenstein 1992). However, for pairwise differences, the signature arising from exponential growth at a constant rate is indistinguishable from that due to rapid expansion followed by a period of stasis (or by a selective sweep) (Rand *et al.* 1994). This approach was extended by Nee *et al.* (1995, and Chapter 5) who, by analysing the number of lineages through time, were able to distinguish qualitatively different forms of population growth.

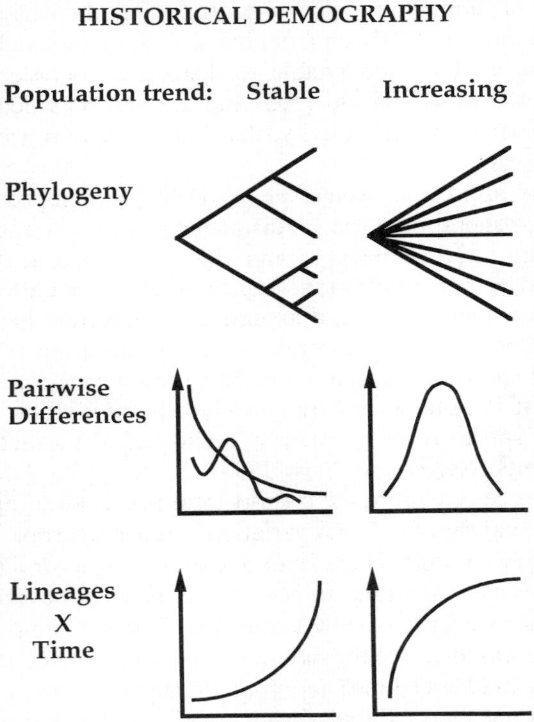

Fig. 13.2 Diagramatic summary of inferences from DNA sequences about trends in population size (see text). The lineages axis of the lineages × time plot is logarithmically scaled.

It is not yet clear how these measures respond to large fluctuations in population size in the short term, i.e. tens to hundreds of generations, but, being based on the dynamics of mutations, they are most likely to be relevant to long-term processes. Rogers and Harpending (1992) found that the distribution of pairwise differences responds relatively rapidly to sharp population declines, but 'rapid' was still measured in thousands of generations (i.e. of the order of the inverse of the mutation rate), possibly because this method is over-sensitive to deep branches in the genealogy (Felsenstein 1992).

A striking example of the disparity between population trends inferred from molecular phylogeny and current status concerns the coconut crab, *Birgus latro*. A study of mtDNA variation (Lavery *et al.*, submitted) revealed extremely high allelic diversity, but mostly low sequence divergences. Populations from the Pacific Ocean revealed the genetic signature of an exponentially expanding population: the phylogeny was star-like, the distribution of pairwise differences fitted a Poisson, and the lineage versus time plots suggest exponential growth at a constant or, more recently, an accelerating rate. In fact, the species has recently declined in numbers to the point where it is now extinct from most of its former range. Lavery *et al.* (submitted) suggested that the current phylogenetic signature reflects rapid population expansion during periods of lower sea level in the Pleiostocene, rather than the recent population declines.

13.4 Conservation of processes instead of entities

From the theory and examples reviewed above, it appears that the most robust contributions of molecular phylogenies to conservation are in defining conservation units, specifically ESUs, and, perhaps, in making inferences about population processes over evolutionary time. Conversely, analyses of molecular phylogenies could be misleading about current or very recent population processes if the species concerned have undergone dramatic declines, as is often the case for those requiring active conservation management.

One option is to use phylogenies just to define entities for conservation and avoid making inferences about process. This would be reasonable if the goal was simply to identify areas containing substantial evolutionary diversity and place them into reserves. However, adaptive management of such reserves and of their surrounds requires information on population size and connectivity, both to assess current status and to predict outcomes of specific management actions. Conceivably, molecular phylogenies could contribute here by providing insights into long-term population trends and patterns of gene flow, against which current behaviour can be compared (Milligan *et al.* 1994). Dramatic differences between long-term and current processes that can be attributed to human modification of the landscape may signal the need for intervention. Of course, common sense should dictate the extent to which historical processes can be maintained in the current, human-modified landscape.

For example, around the world a large number of amphibian species appear to be declining precipitously (Blaustein and Wake 1990; Richards *et al.* 1993). While most ecologists view this with alarm, another view is that this could be part of a natural cycle of population contractions and expansions (Pechmann *et al.* 1991), rather than a set of anthropogenic extinctions. This hypothesis is difficult to assess without information on long-term population sizes, data that are virtually non-existent; molecular phylogenies may be informative.

A second issue for which information on long-term versus current processes may be useful relates to translocations and the related questions of managing gene flow or hybridization. On one hand, it is often suggested that a low level of managed migration within a metapopulation may prevent inbreeding and loss of genetic variation through drift (for example Lacy 1987). On the other hand, there are concerns that introduction of foreign genes could result in the loss of local adaptation and cause 'outbreeding depression' (Templeton 1986). Unfortunately, with the exception of dramatic differences in karyotype, our ability to predict the outcomes from the extent of genetic differences among populations is limited (Avise and Aquadro 1982; Lynch 1991; Vogler and DeSalle 1994).

The approach often adopted by wildlife managers—don't cross genetically distinct stocks—appears conservative, at least from the perspective of maintaining entities. However, this approach, which at the extreme can border on 'genetic typology', may be inappropriate when compared to the long-term processes that operated in the species concerned (Wayne *et al.* 1994). From the perspective of long-term process, translocation of individuals within ESUs is unlikely to be detrimental and may well be an advantage, whereas deliberate translocation of individuals between ESUs should be avoided (see also Vrijenhoek 1989; Woodruff 1989). The focus should be on maintaining the overall process, i.e. historical levels of gene flow, rather than the specific entities, for example MUs within ESUs.

An interesting situation arises where there is evidence from molecular studies for occasional hybridization between otherwise distinct species. For example, Degnan (1993) found paraphyly of mtDNA between two species of silvereye (*Zosterops lateralis*) with monophyletic nuclear alleles, a combination strongly suggestive of one or more episodes of historical hybridization. Such hybridization is commonly considered aberrant and to be avoided in managing species. However, several recent studies have suggested that, far from being detrimental, occasional hybridization may be an important part of the evolutionary process (for example DeMarias *et al.* 1992), injecting genetic variance back into populations that would otherwise lose their diversity through genetic drift (Grant and Grant 1992, 1994).

13.5 Conclusion

Information on gene phylogeny can make a significant contribution to conservation through the more rigorous definition of entities and by contributing to a better understanding of historical population processes. It is important that the potential limitations of phylogenetic information in making inferences about current processes be explored and recognized. There is need for further development of theory, particularly in relation to the response of measures of gene flow or population trends to rapid fluctuations in size and connectivity of populations. This area is also ripe for experiments to determine whether management recommendations based on inferences from molecular data about long-term processes are valid or general. Such experiments are needed for a variety of species, vertebrate and invertebrate, and can be conducted in the laboratory or in the field, perhaps taking advantage of reintroductions under way as part of existing conservation programs.

Acknowledgements

Thanks to M. Bruford, M. Cunningham, S. Degnan, S. Lavery, R. Slade, C. Schneider, and J. Worthington Wilmer for critiques, discussion and/or permission to discuss unpublished data. The research described was funded by the Australian Research Council, the Queensland Department of Environment and Heritage, and the Collaborative Research Centre for Tropical Rainforest Ecology and Management.

References

Avise, J. C. (1989). Gene trees and organismal histories: a phylogenetic approach to population biology. *Evolution*, **43**, 1192–208.
Avise, J. C. (1992). Molecular population structure and the biogeographic history of a regional fauna: a case history with lessons for conservation biology. *Oikos*, **63**, 62–76.
Avise, J. C. (1994). *Molecular markers, natural history and evolution*. Chapman and Hall, New York.
Avise, J. C. (1996). Towards a regional conservation genetics perspective. In *Conservation genetics: case histories from nature*, (ed. J. C. Avise and J. Hamrick) Chapman and Hall, New York. (In press.)
Avise, J. C. and Aquadro, C. F. (1982). A comparative summary of genetic distances in the vertebrates. *Evol. Biol.*, **15**, 151–85.
Avise, J. C. and Ball, R. M. (1990). Principles of genealogical concordance in species concepts and biological taxonomy. *Oxford Surv. Evol. Biol.*, **7**, 45–68.
Avise, J. C., Neigel, J. E., and Arnold, J. (1984). Demographic influences on mitochondrial DNA lineage survivorship in animal populations. *J. Mol. Evol.*, **20**, 99–105.
Baker, C. S., Parry, A., Bannister, J. L. *et al.* (1993). Abundant mitochondrial DNA variation and world-wide population structure in humpback whales. *Proc. Natl. Acad. Sci. USA*, **90**, 8239–43.
Ball, R. M. J., Neigel, J. E., and Avise, J. C. (1990). Gene geneologies within the organismal pedigree of random-mating populations. *Evolution*, **44**, 360–70.
Blaustein, A. R. and Wake, D. B. (1990). Declining amphibian populations: a global phenomenon? *TREE*, **5**, 203–4.
Brooks, D. R. and McLellan, D. A. (1991). *Phylogeny, ecology and behaviour*. University of Chicago Press.
Brooks, D. R., Mayden, R. L., and McLennan, D. A. (1992). Phylogeny and biodiversity: conserving our evolutionary legacy. *TREE*, **7**, 55–9.
Clegg, S., Hale, P. and Moritz, C. Extensive genetic diversity and gene flow in the red kangaroo, *Macropus rufus*. *Molecular Ecology*. (Submitted.)
Crozier, R. H. (1992). Genetic diversity and the agony of choice. *Biol. Conserv.*, **61**, 11–15.

Crozier, R. H. and Kusmierski, R. M. (1994). Genetic distances and the setting of conservation priorities. In *Conservation genetics*, (ed. V. Loeschcke, J. Tomiuk, and S. K. Jain), pp. 227–37. Birkhäuser, Basel.

Daugherty, C. H., Cree, A., Hay, J. M., and Thompson, M. B. (1990). Neglected taxonomy and continuing extinctions of the tuatara (*Sphenodon*). *Nature*, **347**, 177–9.

Degnan, S. D. (1993). The perils of single gene trees—mitochondrial versus single-copy nuclear DNA variation in white-eyes (Aves: Zosteropidae). *Mol. Ecol.*, **2**, 219–25.

DeMarias, B. D., Dowling, T. E., Douglas, M. E., Minckley, W. L. and Marsh, P. C. (1992). Origin of *Gila seminuda* (Teleostei: cyprinidae) through introgressive hybridisation: implications for evolution and conservation. *Proc. Natl. Acad. Sci. USA*, **89**, 2747–51.

Dizon, A. E., Lockyer, C., Perrin, W. F., Demaster, D. P., and Sisson, J. (1992). Rethinking the stock concept: a phylogeographic approach. *Cons. Biol.*, **6**, 24–36.

Erwin, T. L. (1991). An evolutionary basis for conservation strategies. *Science*, **253**, 75–2.

Faith, D. P. (1992). Conservation evaluation and phylogenetic diversity. *Biol. Conserv.*, **61**, 1–10.

Felsenstein, J. (1992). Estimating effective population size from samples of sequences: inefficiency of pairwise and segregating sites as compared to phylogenetic estimates. *Genet. Res. (Camb.)*, **59**, 139–47.

Fjeldså, J. (1994). Geographical patterns for relict and young species of birds in Africa and South America and implications for conservation priorities. *Biodiv. Conserv.*, **3**, 207–26.

Gillespie, J. H. (1986). Rates of molecular evolution. *Ann. Rev. Ecol. Syst.*, **17**, 637–65.

Grant, P. R. and Grant, B. R. (1992). Hybridization of bird species. *Science*, **256**, 193–7.

Grant, P. R. and Grant, B. R. (1994). Phenotypic and genetic effects of hybridization in Darwin's finches. *Evolution*, **48**, 297–316.

Harvey, P. H. and Pagel, M. D. (1991). *The comparative method in evolutionary biology*. Oxford University Press.

Hillis, D. M. and Moritz, C. (ed.) (1990). *Molecular systematics*. Sinauer, Sunderland, MA.

Hudson, R. R. (1990). Gene genealogies and the coalescent process. *Oxford Surv. Evol. Biol.*, **7**, 1–44.

Hudson, R. R., Boos, D. D., and Kaplan, N. L. (1992). A statistical test for detecting geographic subdivision. *Mol. Biol. Evol.*, **9**, 138–151.

Joseph, L., Moritz, C., and Hugall, A. (1995). Molecular support for vicariance as a source of diversity in rainforest. *Proc. R. Soc.*, **B260**, 177–182.

Kennedy, M. (ed.) (1992). *Australasian marsupials and monotremes: an action plan for their conservation*. IUCN, Gland, Switzerland.

Lacy, R. C. (1987). Loss of genetic variation from managed populations: interacting effects of drift, mutation, immigration, selection and population subdivison. *Cons. Biol.*, **1**, 143–58.

Lavery, S., Moritz, C., and Fielder, D. R. (1996). Genetic patterns suggest exponential population growth in a declining species. *Mol. Biol. Evol.* (Submitted.)

Leberg, P. L. (1992). Effects of population bottlenecks on genetic diversity as measured by allozyme electrophoresis. *Evolution*, **46**, 477–94.

Lynch, M. (1991). The genetic interpretation of inbreeding and outbreeding depression. *Evolution*, **45**, 622–9.

Maddison, W. P. and Slatkin, M. (1990). Null models for the number of evolutionary steps in a character on a phylogenetic tree. *Evolution*, **45**, 1184–97.

Milligan, B. G., Leebens-Mack, J., and Strand, A. E. (1994). Conservation genetics: beyond the maintenance of marker diversity. *Mol. Ecol.*, **3**, 423–35.

Moritz, C. (1994*a*). Applications of mitochondrial DNA analysis in conservation: a critical review. *Mol. Ecol.*, **3**, 401–11.

Moritz, C. (1994*b*). Defining evolutionarily significant units for conservation. *TREE*, **9**, 373–5.

Nee, S., Holmes, E. C., May, R. M. and Harvey, P. H. (1994). Extinction rates can be estimated from molecular phylogenies. *Phil. Trans. R. Soc.*, **B344**, 77–82.

Nee, S., Holmes, E. C. and Harvey, P. H. (1995). Inferring population history from molecular phylogenies. *Phil. Trans. Roy. Soc. Lond.*, **B349**, 25–31.

Nei, M., Maruyama, T., and Chakraborty, R. (1975). The bottleneck effect and genetic variability in populations. *Evolution*, **29**, 1–10.

Neigel, J. and Avise, J. C. (1986). Phylogenetic relationships of mitochondrial DNA under various demographic models of speciation. In *Evolutionary processes and theory* (ed. E. Nevo and S. Karlin), pp. 515–34. Academic, New York.

Neigel, J. E. and Avise, J. C. (1993). Application of a random walk model to geographic distributions of animal mitochondrial DNA variation. *Genetics*, **135**, 1209–20.

Neigel, J. E., Ball, R. M., and Avise, J. C. (1991). Estimation of single generation migration distances from geographic variation in animal mitochondrial DNA. *Evolution*, **45**, 423–32.

Pamilo, P. and Nei, M. (1988). Relationships between gene trees and species trees. *Mol. Biol. Evol.*, **5**, 568–83.

Pechmann, J. F. K., Scott, D. E., Semlitsch, R. D., Caldwell, J. P., Vitt, L. J., and Gibbons, J. W. (1991). Declining amphibian populations: the problems of separating human impacts from natural fluctuations. *Science*, **253**, 892–5.

Rand, D. M., Dorfsman, M. and Kann, L. M. (1994). Neutral and non-neutral evolution of *Drosophila* mitochondrial DNA. *Genetics*, **138**, 741–56.

Richards, S. R., McDonald, K. R. and Alford, R. A. (1993). Declines in populations of Australia's endemic tropical rainforest frogs. *Pacific Cons. Biol.*, **1**, 66–77.

Rogers, A. R. and Harpending, H. (1992). Population growth makes waves in the distribution of pairwise genetic differences. *Mol. Biol. Evol.*, **9**, 552–69.

Sibley, C. G. and Ahlquist, J. E. (1990). *Phylogeny and classification of birds. A study in molecular evolution.* Yale University Press, New Haven, CT.

Slade, R. W., Moritz, C. and Heidemann, A. (1994). Multiple nuclear gene phylogenies: applications to pinnipeds and a comparison with a mtDNA gene phylogeny. *Mol. Biol. Evol.*, **6**, 341–56.

Slatkin, M. (1987). Gene flow and the geographic structure of animal populations. *Science*, **236**, 787–92.

Slatkin, M. (1993). Isolation by distance in equilibrium and non-equilibrium populations. *Evolution*, **47**, 264–79.

Slatkin, M. and Hudson, R. R. (1991). Pairwise comparisons of mitochondrial DNA sequences in stable and exponentially growing populations. *Genetics*, **129**, 555–62.

Slatkin, M. and Maddison, W. P. (1989). A cladistic measure of gene flow inferred from phylogenies of alleles. *Genetics*, **123**, 603–13.

Slatkin, M. and Maddison, W. P. (1990). Detecting isolation by distance using phylogenies of genes. *Genetics*, **126**, 249–60.

Templeton, A. R. (1986). Coadaptation and outbreeding depression. In *Conservation biology: the science of scarcity and diversity*, (ed. M. E. Soule), pp. 105–16. Sinauer, Sunderland, MA.

Vane-Wright, R. I., Humphries, C. J., and Williams, P. H. (1991). What to protect—systematics and the agony of choice. *Biol. Conserv.*, **55**, 235–54.

Vogler, A. P. and DeSalle, R. (1994). Diagnosing units of conservation management. *Cons. Biol.*, **8**, 354–63.

Vrijenhoek, R. C. (1989). Population genetics and conservation. In *Conservation for the twenty first century*, (ed. D. Western and M. Pearl), pp. 89–98. Oxford University Press, New York.

Wayne, R. K., Bruford, M. W., Girman, D., Rebholz, W. E. R., Sunnucks, P., and Taylor, A. C. (1994). Molecular genetics of endangered species. In *Creative conservation. Interactive management of wild and captive animals*, (ed. P. J. S. Olney, G. M. Mace, and A. T. C. Feistner), pp. 92–117. Chapman and Hall, London.

Woodruff, D. S. (1989). The problems of conserving genes and species. In *Conservation for the twenty first century*, (ed. D. Western and M. Pearl), pp. 76–88. Oxford University Press, New York.

Worthington Wilmer, J., Moritz, C., Hall, L., and Toop, J. (1994). Extreme population structuring in the threatened Ghost Bat, *Macroderma gigas*: evidence from mitochondrial DNA. *Proc. R. Soc.*, **B257**, 193–8.

Combining phylogenetic evidence

14

Testing the time axis of phylogenies

M. J. Benton

14.1 Introduction

Phylogenetic trees produced from morphological or molecular information are generally represented with a horizontal axis that represents clade diversity and a vertical axis that represents character-state change. The nodal order of branching may be calibrated as a time axis by the addition of stratigraphic information about the occurrence of fossils in geological time. There is no other, independent, source of information for calibrating the time of cladogenetic events in phylogenies, and yet many doubts have been expressed about the trustworthiness of the fossil record. This subject will be explored further, and a case will be made for the use of stratigraphic and palaeontological data in producing and in testing phylogenies.

Two temporal aspects of phylogenies are of interest: the implied order of nodes in a phylogenetic tree, and the absolute time represented. Both of these aspects of phylogenies may be tested by direct comparison between cladograms, or molecular trees, and stratigraphic data. In particular, it is shown that: (1) stratigraphic data on group appearances match cladistic data on branching patterns; (2) palaeontological data have improved by about 5 per cent over the past 26 years; and (3) marine and continental, and invertebrate and vertebrate, fossil records may be equally complete.

14.2 The value of palaeontological data

Incompleteness of the fossil record

The fossil record of the history of life is incomplete (Darwin 1859; Raup 1972; Simpson 1983; Allison and Briggs 1991). Clearly, very few individual organisms that have ever existed produce a fossil. A number of substantial filters must be crossed between the death of an organism and its identification as a fossil: (1) habitat and ecology (short-lived benthic marine organisms with hard skeletons are more likely to be buried than long-lived arboreal organisms which lack skeletons); (2) sedimentation patterns (many organisms live in areas where deposition is not taking place); (3) subsequent geological history (older rocks may be metamorphosed, subducted, eroded, and suffer all kinds of similar indignities); (4) serendipity and human factors (fossils do not exist as unique taxa until they have been collected by an interested person, studied, and described).

Some of these filters against fossilization may be tested quantitatively. Schopf (1978) found that 60 per cent of organisms living in the coastal habitats around Friday Harbor, Washington would not be preserved since they lack skeletons, they live on rocks and in other environments which are not preserved, or they are highly mobile and may escape burial. Other studies have yielded similar, or less favourable, figures.

The conclusion from such gloomy studies could be that there is little chance that palaeontologists can offer more than a tiny and highly distorted picture of the former diversity of life. It could be asserted that the number of data lost through incomplete preservation and sampling are so vast as to render the remaining fossil record virtually worthless. This claim has rarely been made in such extreme terms, but clearly opinions are coloured by considerations of the sort described. For example, in an attempt to avoid the incompleteness of palaeontological data, Harvey *et al.* (1994) used 'fossil-free trees' to study macroevolution; they assumed that molecular phylogenies will, in the future, become much more reliable than morphological cladistic phylogenies (and especially those that incorporate fossils).

Two specific lower-order criticisms of the fossil record have been made: (1) fossils provide far less character information than modern taxa, and so should take only a secondary role in phylogeny reconstruction (Patterson 1981; Ax 1987), and (2) the order of occurrence of fossils is so mixed up by the patchiness of representation that the fossil record reveals only a crude outline of what happened in the past. These criticisms, focusing on (1) missing characters and (2) missing taxa deserve consideration and testing.

Missing characters?

A number of systematists have made the assertion that fossil taxa are much inferior to living taxa for the reconstruction of phylogeny, and that they should be ignored, or accorded much lower value in the assessment of branching patterns (Hennig 1966, 1981; Nelson 1969; Nelson and Platnick 1981; Patterson 1981; Ax 1987; Goodman 1989; Meyer and Wilson 1990; Meyer and Dolven 1992). Specific criticisms of fossils were: (1) much character information is missing because of the loss of soft tissues; and (2) it is impossible to include fossil taxa in molecular phylogenies.

No-one can deny that fossils often lack soft-part data, and it is possible to show that fossil taxa are typically associated with more missing data in taxon/character data matrices than are extant taxa. However, it is important to recall that incompleteness is not the same as an absence of information (Donoghue *et al.* 1989; Wilkinson 1994). Even if only one or two characters are available for a particular taxon, these may be the attributes that determine the nature of a particular node in the cladogram. More completely coded taxa may offer little phylogenetically informative data (Wilkinson 1994).

Events have now overtaken the second criticism. Proteins and DNA have been recovered from a variety of fossils, and molecular phylogenetic trees, incorporating fossil taxa, have been published (for example Golenberg *et al.* 1990; DeSalle *et al.* 1992; Cano *et al.* 1993). When problems of analysis and interpretation are resolved, such studies may become commonplace. Fossils may yield molecular sequences that assist in phylogeny reconstruction, especially when long terminal branches are present (Smith and Littlewood 1994).

Numerous studies (Doyle and Donoghue 1987; Gauthier *et al.* 1988; Donoghue *et al.* 1989; Huelsenbeck 1991; Novacek 1992*a*, *b*; Wilson 1992; Smith and Littlewood 1994) have now contradicted the view that fossils are of secondary value in phylogeny reconstruction. These studies have used a variety of empirical and modelling studies to show that fossil taxa contribute positively to phylogeny reconstruction as a result of their unique properties (Smith 1994). Advantages of fossils are that (1) they give the only direct evidence of the order and precise date of the acquisition of characters and character complexes, (2) they may allow the coding of characters that have been overwritten by subsequent evolution within a clade (such as the teeth of turtles and birds, which are known only in fossil forms), and (3) they frequently present character combinations not found in modern forms.

The last point, that fossils may provide unique character combinations, is critical. To ignore fossils is to ignore real terminal taxa that form independent branches in cladograms. Even when fossil taxa are rare or incomplete, they may offer character data critical for resolving particular nodes. Fossil taxa may assist in distinguishing synapomorphy (see the glossary for meaning of this and other technical terms) from homoplasy, and this is particularly so in cases where there are long terminal branches; the fossil taxa may divide these long branches up, and provide positive evidence of convergence of wrongly assessed postulated synapomorphies. Fossils may also assist in polarity determination in cases where there has been a change of character state within a clade (a postulated synapomorphy may be found to occur in early members of the supposed outgroup, which shows that its distribution was once more general, and that it is in fact a symplesiomorphy).

Missing taxa?

The question of missing taxa in the fossil record would seem to be insurmountable, and hard to test. Darwin (1859) predicted that phylogenetic patterns would emerge as palaeontologists collected more and more fossils, but this seems less achievable now than he had hoped because of the patchiness of the fossil record. When fossils occur in vast numbers in closely spaced stratigraphic horizons, it may be possible to reconstruct phylogenies by the stratophenetic method (Gingerich 1979; Cheetham and Hayek 1988), which consists of linking sequences of populations through time into lineages. However, as the density of sampling on a phylogenetic tree diminishes, the chances of encountering fossil taxa out of sequence increases (Fortey and Jefferies 1982).

Glossary

Explanation of some terminology relevant to cladistic analysis.

Apomorphy, a derived character.
Clade, a monophyletic group.
Cladistics, the search for patterns of relationships based on the identification of truly homologous attributes.
Homology, similarity of features owing to common descent.
Homoplasy, a feature that is not a homologue, but may be a shared primitive resemblance, or something that has evolved convergently or in parallel.
Monophyletic group, a group that includes all the descendants from a single common ancestor.
Node, a branching point in a cladogram.
Outgroup, a group of close relatives of the organisms whose relationships are under study (the in group), and which may give indications of character polarity.
Plesiomorphy, a primitive, or general, character that does not indicate relationships.
Polarity, the direction from general (primitive) to specialized (apomorphic or derived).
Rooting, the search for the root of a cladogram, which depends on polarity determination of characters
Symplesiomorphy, a shared primitive character.
Synapomorphy, a shared derived character, a character shared uniquely by the decendants of the first organism to have acquired the feature.
Terminal taxon (pl. *taxa*), a species or other clade that occurs at the end of one of the branches of a cladogram.

Some palaeontologists (for example Harper 1976; Gingerich and Schoeninger 1977; Szalay 1977) have argued that stratigraphic occurrence may be used as a test of character polarity, that the character state found in the oldest fossil is the ancestral state. However, this simplistic assumption that early = primitive has been criticized by Eldredge and Cracraft (1980), Nelson and Platnick (1981), and many others. The equation is probably valid in many cases, but there is no guarantee that, even in well-sampled groups, the order of occurrence of fossils is correct. A cladogram must be constructed from character data in order to establish the hierarchical branching pattern.

This debate, about the use of fossils in polarity determination, as well as the wider debates between cladists and traditional systematists, has left an impression that fossil taxa are generally so sparsely represented in time that little may be gained from palaeontological data. Until recently, the only replies that could be made by palaeontologists were qualitative: simple assertions that the nature of the fossil record had been misrepresented, and that it was not so bad as people had suggested. There are now two simple quantitative techniques that give absolute measures of the quality of the fossil record, and their results are favourable to palaeontological data. These techniques depend upon an assumption that cladograms and molecular trees are independent of stratigraphic data.

14.3 Three independent sources of evidence about phylogeny

There are three independent sources of data on series of events in phylogeny: cladistic morphological, molecular, and stratigraphic. The first two, of course, provide branching patterns, while the third, stratigraphy, offers only a sequence of dated events. There are probably more than three, if it can be maintained that the nucleic acids evolve independently of proteins such as myoglobin, haemoglobin, and cyctochrome c, and indeed it may be the case that unrelated proteins offer a further selection of independent sources of data on the order of phylogenetic events. Likewise, morphological data from larvae, as well as other forms of organism-level data, such as physiological and behavioural characters, might offer further broadly independent datasets.

Cladistic analyses are generally based on morphological characters alone. The hierarchical structure of the cladogram is discovered by analysing a matrix of characters coded across a number of taxa. At times, fossils have been proposed as the arbiters of two equivalent techniques in cladistics: (1) character polarity and (2) tree rooting, and if these methods relied upon the age of the fossils, then cladistic methods would not be independent of stratigraphic data.

Fossils have been used on occasion to determine character polarity, but that method is discredited (see above). The technique of choice for polarity determination is outgroup comparison, and this method treats all organisms under consideration, whether living or extinct, as equivalent terminal taxa. In any case, most current analytical methods require no assumptions about character polarity, and the issue may be avoided completely. The modern techniques identify homologies, but they do not indicate the direction of change. This is discovered by the process of tree rooting.

Tree rooting is carried out by identifying one or more outgroups, which fix the shape of the cladogram, and thereby determine the direction of character change. This is perhaps a more contentious issue, since fossil taxa are often chosen as the outgroups. There is potential confusion here, however, about why the fossil taxa were chosen for tree rooting: it was not because they occurred earlier in time than the extant taxa (a mixing of stratigraphic and cladistic data), but because they illustrate a more plesiomorphous set of characters (Huelsenbeck 1991; Wilson 1992; Smith 1994). The outgroup taxon, whether fossil or Recent, is chosen because it is minimally divergent from the ingroup taxa, and hence can be more informative in determining the directions of character change. As an example, in attempting to resolve a cladogram of lungfish, coelacanth, and tetrapods, which diverged about 400 million years ago, it is more helpful to use one or two Devonian fishes as outgroups than to select a modern cod or seahorse, since they have built up a 400 million year overprint of their own character transformations, which obscures many of the relevant character states.

Molecular phylogenies are compiled either from comparisons of relative similarity between proteins or nucleic acids (DNA, RNA) of different species, or by comparisons of sequences of amino acids in proteins, or of bases in nucleic acids. The quantitative data on similarities among taxa, or on precise differences between sequences, are converted into trees by a variety of multivariate techniques. The only point at which fossils enter the technique is in calibrating the time-scale of the trees. The fossils do not affect the shape of the tree, and hence there is no mixing of molecular and stratigraphic data in the identification of the hierarchy of branching points.

If stratigraphic, morphological cladistic, and molecular techniques provide independent evidence about branching patterns, then those patterns may be mutually tested in an attempt to check the performance of each technique. In addition, such cross-testing of historical patterns of phylogeny acts as a test of evolution in general. Evolution, meaning simply the patterns of organic change through time (Mayr 1982), has been seen as a theory founded on such a broad accumulation of observations and hypotheses that it cannot be tested in any simple way. And yet, classic evolutionists (Darwin 1859; Ridley 1993) have indicated that the kind of overwhelming fact that would confound their acceptance of evolution would be the discovery of a fossil in a wholly anomalous stratigraphic situation. J. B. S. Haldane once said that his faith in evolution would be destroyed by the discovery of a fossil rabbit in Precambrian rocks. Creationists have striven to supply such anomalous fossils: finds such as trilobites preserved in the base of human sandal-prints and dinosaur and human footprints occurring together have turned out to be hoaxes.

It is assumed in the above discussion that stratigraphic, morphological cladistic, and molecular data all offer equally valid approaches to the discovery of the order of events in phylogeny. The old debate about whether molecular or morphological data are better than each other is sterile. There is no evidence for the assertions by some biologists (for example Goodman 1989; Meyer and Wilson 1990; Graur 1993) that molecular phylogeny reconstruction is superior to morphological cladistic analysis of living and fossil taxa. Experience shows that neither approach consistently produces unequivocal results, and it is best to accept the validity of both methods and to enrich phylogenetic studies by informed comparisons between the two (Marshall and Schultze 1992; Smith and Littlewood 1994).

A recent suggestion has been, not to compare trees founded on molecular and morphological data, but to combine them—the total evidence approach of Kluge (1989) and Eernisse and Kluge (1993). The total evidence method is based on the idea that all available data should be drawn upon in an attempt to distinguish homology from homoplasy. Parsimony methods of phylogeny reconstruction depend on minimizing character incongruence, and, the more characters that are available, the truer will be the most parsimonious (= most congruent) tree that is discovered. However, it is not evident that a great quantity of data will always strengthen the phylogenetic signal: indeed, addition of diverse character sets may obscure the node support given by a few reliable characters beneath a cacophony of statistical noise (Bull *et al.* 1993). There are also technical problems with the total evidence approach, in assessing that the data are truly homogeneous, and in determining how to code morphological and molecular characters, and how to weight them so that they are considered in a way that is theory independent. Nevertheless, testing of evolution, and testing of datasets and methods of phylogeny reconstruction, requires that stratigraphic, morphological cladistic, and molecular data are kept separate.

14.4 Testing the quality of the fossil record

Order of branching

The order of branching may be compared readily among different hypotheses for the phylogeny of a particular group. The basic approach (Fig. 14.1) is to enumerate the order of nodes on each phylogenetic tree and to compare them with stratigraphic data for mismatches of ordering. An appropriate simple statistical test is Spearman rank correlation (SRC), which assesses the probability that one predicts the other better than would be expected by chance. The method works only for samples of more than four taxa, since there must be more than three nodes (number of nodes = number of taxa − 1) for the test to have any meaning.

The method must be amplified in order to deal with real cases (Norell 1992, 1993; Norell and Novacek 1992*a*, *b*; Benton 1994; Benton and Storrs 1994, 1995). It is only possible to assess node order in a pectinate (unbalanced, Hennigian comb) type of cladogram (Fig. 14.1B), since there is no evidence in a more balanced cladogram (Fig. 14.1A) for the relative ordering of nodes on both major branches: should the branching point between A and B be numbered 2, 3, or 4? Hence, one branching stem must be collapsed at each equivocal point, and that stem is then treated as a single taxon, equivalent to all the others (Fig. 14.1B). It is possible to collapse balanced cladograms in different ways, and it is sometimes informative to test one version with the left-hand half collapsed, and then to test a second version with the right-hand half collapsed.

Tests of relative ordering of branching points in morphological cladograms with known stratigraphic ordering of first occurrences of the same taxa have demonstrated three facts: (1) clade data and age data agree; (2) improvements in palaeontological knowledge do not necessarily improve the fit between clade and age data; and (3) continental vertebrates show better levels of matching between clade and age data than do (marine) echinoderms.

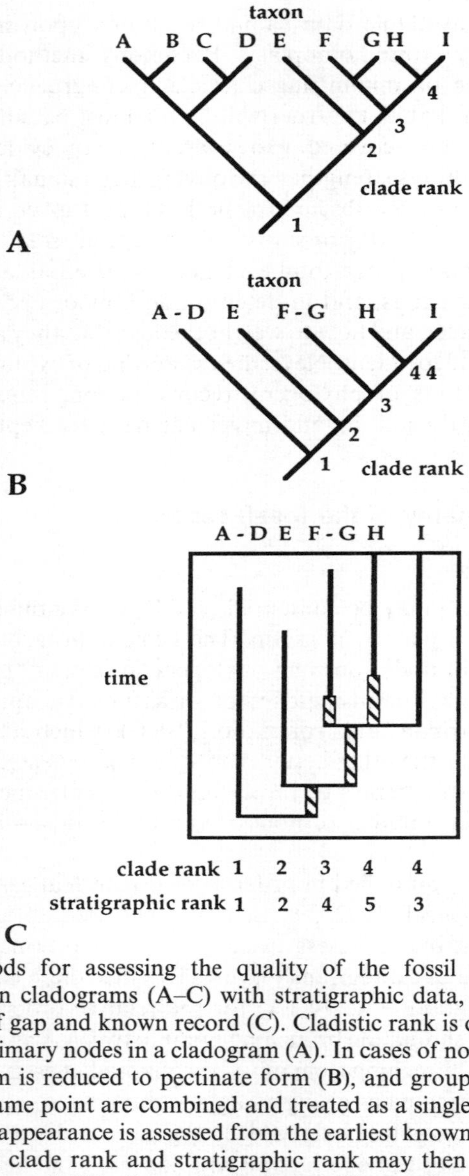

Fig. 14.1 Methods for assessing the quality of the fossil record, by comparing branching order in cladograms (A–C) with stratigraphic data, and by comparing the relative amount of gap and known record (C). Cladistic rank is determined by counting the sequence of primary nodes in a cladogram (A). In cases of non-pectinate cladograms (A), the cladogram is reduced to pectinate form (B), and groups of taxa that meet the main axis at the same point are combined and treated as a single unit. The stratigraphic sequence of clade appearance is assessed from the earliest known fossil representative of sister groups, and clade rank and stratigraphic rank may then be compared (C). The minimum implied gap (MIG, diagonal rule) is the difference between the age of the first representative of a lineage and that of its sister, as oldest known fossils of sister groups are rarely of the same age. MIG is a minimum estimate of stratigraphic gap, as the true age of lineage divergence may lie well before the oldest known fossil.

Clade data and age data generally agree (Fig. 14.2A). Norell and Novacek (1992*a*) found that 18 out of 24 test cases of cladograms of vertebrates (75 per cent) gave statistically significant ($P<0.05$) correlations of clade and age data, using the SRC test. In larger samples, Norell and Novacek (1992*b*) found significant correlation in 24 of 33 test cases (73 per cent), while Benton and Storrs (1994) found significant correlation in 41 of 74 test cases (55 per cent).

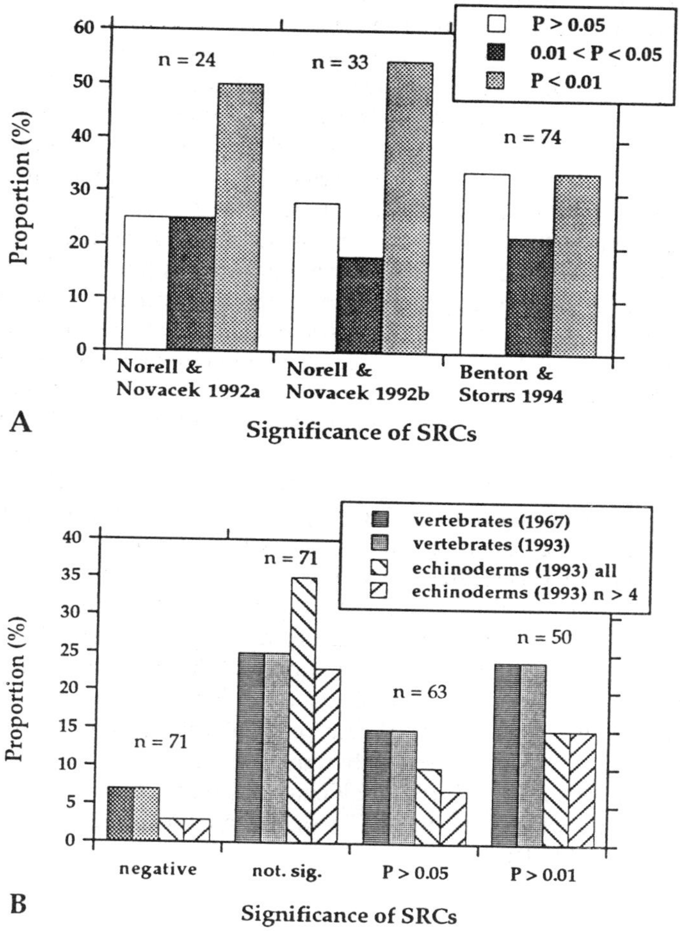

Fig. 14.2 Comparison of measures of completeness of the fossil record of vertebrates and echinoderms. (A) Assessments of the statistical significance of Spearman rank correlation (SRC) tests for three studies of cladograms of vertebrates (Norell and Novacek 1992*a*, *b*; Benton and Storrs 1994). In all three cases, most cladograms show statistically significant ($P<0.05$) correlation of clade order and age order, but the pass rate declines with larger datasets. (B) Comparison of the statistical significance of SRC tests for four sets of fossil data, two of vertebrates (Harland *et al.* 1967; Benton 1993) and two of echinoderms (both with stratigraphic data from Benton (1993)), one the set of all echinoderm cladograms, and one the set of echinoderm cladograms with more than four terminal taxa. The statistical significance of SRC matching does not change from 1967 to 1993 for vertebrates, but vertebrates show more statistically significant matches ($P<0.05$ and $P<0.01$) than does either of the echinoderm data sets.

Both pairs of investigators found the same results in independent analyses of the same cladograms, and the larger sample of Benton and Storrs (1994) shows lower levels of correlation, possibly because of the addition of some less well resolved cladograms. All these studies, based only on vertebrates, confirm none the less that clade rank and age rank match.

It might be expected that the addition of new fossil finds and reanalysis of older ones would improve the fit of age data to a fixed sample of cladograms, by the filling of gaps and corrections of former taxonomic assignments. However, in a comparison of a 1967 dataset (Harland *et al.* 1967) and one from 1993 (Benton 1993), Benton and Storrs (1994) found no change at all in the proportions of cladograms that showed statistically significant ($P < 0.05$ and $P < 0.01$) matching of clade and age order (Fig. 14.2B), although there had been a change in the status of 28 of the 71 cladograms compared (39 per cent). In other words, as a result of 26 years of work, new discoveries and reassignments had improved the fit in 20 per cent of cases, but caused mismatches of clade and age data in a further 20 per cent of cases. Sometimes a new fossil does not fill a gap but creates additional gaps on other branches of a cladogram.

This discovery of a lack of improvement in the congruence of clade versus age rank order is important, since it highlights the fact that mismatches may arise from subtle changes in knowledge. Non-correlation may result from minor variations in fossil dating, and may not imply wildly different evidence about the history of life from cladograms and from fossil occurrences.

If cladograms of vertebrates are generally matched by the correct order of appearance of the fossils, it might be expected that marine animals, with potentially richer fossil records, might show better results. However, in a study of cladograms of echinoderms, Benton and Simms (1995) found correlation in only 24 out of 63 test cases (38 per cent). This disappointing result may arise from the fact that many (21 per cent) of the echinoderm cladograms consisted of only four taxa. When these were excluded, 23 out of 50 (46 per cent) cladograms showed a statistically significant ($P < 0.05$) correlation of clade and age rank; somewhat better, but still worse than the results for continental vertebrates (Fig. 14.2B).

Relative completeness

Tests of the order of branching provide a great deal of information, but they are highly sensitive to perturbation, especially when the time intervals over which several branching events occurred is small. In such cases, a trivial change in the relative dates of some basal taxa could entirely overturn the stratigraphic rank order. In addition, the significance of SRC values may correlate with cladogram size, for vertebrates at least (Benton and Storrs 1994), although this was not found for echinoderms (Benton and Simms 1995). The tests of rank-order correlation take no account of the amount of time involved, and an additional test is required.

A number of statistics have been proposed which measure the relative amount of time represented by fossils and unrepresented by fossils. Clearly, if a particular fossil sequence is represented by densely packed specimens throughout, that sequence can be said to be more complete than one represented by only a few specimens scattered through the entire duration. Paul (1982) suggested a simple test of relative completeness of fossil ranges, termed by Benton (1987) the simple completeness metric (SCM), where

$$\mathrm{SCM} = \frac{\text{number of time units with fossils}}{\text{total number of time units in fossil ranges}} \times 100\%.$$

Values of SCM range from 0 per cent, where no fossils are preserved, although other evidence indicates that the taxa were present, to 100 per cent, where every time interval contains actual fossil evidence. Values fall below 100 per cent because usually some time intervals included within the stratigraphic range of a taxon lack fossils, and the former existence of the group during that time is inferred from the discovery of specimens of the taxon below and above the gap.

The SCM measure is an assessment of *relative* completeness, and it takes no account of possible unknown range extensions above and below the currently known stratigraphic ranges of taxa. A basic property of cladograms, the fact that sister taxa originate from a node, which represents a single point in time, allows the calculation of a different statistic that gives an *absolute* measure of minimum range extensions downwards.

The cladistic measure of absolute completeness is based on the identification of minimum implied gaps (MIGs), or ghost ranges, for each cladogram (Fig. 14.1C) (Norell 1992, 1993; Norell and Novacek 1992*a*, *b*; Weishampel and Heinrich 1992; Benton, 1994; Benton and Storrs 1994, 1995). MIGs may be measured and summed for any phylogenetic tree (a cladogram plotted against known stratigraphic taxon ranges) giving total MIG value. This is best cited as a measure of *relative* completeness of the tree, as the quality of any particular fossil record will depend upon the numbers of taxa involved and their known stratigraphic ranges.

Two similar statistics have been proposed that seek to measure the relative amount of MIG in cladograms. Norell (1992) developed a statistic, his index Z, defined as

$$Z = \frac{\sum(\mathrm{MIG})/n}{\text{maximum known clade age}},$$

while Benton and Storrs (1994) presented an alternative relative completeness index (RCI), calculated by comparing the amount of gap in a particular fossil record (assessed as MIG) to the part of the record represented by fossils

$$\mathrm{RCI} = \left(1 - \frac{\sum(\mathrm{MIG})}{\sum(\mathrm{SRL})}\right) \times 100\%.$$

In these formulae n is the number of taxa, and SRL is the simple range length for each taxon (the total time between first and last known appearance). The statistics differ in that MIG is scaled to the number of taxa in the Z index, and this proportional figure is divided by the maximum known age of the clade under study (i.e. the age of the oldest fossil in the test case). In the RCI index, known and cladistically implied ranges are summed for all lineages in the test study, and the ratio of the two sums is then calculated. The Z index is highly sensitive to the choice of outgroups, since the addition or deletion of a single basal outgroup may greatly affect the denominator. Hence, the RCI index is preferred here, but full comparisons of the behaviour of the two measures are required.

RCI, being a measure of total inherent gap (based upon a predictive phylogeny) to known lineage duration based upon fossil appearances, theoretically may vary from a negative value, where the amount of expected gap exceeds the sum total of proven stratigraphic range lengths, to 100 per cent, where no gaps are evident. Examples of the total gap being greater than the total known record appear to be rare for analyses involving numerous taxa, although Benton and Storrs (1994) found one example among the vertebrates, and Benton and Simms (1995) found one among the echinoderms.

In a study of 74 cladograms of vertebrates, Benton and Storrs (1994, 1995) found that 71 (96 per cent) had RCI values in excess of 50 per cent (Fig. 14.3A). In other words, all but three of the 74 cladograms represented phylogenies in which more than half the range is represented by fossil specimens. Indeed, the mean RCI value is 72.3 per cent, indicating that ghost ranges make up just over one-quarter of all the total stratigraphic ranges.

The RCI tests on cladograms of vertebrates also showed a statistically significant improvement over the past 26 years of research. In a comparison of the RCI values implied by the 1967 and 1993 datasets, assessed across a fixed sample of 73 cladograms, the mean RCI value shifted from 67.9 to 72.3 per cent, a statistically significant difference, according to a paired t test ($P = 0.045$). A comparison between the two datasets (Fig. 14.3A), using a Wilcoxon signed ranks test, a non-parametric analogue of the t test, also indicated a statistically significant ($P = 0.026$) difference between 1967 and 1993 values. In other words, comparisons of the relative completeness of cladograms shows a significant *improvement*, by about 5 per cent, in knowledge of the fossil record over the past 26 years of research. Hence, new fossil discoveries, and reassignments of older ones, do positively affect the amount of ghost range, although such changes in knowledge did not apparently affect the match of clade and age rank order (see above).

The comparison of rank-order correlation of clade and age data on cladograms of echinoderms and vertebrates (Benton and Simms 1995) suggested that vertebrates showed better matching (see above). The RCI comparisons confirm this finding when all echinoderm cladograms are considered: 78 per cent of echinoderm cladograms have RCI values of over 50 per cent, whereas the figure is 95 per cent for cladograms of continental vertebrates (Fig. 14.3B). Mean RCI

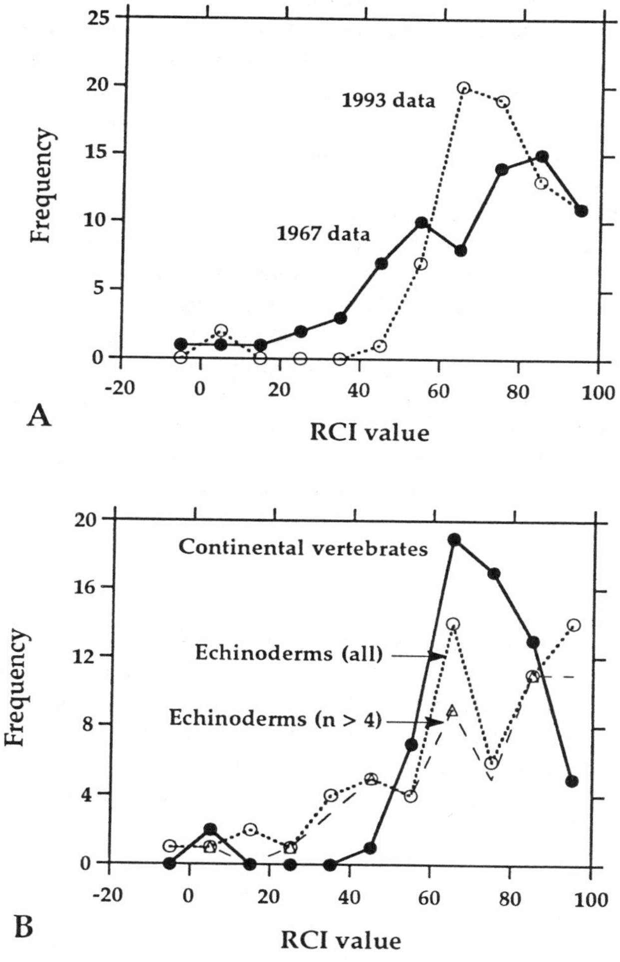

Fig. 14.3 Comparisons of relative completeness of different fossil records, measured by the relative completeness index (RCI). (A) Relative improvement in fossil record quality as determined from Harland *et al.* (1967) and Benton (1993). There is a statistically significant 5 per cent shift of the distribution of RCI values to the right from 1967 to 1993, indicating improvement in palaeontological knowledge. (B) Comparison of the fossil records of continental vertebrates and of echinoderms (all cladograms, and all cladograms with more than four terminal taxa). The sample of cladograms of all echinoderms ($n=63$) has a mean RCI = 66 per cent, for echinoderm cladograms with more than four terminal taxa ($n=50$), the mean RCI = 70 per cent and for continental vertebrate cladograms ($n=63$), the mean = 70 per cent. Continental vertebrates have a more complete fossil record than do echinoderms, based on the all-echinoderm distribution (Kolmogorov-Smirnov test, $P<0.01$), but the distributions are the same for continental vertebrates and echinoderms, based on the larger ($n>4$) echinoderm cladograms (Kolmogorov-Smirnov test, $P<0.01$).

values are 66 per cent for echinoderms and 70 per cent for continental vertebrates. When cladograms of echinoderms with four taxa are excluded from the comparisons, 80 per cent have RCI values over 50 per cent, and the mean RCI value increases to 70 per cent, the same as for continental vertebrates. Hence, comparisons of relative completeness indicate that echinoderms have similarly well-represented, or somewhat poorer, fossil records than do continental vertebrates.

The finding that continental vertebrates have a *better* fossil record than echinoderms suggests two observations. (1) The relative abundance of specimens at individual fossil localities is no indicator of the completeness of their fossil record on a large scale: this depends on the number of stratigraphic horizons that have yielded fossils, and on the packing of those horizons in time. (2) The fossil record of vertebrates has probably been more intensively studied than has that of echinoderms. Hence, our knowledge of the vertebrate fossil record is placed higher on the collector curve (numbers of taxa versus effort), and may be assumed to approach closer to the level of complete sampling and full knowledge of all fossil taxa that exist in the rocks.

14.5 Discussion

The time axis of phylogenies is important, and it can be calibrated, with increasing confidence, by the use of stratigraphic data about geological dates of occurrence of basal taxa within clades. Recent studies have shown a good concordance between stratigraphic information about the fossil record and morphological cladistic data on branching patterns. In addition, it has also been shown that knowledge of the fossil record of vertebrates at least has improved by 5 per cent in the past 26 years of research. There is no reason to doubt that such a measure of improvement might apply to other segments of the fossil record, nor that such improvements might continue into the future. Additional analyses, both of additional groups of organisms, and after further increments of research input, may confirm or refute these findings. In addition, further revisions of cladograms and molecular trees may have a bearing on the quality of matching between clade and age data.

Measures of the absolute time-scale of phylogeny are necessary in testing for variation in the rate of molecular evolution (the molecular clock), the nature of character acquisition during times of rapid branching (radiations), the meaning of disparity in the origins of higher taxa, and the quality of different parts of the fossil record. The work outlined above should dispel some of the fears that have been expressed recently about the value of fossils in phylogeny reconstruction, both as repositories of useful character information and as indicators of the minimum ages of branching points.

In the above discussions, mutual testing among the independent sources of data on the order of phylogenetic events, from morphological cladistics, molecular phylogenetics, and the stratigraphic distribution of fossils, has been aimed at testing the quality of the fossil record. However, since none of these three forms of information can be said *a priori* to be more reliable in all cases than the other two, mutual testing can extend to assessing the validity of cladograms derived from molecular and morphological data. In a case where a number of equally most parsimonious trees (MPTs) are identified, the validity of one scheme over the others could be tested by a probabilistic statement based on calculation of the correlation between clade and age data (SRC), and of the RCI implied by each cladogram. The cladogram that gave the most significant SRC metric of correlation, and the highest RCI value (i.e. smallest amount of ghost range) has the greatest probability of being correct, all other things being equal.

Acknowledgements

Thanks to Derek Briggs. Andrew Smith, and Matthew Wills for helpful comments on earlier drafts of this paper. This work was funded in part by Grant F/182/AK from the Leverhulme Trust.

References

Allison, P. A. and Briggs, D. E. G. (1991). *Taphonomy: releasing the data locked in the fossil record.* Plenum, New York.

Ax, P. (1987). *The phylogenetic system.* Wiley, New York.

Benton, M. J. (1987). Mass extinctions among families of non-marine tetrapods: the data. *Mém. Soc. Géol. Fr.*, **150**, 21–32.

Benton, M. J. (1993). *The Fossil Record 2.* Chapman and Hall, London.

Benton, M. J. (1994). Palaeontological data, and identifying mass extinctions. *Trends. Ecol. Evol.*, **9**, 181–5.

Benton, M. J. and Simms, M. J. (1995). Testing the marine and continental fossil records. *Geology*, **23**, 601–4.

Benton, M. J. and Storrs, G. W. (1994). Testing the quality of the fossil record: paleontological knowledge is improving. *Geology*, **22**, 111–14.

Benton, M. J. and Storrs, G. W. (1995). Diversity in the past: comparing cladistic phylogenies and stratigraphy. in press. In *Phylogeny and biodiversity* (ed. M. Hochberg), pp. 19–40. Oxford University Press.

Bull, J. J., Huelsenbeck, J. P., Cunningham, C. W., Swofford, D. L., and Waddell, P. J. (1993) Partitioning and combining data in phylogenetic analysis. *Syst. Biol.*, **42,** 384–97.

Cano, R. J., Poinar, H. N., Pieniazek, N. J., Acra, A., and Poinar, G. O. (1993). Amplification and sequencing of DNA from a 120–135 million-year-old weevil. *Nature*, **363**, 536–8.

Cheetham, A. H. and Hayek, L. C. (1988). Phylogeny reconstruction in the Neogene bryozoan *Metrarabdotos*: a paleontologic evaluation of methodology. *Hist. Biol.*, **1**, 65–83.

Darwin, C. (1859). *On the origin of species by means of natural selection.* John Murray, London.

DeSalle, R. J., Gatesy, J., Wheeler, W., and Grimaldi, D. (1992). DNA sequences from a fossil termite in Oligo-Miocene amber and their phylogenetic implications. *Science*, **257**, 1933–6.

Donoghue, M. J., Doyle, J. A., Gauthier, J., and Kluge, A. G. (1989). The importance of fossils in phylogeny reconstruction. *Ann. Rev. Ecol. Syst.*, **20**, 431–60.

Doyle, J. and Donoghue, M. (1987). Seed plant phylogeny and the origin of angiosperms: an experimental cladistic approach. *Bot. Rev.*, **52**, 321–31.

Eernisse, D. J. and Kluge, A. G. (1993). Taxonomic congruence versus total evidence, and amniote phylogeny inferred from fossils, molecules, and morphology. *Mol. Biol. Evol.*, **10**, 1170–95.

Eldredge, N. and Cracraft, J. (1980). *Phylogenetic patterns and the evolutionary process.* Columbia University Press, New York.

Fortey, R. A. and Jefferies, R. P. S. (1982). Fossils and phylogeny—a compromise approach. In *Problems of phylogenetic reconstruction*, Systematics Association Special Volume, (ed. K. A. Joysey and A. E. Friday), pp. 197–234. Academic, London.

Gauthier, J., Kluge, A. G., and Rowe, T. (1988). Amniote phylogeny and the importance of fossils. *Cladistics*, **4**, 105–209.

Gingerich, P. D. (1979). The stratigraphic approach to phylogeny reconstruction in vertebrate paleontology. In *Phylogenetic analysis and paleontology*, (ed. J. Cracraft and N. Eldredge), pp. 41–77. Columbia University Press, New York.

Gingerich, P. D. and Schoeninger, M. (1977). The fossil record and primate phylogeny. *J. Hum. Evol.*, **6**, 484–505.

Golenberg, E. M., Giannassi, D. E., Clegg, M. T., Smiley, C. J., Durbin, M., Henderson, D. *et al.* (1990). Chloroplast DNA sequence from a Miocene *Magnolia* species. *Nature*, **344**, 656–8.

Goodman, M. (1989). Emerging alliance of phylogenetic systematics and molecular biology: a new age of exploration. In *The hierarchy of life*, (ed. B. Fernholm, K. Bremer, and H. Jornvall), pp. 43–61. Elsevier, New York.

Graur, D. (1993). Molecular phylogeny and the higher classification of eutherian mammals. *Trends Ecol. Evol.*, **8**, 141–7.

Harland, W. B. *et al.* (1967). *The fossil record; a symposium with documentation.* Geological Society of London, London.

Harper, C. W. (1976). Phylogenetic inference in paleontology. *J. Paleontol.*, **50**, 180–93.

Harvey, P. H., Holmes, E. C., Mooers, A. Ø., and Nee, S. (1994). Inferring evolutionary processes from molecular phylogenies. In *Models in phylogeny reconstruction*, Systematics Association Special Volume 52, (ed. R. W. Scotland, D. J. Siebert, and D. M. Williams), pp. 315–33. Academic, London.

Hennig, W. (1966). *Phylogenetic systematics.* University of Illinois Press, Urbana.

Hennig, W. (1981). Insect phylogeny. Wiley, New York.

Huelsenbeck, J. P. (1991). Tree length distribution skewness: an indicator of phylogenetic information. *Syst. Zool.*, **40**, 257–70.

Kluge, A. G. (1989). A concern for evidence and a phylogenetic hypothesis of relationships among *Epicrates* (Boidae, Serpentes). *Syst. Zool.*, **38**, 7–25.

Marshall, C. R. and Schultze, H.-P. (1992). Relative importance of molecular, neontological, and paleontological data in understanding the biology of the vertebrate invasion of land. *J. Mol. Evol.*, **35**, 93–101.

Mayr, E. (1982). *The growth of biological thought: diversity, evolution and inheritance.* Harvard University Press, Cambridge, MA.

Meyer, A. and Dolven, S. I. (1992). Molecules, fossils, and the origin of tetrapods. *J. Mol. Evol.*, **35**, 102–13.

Meyer, A. and Wilson, A. C. (1990). Origin of tetrapods inferred from their mitochondrial DNA affiliation to lungfish. *J. Mol. Evol.*, **31**, 359–64.

Nelson, G. J. (1969). Origin and diversification of teleostean fishes. *Ann. N. Y. Acad. Sci.*, **167**, 18–30.

Nelson, J. S. and Platnick, N. I. (1981). *Systematics and biogeography. Cladistics and vicariance*. Columbia University Press, New York.

Norell, M. A. (1992). Taxic origin and temporal diversity: the effect of phylogeny. In *Extinction and phylogeny*, (ed. M. J. Novacek and Q. D. Wheeler), pp. 89–118. Columbia University Press, New York.

Norell, M. A. (1993). Tree-based approaches to understanding history: comments on ranks, rules, and the quality of the fossil record. *Am. J. Sci.*, **293A**, 407–17.

Norell, M. A. and Novacek, M. J. (1992*a*). The fossil record and evolution: comparing cladistic and paleontologic evidence for vertebrate history. *Science*, **255**, 1690–3.

Norell, M. A. and Novacek, M. J. (1992*b*). Congruence between superpositional and phylogenetic patterns: comparing cladistic patterns with fossil records. *Cladistics*, **8**, 319–37.

Novacek, M. J. (1992*a*). Fossils, topologies, missing data and the higher-level phylogeny of eutherian mammals. *Syst. Biol.*, **41**, 58–73.

Novacek, M. J. (1992*b*). Fossils as critical data for phylogeny. In *Extinction and phylogeny*, (ed. M. J. Novacek and Q. D. Wheeler), pp. 46–88. Columbia University Press, New York.

Patterson, C. (1981). Significance of fossils in determining evolutionary relationships. *Ann. Rev. Ecol. Syst.*, **12,** 195–223.

Paul, C. R. C. (1982). The adequacy of the fossil record. In *Problems of phylogenetic reconstruction*, Systematics Association Special Volume, and (ed. K. A. Joysey and A. E. Friday), pp. 75–117. Academic, London.

Raup, D. M. (1972). Taxonomic diversity during the Phanerozoic. *Science*, **215**, 1065–71.

Ridley, M. (1993). *Evolution*. Blackwell Scientific, Oxford.

Schopf, T. J. M. (1978). Fossilization potential of an intertidal fauna: Friday Harbor, Washington. *Palaeobiology*, **4**, 261–70.

Simpson, G. G. (1983). *Fossils and the history of life*. Freeman, San Francisco.

Smith, A. B. (1994). *Systematics and the fossil record.* Blackwell Scientific, Oxford.

Smith, A. B. and Littlewood, D. T. J. (1994). Paleontological data and molecular phylogenetic analysis. *Paleobiology*, **20**, 259–73.

Szalay, F. S. (1977). Ancestors, descendants, sister groups and testing of phylogenetic hypotheses. *Syst. Zool.*, **26**, 12–18.

Weishampel, D. B. and Heinrich, R. E. (1992). Systematics of Hypsilophodontidae and basal Iguanodontia (Dinosauria: Ornithopoda). *Hist. Biol.*, **6**, 159–84.

Wilkinson, M. (1994). Common cladistic information and its consensus representation: reduced Adams and reduced cladistic consensus trees and profiles. *Syst. Biol.*, **43**, 343–68.

Wilson, M. V. H. (1992). Importance for phylogeny of single and multiple stem-group fossil species, with examples from freshwater fishes. *Syst. Biol.*, **41**, 462–70.

15

Comparative evolution of larval and adult life-history stages and small subunit ribosomal RNA amongst post-Palaeozoic echinoids

A. B. Smith, D. T. J. Littlewood, and G.A. Wray

15.1 Introduction

With the construction of robust phylogenies based on multiple corroborative datasets, comes the opportunity to compare the phylogenetic histories of different subsets of characters. Until recently workers have focused largely on comparing the performance of morphological and molecular data in tree construction, with the aim of showing that one or other is superior. However, where two or more independent datasets give broadly comparable topologies, a comparative approach to evolutionary patterns can be adopted. Subsets of characters that can be considered as independent of one another can be optimized onto the preferred phylogeny and compared to investigate how different aspects of a clade's biology have evolved over time. This approach is not confined simply to the morphology versus molecular dichotomy, but can be expanded to include comparisons of different kinds of molecular data (for example coding versus non-coding regions; ribosomal loop versus stem regions, etc.) or different stages in the life history of an organism (for example larval versus adult evolution). Although there has been substantial work done on the comparison of different kinds of molecular data, no one has previously attempted to compare the evolution of different life-history stages within a clade.

The Echinoidea are an excellent group for evolutionary studies because of their skeletal complexity and relatively good fossil record. A direct reading of their phylogenetic history, as deduced from the fossil record, indicates a complex history that includes periods of rapid morphological radiation as well as periods of extreme conservativism. However, all that we glean from the fossil record is the history of adult morphology; there is no record of how larval morphology has changed (other than our ability to distinguish fossils with planktotrophic from non-planktotrophic life histories on the basis of adult apical disc crystallography (Emlet 1985, 1989)), and no fossil biomolecules have yet been recovered from geologically ancient echinoids. Thus we have only a partial and highly biased understanding of how evolution has proceeded in the Echinoidea.

Echinoids, like most marine invertebrates, pass through a complex life-history cycle composed of two morphologically very contrasting phases; the adult benthic phase and the larval planktonic phase (echinopluteus) (Fig. 15.1). Furthermore, most of the complex morphological features of the larval stage are resorbed at metamorphosis, and adult structures form *de novo* from a rudiment on the larval body, so there is little obvious connection between larval and adult morphological characters.

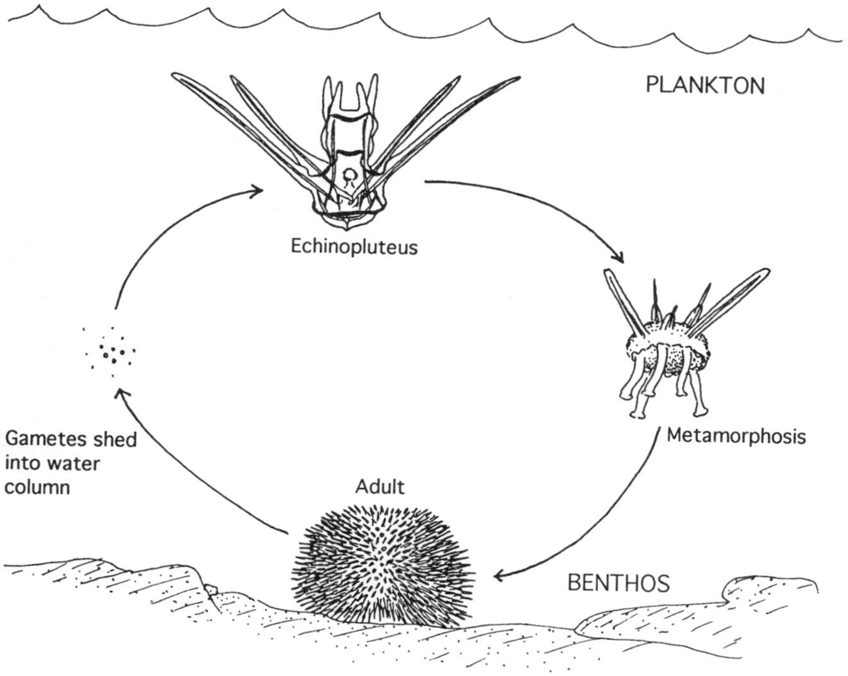

Fig. 15.1 Schematic life-history cycle of an echinoid, with a benthic adult phase alternating with a pelagic planktonic phase. Not to scale.

We knew virtually nothing about the evolutionary history of echinoid larval forms until one of us (Wray 1992) compiled a 75-character data matrix for the echinoplutei of extant echinoids, optimizing character changes onto a cladogram constructed from adult morphology. Using this Wray investigated several aspects of the evolution of echinopluteus morphology concluding, amongst other things, that a number of important characters probably evolved very early on during the echinoid crown-group radiation. He argued that the persistence of major traits over time was evidence for strong selective pressure, and provided several examples of parallel evolutionary transformations that could be interpreted as having functional importance. Finally Wray observed that larval and adult characters probably displayed mosaic evolution, noting that larval morphology was very highly conserved within Clypeasteroidea but highly varied in Echinoida, whereas the opposite was true for adult morphologies.

We wish to explore comparative patterns of evolution in a more rigorous way, to test whether larval and adult morphologies are evolving in tandem or independently, and to explore the relationship between morphological and molecular evolution. The earlier study (Wray 1992) was largely anecdotal and flawed by having optimized larval characters onto a phylogenetic tree constructed solely from adult morphology, Both larval characters and adult characters provide phylogenetic information about the relationships of echinoid taxa, and there is no *a priori* reason for believing one to be more correct than the other. In particular, if we wish to ask whether larval characters show more convergence than adult characters, then it is important that we have a phylogeny that is not based solely on just one set of data, since this will minimize the homoplasy in that dataset at the expense of the other.

Total evidence has been much discussed recently (Eernisse and Kluge 1993; Kluge and Wolf 1993; De Queiroz 1993; Larson 1994). We use total evidence because it derives the hypothesis of phylogenetic relationships that is best supported overall, and because it allows us to examine and compare evolutionary patterns from diverse datasets directly. Calibration of cladograms from the fossil record can come only from the analysis of adult morphology but, by constructing a phylogeny based on larval and molecular data in addition to adult morphology, we can also calibrate molecular and larval morphological evolution.

The questions that we wish to tackle in this chapter are as follows: (i) Have larval and adult morphological traits evolved independently of one another or in concert? (ii) Are larval characters more prone to homoplasy than adult characters or are homoplasy levels broadly similar? (iii) Are rates of change in the ribosomal RNA gene correlated with morphological change? Do they correlate with elapsed time more closely than morphological characters?

15.2 Methods and materials

Taxa and characters used

We began with 29 species of Recent echinoid (Table 15.1), which we have scored for 163 morphological characters of the adult (136 phylogenetically informative characters; listed in Littlewood and Smith (1995)). For 26 of these species we have compiled complimentary information on their planktotrophic larval morphology (48 characters, of which 40 are phylogenetically informative: Appendix 1). For *Fellaster zelandiae* we have used the larval characters of a closely related species (*Echinarachnius parma*). Of the other two species, one has lecithotrophic development and thus lacks a planktotrophic larva, and larval development of the other has never been reported. Larval characters were compiled from the original 75 characters used by one of us previously (Wray 1992), removing characters that were invariant amongst our 28 species, and deleting characters now known to be variable within species. The data matrix of larval characters is given in Appendix 1. To complement these morphological data we have complete small subunit ribosomal RNA (SSU rRNA) gene sequences for 22 of the species (278 characters, of which 85 are phylogenetically informative), and partial large subunit ribosomal RNA (LSU rRNA) sequences for 12 of these species (91 characters, 36 of which are phylogenetically informative) (data from Littlewood and Smith 1995). Finally, we also included an additional 13 species of fossil taxa for which we can score only adult morphological traits, but which belong to basal parts of long branches leading to extant taxa. Details of all taxa and their sources are given in Littlewood and Smith (1995).

Table 15.1 Recent species used in this study and the character sets available. Full details of source for these echinoids is given in Littlewood and Smith (1995)

Taxon	Morphology		Gene sequence	
	Larva	Adult	SSU rRNA	LSU rRNA
Cidaris cidaris	✓	✓	—	✓
Eucidaris tribuloides	✓	✓	✓	—
Asthenosoma owstoni	✓	✓	✓	✓
Centrostephanus coronatus	—	✓	✓	—
Diadema setosum	✓	✓	—	—
Cassidulus mitis	—	✓	✓	—
Echinolampas crassa	✓	✓	—	—
Fellaster zelandiae/Archanoides placenta	✓	✓	✓	—
Echinocyamus pusillus	✓	✓	—	✓
Encope aberrans	✓	✓	✓	✓
Echinodiscus bisperforatus	✓	✓	✓	—
Echinocardium cordatum	✓	✓	✓	✓
Meoma ventricosa	✓	✓	✓	—
Spatangus purpureus	✓	✓	—	✓
Brissopsis lyrifera	✓	✓	✓	—
Arbacia lixula	✓	✓	✓	✓
Stomopneustes variolaris	✓	✓	✓	—
Glytocidaris crenularis	✓	✓	—	—
Temnopleurus hardwickii	✓	✓	✓	—
Salmacis sphaeroides	✓	✓	✓	—
Mespilia globulus	✓	✓	✓	—
Echinus esculentus	✓	✓	✓	✓
Psammechinus miliaris	✓	✓	✓	✓
Paracentrotus lividus	✓	✓	—	✓
Strongylocentrotus intermedius	✓	✓	✓	—
Colobocentrotus atratus	✓	✓	✓	—
Tripneustes gratilla	✓	✓	✓	—
Sphaerechinus granularis	✓	✓	✓	✓
Lytechinus variegatus	✓	✓	—	✓

Whereas we can objectively quantify the numbers of fixed-point mutations that distinguish two or more molecular sequences, the definition of morphological characters and their transformational states is much more subjective. Morphological attributes cannot be codified objectively in the same way that nucleotide bases can, and the numbers and states of the characters defined will vary to some extent from observer to observer. However, we have tried to make our morphological databases as comprehensive as possible, including all characters that have been used as differential characters by systematists in the past. Our assumption throughout this chapter is that our morphological sets of characters, though neither totally objective nor exhaustive, are a reasonable and unbiased subset of all morphological characters that can be defined.

Although the datasets are of different sizes, in terms of the numbers of phylogenetically informative characters that each includes, this is largely independent of the relative strength of phylogenetic signal that each dataset contributes. A large character matrix of near random data will have many phylogenetically informative characters, but very weak phylogenetic signal, whereas a small character matrix with a strong hierarchical structure to its character distribution will provide a very strong phylogenetic signal. Thus there is no *a priori* reason for weighting character sets based according to the number of characters included, and all characters are given equal weight.

Echinoid phylogenetic relationships: the working hypothesis

Individual datasets (adult morphology, larval morphology, large and small subunit ribosomal RNA gene sequences) each provide independent evidence from which to estimate the phylogenetic relationships of the taxa under study. The true phylogenetic relationships of these taxa are unknowable, but parsimony analysis of each dataset alone gives an independent estimate (with error) of this phylogeny. With the exception of the larval data, these separate analyses produce cladograms that have many branches in common, or represent more or less resolved topologies that are compatible with one another (Fig. 15.2). However, there are some significant differences in the position of a few taxa (notably the placement of *Arbacia* with the spatangoids in larval data and as sister group to Echinacea in SSU rRNA and adult morphology data). A strict consensus tree of all trees from these four analyses is largely uninformative, with many branches collapsed into large polychotomies. Since we have no way of telling, *a priori*, which of our four analyses is likely to be closest to the truth, we have combined all our data into a single large matrix to determine where strong signal is to be found.

We ran a parsimony analysis on the data matrix of all data combined for the 29 recent and 13 fossil taxa using the computer program PAUP (Swofford (1993): heuristic search with all characters treated as unordered and of equal weight). This found 28 equally parsimonious trees with most of the variation generated by instability of purely fossil taxa, which include a large number of missing data. When fossil forms are pruned from this tree only two equally parsimonious solutions for the remaining taxa result. These are not precisely the same as the trees obtained from parsimony analysis of extant taxa alone; the inclusion of fossil taxa in the parsimony analysis alters relationships for one or two extant taxa compared to analyses from which they are omitted. The inclusion of fossils helps to discriminate homoplasy from homology because it provides a denser sampling of all character associations that have existed within the clade (Smith 1994; Littlewood and Smith 1995).

The two rival cladograms differ only in the relative placement of *Psammechinus*, *Paracentrotus*, and *Strongylocentrotus*, and this is due largely to conflict between adult pedicellarial characters and larval characters. We resolved this conflict by pruning *Psammechinus* from the cladogram. In this way we were left with a single topology that represents our best working hypothesis of the relationships amongst our 28 extant taxa. Bootstrap values, based on 1000 replicates, are greater than 70 per cent for all but two branches of this cladogram (Fig. 15.3a) and we use this topology in all subsequent analyses.

Character optimization and cladogram calibration

Each set of morphological or molecular data was optimized onto the total evidence cladogram and individual character transformations were assigned to a single branch. For morphological data pertaining to adults, a few trivial corrections were required to the computer-based optimization, because there were instances where character convergence between two extant taxa, demonstrable as homoplasy in analyses where all taxa were included, were mistakenly listed as homologies in our 'living taxa only' cladogram. We could not make similar corrections for other datasets, but simply accepted the optimization as derived from character distributions of extant taxa alone. Although this will have introduced some error, we do not believe that it is either significant or systematic in its bias. All character reconstructions were checked for consistency and, where doubt existed, characters were partitioned between two or more branches. For each branch of our total evidence cladogram we therefore derived an estimate for the number of character transformations for larval and adult characters and for rRNA fixed point mutations.

Our rRNA data show that there have been major differences in the rate of molecular evolution amongst echinoids, with three clades with much longer branches than the rest. Consequently we could not use a molecular clock hypothesis to calibrate our cladogram. In any case, one of our goals was to investigate how clock-like evolution has been amongst our different character sets, and this requires an independent means of calibration. We therefore calibrated the total evidence cladogram by reference to the fossil record of adult morphologies. The first appearance in the stratigraphic record of a fossil with one or more apomorphies of a clade defines the latest divergence time of that clade. For each pair of sister taxa we took the earlier first appearance to date the time of sister group divergence, and hence date the node on the

(a) Larval morphology

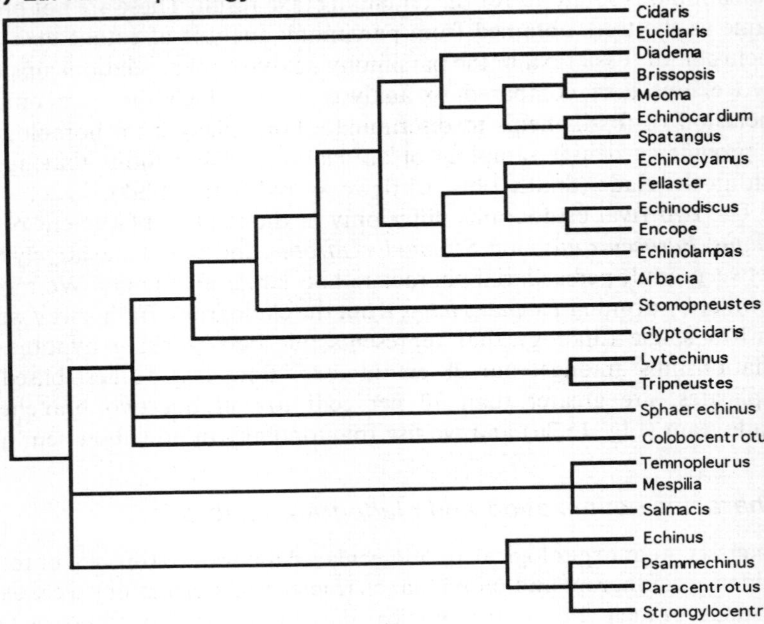

(b) Adult morphology

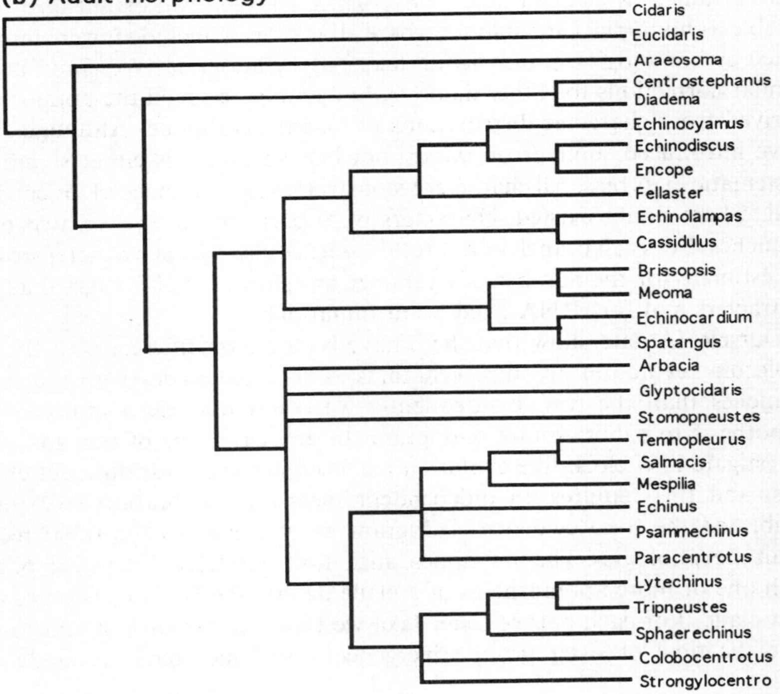

(c) SSU rRNA sequence data

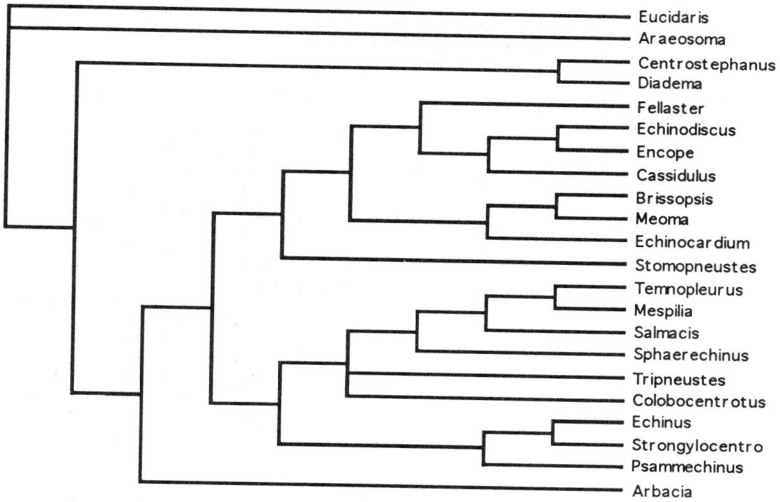

(d) SSU rRNA sequence data

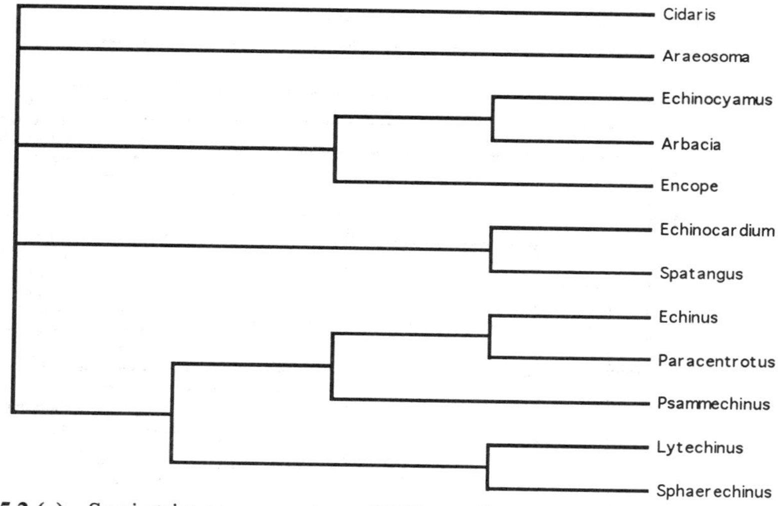

Fig. 15.2 **(a)** Semi-strict consensus tree of 208 equally most parsimonious cladograms for 27 extant taxa derived from 48 larval morphological characters (data in Appendix). Tree length = 91 steps, CI (exluding uninformative positions) = 0.55; RI = 0.80; RC = 0.48. (**b**) Semi-strict consensus of 13 equally most parsimonious cladograms for 29 extant taxa derived from 163 adult morphological characters (character matrix from Littlewood and Smith (1995)). Tree length = 261 steps, CI (excluding uninformative positions) = 0.70; RI = 0.89; RC = 0.65. (**c**) Strict consensus tree of two equally most parsimonious cladograms for 22 extant taxa derived from complete SSU rRNA gene sequence data (character matrix from Littlewood and Smith (1995)). Tree length 440 steps, CI (excluding uninformative positions) = 0.50; RI = 0.59; RC = 0.44. (**d**) Semi-strict consensus tree of eight equally most parsimonious cladograms from 12 extant taxa derived from partial LSU rRNA gene seqences (character matrix from Littlewood and Smith (1995)). Tree length 132 steps, CI (excluding uninformative positions) = 0.58; RI = 0.79; RC = 0.55.

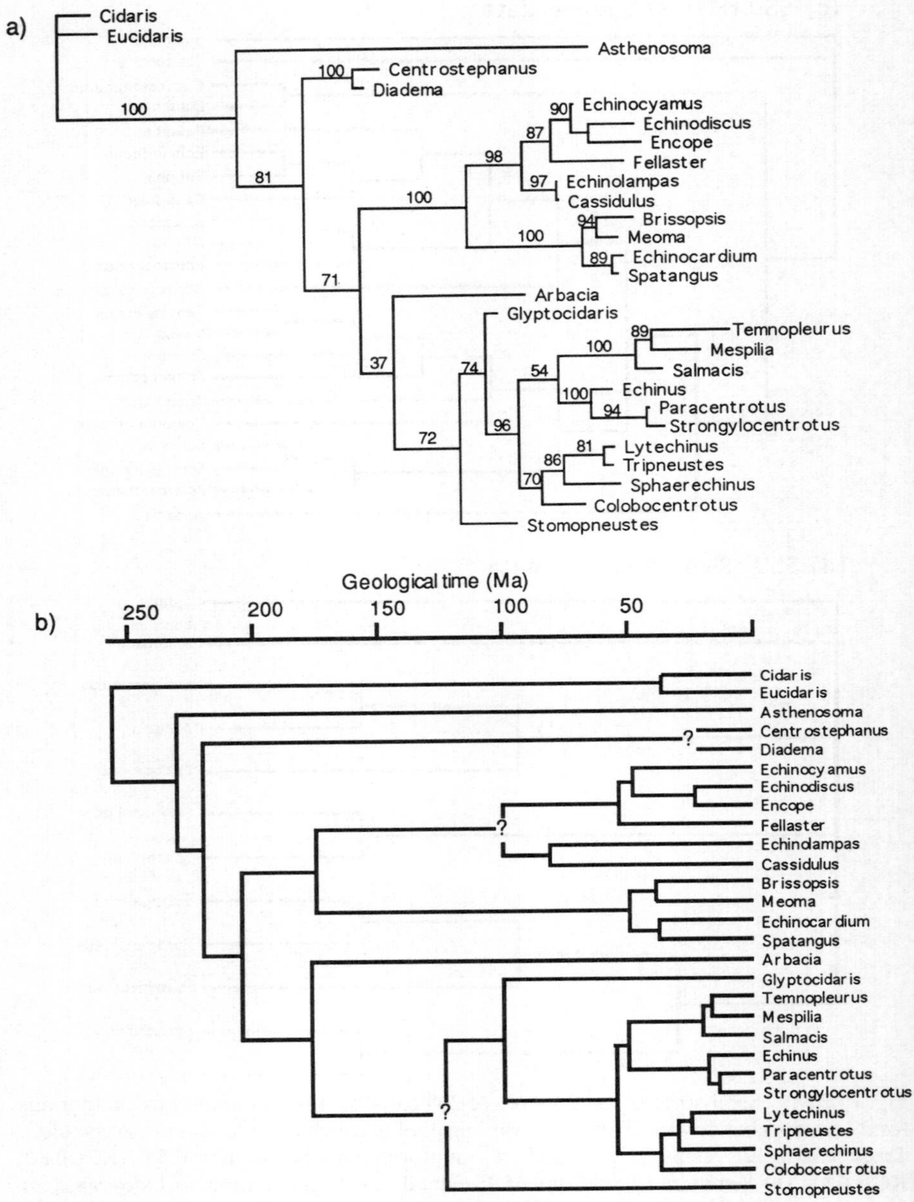

Fig. 15.3 (a) Cladogram of 27 extant taxa used as our best working estimate of echinoid phylogenetic relationships. This was derived by parsimony analysis of the full 46 Recent and fossil taxa, and all 580 morphological and molecular characters, but fossil taxa and *Psammechinus* have been subsequently deleted by pruning. Bootstrap values are shown for each branch based on 1000 replicates. (**b**) Calibrated evolutionary tree derived by using the fossil record of first appearances to date nodes identified in the combined data cladogram. Question marks indicate major branch points where there is little or no fossil evidence to date the event.

cladogram. This provides estimates of the absolute time represented by branches on the cladogram. However, not all lineages have equally dense and well-sampled fossil records. Whereas we can be reasonably confident about dates such as the first appearance of irregular echinoids (Sinemurian, Early Jurassic, about 205 million years ago), other parts of the cladogram are more problematic. For example, it is impossible to be precise about the dating of the split between *Diadema* and *Centrostephanus* since that clade has a very poor fossil record. Nor can we be certain as to the date at which cassiduloids split from clypeasteroids since the sister group of clypeasteroids amongst cassiduloids remains poorly constrained (Suter 1994). However, removing poorly dated nodes leaves 29 branches for which we have reasonably strong fossil evidence to provide estimates of their duration. A calibrated tree is shown in Fig. 15.3b.

15.3 Results and discussion

Has larval and adult morphological evolution proceeded independently or in concert?

Given the very different modes of life that larval and adult echinoids pass through, it is likely that at least some of the selective pressures acting on one phase of the life cycle will be irrelevant to the other phase. Over geological time, therefore, one might expect adult and larval morphological character transformations to have gone on more or less independently. Such mosaic evolution (DeBeer 1958) has often been invoked to explain dissociated patterns of evolution between larval and adult form in echinoids and other groups (for example Jablonski and Lutz 1983; Levinton 1989; Wray 1992). Conversely, though, periods when morphological diversification of adult form was proceeding rapidly might also coincide with periods when larval morphology was also diversifying rapidly. Wray (1992) noted that many changes in the larval form appeared early on in the diversification of crown group echinoids, corresponding with the well-established rapid morphological diversification in adult form.

If adult and larval evolution have proceeded in tandem then branches of the evolutionary tree where adult morphology has been evolving rapidly should correspond to branches when larval character evolution has also been proceeding rapidly. Conversely, if larval and adult morphology have been evolving largely independently then we should see no correlation between periods of enhanced morphological evolution in larval and adult forms.

To address this question we compared the number of inferred larval character transformations with the number of inferred adult character transformations for each branch on the total evidence cladogram. Since our estimates of the number of changes are not precise, due to subjectivity of character definition and the problem of occasional homoplasies mistaken for homology, we have used a non-parametric rank correlation statistic which is dependent upon only the relative amount of change. Our total evidence cladogram provides 46 branches where the numbers of both larval and adult character transformations can be estimated. Applying Spearman's rank correlation test gives a correlation coefficient of 0.35, showing that there is little evidence for concerted evolution between adult and larval life-history stages (see Table 15.3). Although the largest number of character transformations in both cases occurred along the branch separating cidaroid and non-cidaroid taxa, the second largest change in terms of morphology (the origin of irregular echinoids) was associated with very little larval morphological transformation.

Are larval characters more prone to homoplasy than adult characters?

Strathmann (1988) pointed out that amongst echinoderm classes, no matter what phylogenetic relationship was preferred, considerable convergent evolution must have occurred in larval form. At that time he was not sure whether this was unusual or whether the levels of homoplasy in adult characters were comparable. Wray (1992) also noted a number of parallel transformations in different lineages of echinoplutei, emphasizing the prevalence of homoplasy amongst larvae as evidence of selective evolutionary pressures. Here we wish to address whether larval characters show more homoplasy than adult characters.

There are two possible approaches that could be adopted: (i) we could take an independent estimate of the phylogeny based on molecular data alone and optimize our adult and larval character sets on to the resultant molecular tree to compare levels of homoplasy or (ii) we could combine all the data to generate our best estimate of the true phylogeny and then compare subsets of data optimized on to this tree. The first approach has been strongly advocated by Coddington (1988) and Brooks and McLennan (1991) amongst others. They have argued that it is wrong to use characters whose phylogenetic histories are to be examined to construct the phylogenetic tree that is to be the reference, because of possible circularity of logic. However, phylogenetic hypotheses based on only partial data do not necessarily provide us with the topology that is best supported overall (Kluge and Wolf 1993; Eernisse and Kluge 1993). Furthermore, Deleporte (1993) has shown that no circularity is involved when total evidence is used to construct a phylogeny that is subsequently used as the basis for mapping the evolution of subsets of characters, so long as outgroup rooting is used to polarize the resulting topology. Because we believe that obtaining the correct phylogenetic hypothesis is much more important than having independent estimates of phylogenetic relationships, each of which may represent more or less suboptimal topologies, we have adopted the second approach. If we do not derive our biological interpretations from the most strongly supported phylogeny (i.e. the one most likely to be correct), then we may end up making spurious deductions about the biological processes under investigation because of a wrong phylogenetic hypothesis.

In fact, there is really no problem in our case, since the topology derived from combined data is statistically indistinguishable from topologies derived from either adult morphology or SSU rRNA sequences (using Templeton's test

(Larson 1994)). Consequently our total evidence tree is not significantly different from the topology derived from independent molecular data alone, and morphological character sets optimized over either topology have similar levels of homoplasy. Furthermore, the congruence between topologies derived from adult morphology and SSU rRNA sequence data lend added support to the idea that the total evidence tree is largely correct.

The simplest and most direct measure of homoplasy is the consistency index (CI) (Swofford 1991), which describes the number of observed character transformations divided by the minimum number of possible character transformations, summed over all characters. A CI of 1 implies no homplasy and homoplasy levels increase in the data matrix as the CI index approaches 0. However, there are problems with the CI index (Archie 1989; Sanderson and Donoghue 1989; Meier *et al.* 1991) since it is known to be affected by both the numbers of characters and taxa included in the matrix, and by the number of missing data. In comparing the CIs derived from adult and larval morphological data optimized on to the total evidence tree we have removed all uninformative sites and are comparing cladograms based on the same number of taxa and with similar levels of missing data. However, the numbers of informative characters differ considerably (136 adult characters as opposed to 40 larval characters) and may bias our result. Meier *et al.* (1991, figure 3) showed empirically that datasets with fewer characters tend to have higher CIs.

Our 163 adult morphological characters optimized on the total evidence tree imply a total of 247 character transformations, giving a consistency index of 0.70 (excluding uninformative positions), a retention index of 0.88, and a rescaled consistency index of 0.64. By comparison, the 48 larval morphological characters optimized on the total evidence tree imply a total of 99 character transformations, giving a consistency index of 0.50 (excluding uninformative positions), a retention index of 0.75 and a rescaled consistency index of 0.41. Note that the CI is considerably lower for the smaller dataset contrary to expectation. Since the CI tends to increase not decrease with smaller sample size, the observed difference may even be underestimating the difference between adult and larval characters. Very similar results are obtained using other measures of homoplasy, such as the retention index, or the rescaled consistency index (Table 15.2). Thus at this level of analysis there is *prima facie* evidence for larval characters being more prone to homoplasy than adult characters.

However, a factor that may be biasing this analysis is the relative small size and architectural simplicity of many of the larval characters compared to adult characters. Convergence in gross form may be more readily distinguished and differentially scored if there is structural complexity to support or contradict hypotheses of homology. Homoplasy amongst larval characters may therefore simply reflect the small size and simplicity of the elements being compared.

It is important, therefore, for a more accurate assessment of homoplasy levels, that larval characters are compared specifically with structures in the adult of comparable size and complexity. Most larval characters refer to the structure and arrangement of the small skeletal rods that together form a basket-like framework of the echinopluteus. In addition there are a small number of characters that relate to the development of ciliary bands and other soft-tissue features. The most obvious structures in adult echinoids of comparable size and complexity are the pedicellariae. These are small and relatively simple structures found on all echinoids and which have been given great importance in systematic work. Like larvae, most of the characters are derived from the structure of the few simple skeletal elements of which they are composed (the stem and the three beak-like valves). There are in addition a small number of soft-tissue features that are important (for example gland development and distribution). Pedicellariae are appendages on the test and are functionally and structurally effectively independent of other morphological traits.

Table 15.2 Consistency indices excluding invariant positions (CI), retention indices (RI), and rescaled consistency indices (RC) for morphological and molecular characters optimized onto the total evidence tree.

	Number of characters	CI	RI	RC
Larval characters	48	0.50	0.74	0.40
Adult morphology (all)	163	0.70	0.88	0.64
Test construction	8	0.78	0.91	0.74
Apical disc	20	0.62	0.88	0.58
Ambulacral structure	23	0.72	0.88	0.66
Interambulacral structure	11	0.60	0.81	0.57
Peristomial region	11	0.77	0.94	0.72
Spines and tubercles	29	0.62	0.86	0.58
Pedicellariae	18	0.58	0.79	0.51
Perignathic girdle	4	0.80	0.92	0.76
Lantern	29	0.71	0.89	0.66
Internal anatomy	8	0.87	0.95	0.86
SSU rRNA, all variable positions	279	0.50	0.54	0.39

In Table 15.2 the 163 adult characters are partitioned into a number of subsets representing different skeletal and organ systems. Each subset of characters has been optimized over the total evidence tree to calculate a CI. Pedicellarial characters have higher homoplasy (CI = 0.6) than any of the other subsets of adult morphological characters, but still show considerably lower levels of homoplasy than do larval characters (CI of 0.5). Indeed the CI for larval characters alone lies outside two standard deviations from the mean CI of the 10 subsets of adult morphological characters. We therefore conclude that larval echinoid characters may indeed be more prone to homoplasy than adult characters of a similar structure and complexity. Note that levels of homoplasy amongst larval characters are comparable with those displayed by sequence data of the SSU rRNA gene: it is the adult morphological characters that display an anomolously low level of homoplasy. This may be because, historically, experienced systematists working with adult morphology have already sorted and rejected a number of potential characters as too homoplasous for consideration.

Rates of morphological and molecular evolution over geological time

Character transformations accumulate in lineages over time and, so long as there is not extensive character reversal, there will be a general correlation between the number of character transformations accumulated and the length of time between nodes. However, we know that this is not clock-like for adult morphological characters, since we have evidence from the fossil record that there have been short periods of geological time when considerable morphological diversification has occurred. The diversification of irregular echinoids between about 205 and 175 million years ago, for example, was accompanied by a great number of morphological innovations in adult form.

Whereas adult and larval morphology are likely to have responded in an *ad hoc* way to external events, the same is less likely to be true for SSU rRNA gene sequence data. Although the sequence of nucleotide bases for the SSU rRNA gene is also under selective pressure (to maintain the mature molecule's functionality), it is not directly affected by external factors in the way that morphology is thought to be. Consequently there has been an expectation that molecular sequences will evolve in a clock-like manner (for example Li and Graur 1991; Fitch and Ayala 1994). Although few now believe in the reality of a strict molecular clock (Vawter and Brown 1986; Runnegar 1991, p. 392), there is still a generally held belief that molecular sequence evolution proceeds more regularly than does morphological evolution (Olsen and Woese 1993). However, previous work on echinoderm LSU rRNA gene sequences (Smith *et al.* 1992) suggested that this molecule evolved just as irregularly over geological time as did adult morphology. Cumulative plots of number of changes against geological time for single lineages demonstrate that all three datasets (adult and larval morphology, SSU rRNA) show a positive correlation with time (Fig. 15.4), but it is not immediately apparent whether any dataset is significantly better correlated than the others.

To test whether molecular evolution of rRNA has occurred more regularly over geological time than morphological evolution we used the fossil record of first appearances to date nodes on the cladogram. This gave us 29 branches in total where we had estimates of duration (in millions of years), numbers of adult and larval morphological character transformations, and the numbers of fixed point mutations and transversions for the SSU rRNA molecule.

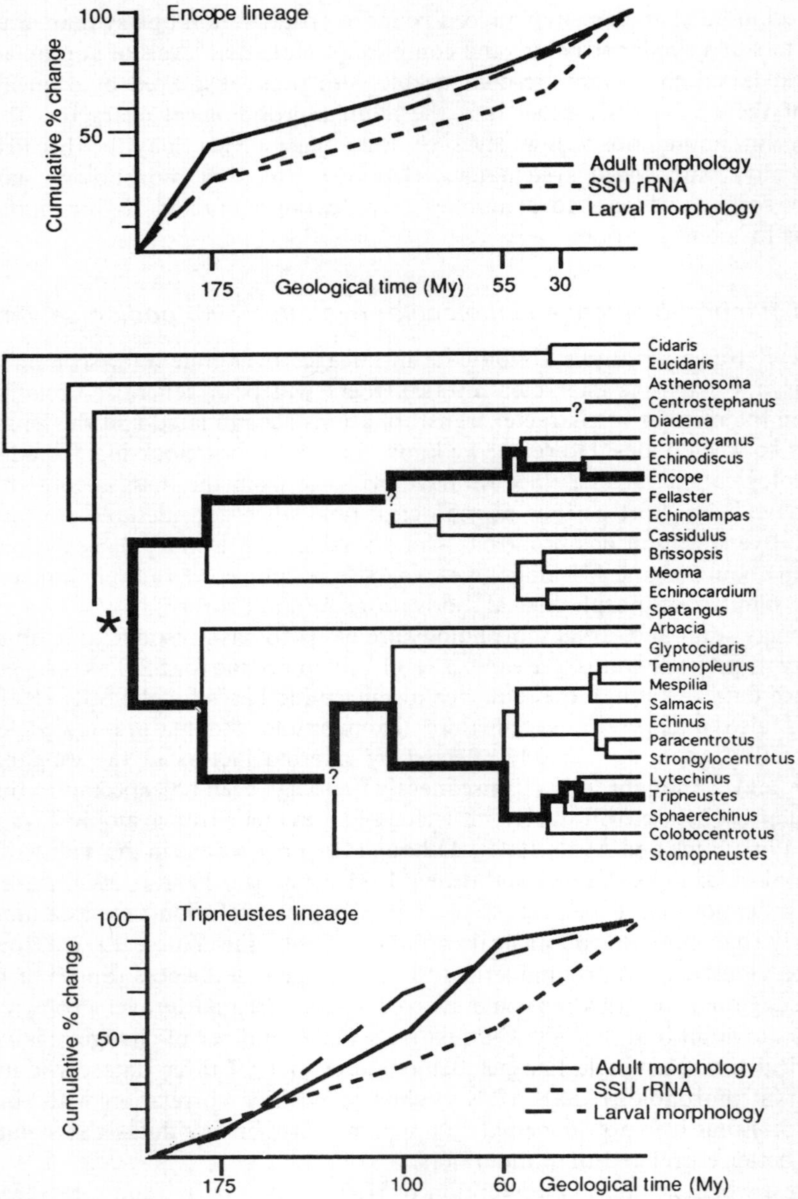

Fig. 15.4 Number of character transformations as a cumulative percentage of the total calculated along two independent branches (diverging from the split between Irregularia and Echinacea 205 million year ago) for larval, adult, and SSU rRNA gene data. Absolute dates are estimated for selected branch points using the first appearance of a fossil member of one or other sister group.

Using Spearman's rank correlation test we found that statistical correlation between time and morphological character transformation is poor (Table 15.3, Fig. 15.5). The correlation coefficient for adult morphology is just 0.42 and for larval characters is even less, at 0.29. However, molecular fixed point mutations of SSU rRNA are no better, and show a correlation coefficient of just 0.39. When transversions alone are considered, the correlation improves slightly, the coefficient rising to 0.47. We conclude from this that morphological and molecular rates of evolution are proceeding more or less equally haphazardly, but that the rate at which transversions become fixed is more regular over time than morphological character transformation. This confirms the findings of Vawter and Brown (1993) that SSU rRNA transversions show the lowest rate of variation amongst different branches and are likely to produce the best 'molecular clock'.

Table 15.3 Spearman rank correlation coefficients (R_s) obtained from pairwise comparison of numbers of apomorphies at each branch on the total evidence tree.

	R_s	Number of pairwise comparisons
Adult morphology versus larval morphology	0.35	46
Adult morphology versus SSU rRNA, all positions	−0.09	41
Larval morphology versus SSU rRNA, all positions	−0.17	37
SSU rRNA loop regions versus stem regions	0.74	40
Estimated branch duration (Ma) versus adult morphology	0.42	29
Estimated branch duration (Ma) versus larval morphology	0.29	28
Estimated branch duration (Ma) versus SSU rRNA, all positions	0.39	29
Estimated branch duration (Ma) versus SSU rRNA, transversions only	0.47	29

We found no evidence for loop or stem differences being important in terms of rates of evolution (Table 15.3, Fig. 15.5). When loop and stem fixed point mutations are treated separately, they show a strong correlation of 0.76. Thus when loop regions are evolving rapidly so too are stem regions, and there is no real reason to favour one over the other for phylogenetic analysis.

Finally, we found that the transition/transversion ratio for the entire SSU rRNA molecule is 1.66:1. For branches that have arisen entirely within the last 60 million years the ratio is 1.98:1, so there may be a tendency for longer or deeper branches to become saturated (and thus underestimate the rate of change along deeper branches). This is most noticable in the long branches (>220 million years) leading to cidaroids and echinothurioids. However, shorter deep branches (>175 million years) show, if anything, higher transition/transversion ratios than recently diverged clades. Vawter and Brown (1993) have found no consistent transition-transversion bias in SSU rRNA genes and we are currently investigating the structural evolution of the SSU rRNA molecule in echinoids in order to clarify the meaning of this pattern.

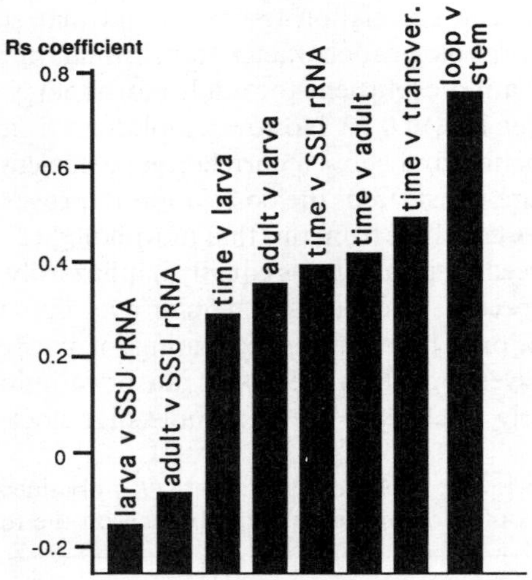

Fig. 15.5 Bar chart showing the relative strengths of correlation for pairwise comparisons of morphological and molecular data as indicated. The vertical axis gives the Spearman rank correlation coefficient.

15.4 Conclusions and prospects

The ability to construct robust phylogenies from the combination of several independent datasets allows comparisons to be drawn concerning the evolutionary behaviour of different kinds of characters over geological time. By including morphological characters as well as molecular characters, not only do we obtain the best possible estimate of a clade's phylogenetic history but we also gain the possibility of calibrating the cladogram and transforming it into an evolutionary tree through recourse to the fossil record. Applied to groups with a reasonably well-established fossil record this approach can provide estimates of branching times, allowing the question of rates of evolution through geological time to be tackled more rigorously. This in turn allows general questions that compare and contrast the way in which different segments of life history, or different kinds of molecular data (for example non-coding versus coding sequences or transitions versus transversions) have evolved in a clade over geological time to be investigated. A great deal has been learned about the structure and evolution of ribosomal RNA genes over geological time (Vawter and Brown 1993; Gutell *et al.* 1994) but we envisage that well-documented phylogenies such as that presented here could help improve our understanding of their structural evolution.

Acknowledgements

We thank the following individuals for help in locating, providing, and transporting echinoid material: S. Amemiya, R. Aronson, A. Baker, A. Bentley,

H. Dixon, S. Donovan, R. Emson, C. Freire, P. Gayle, J. Gage, M. Gibbons, M. Hart, H. Hayashi, G. Hendler, W. Hide, T. Kikuchi, J. Korrubel, B. Lafay, D. McKenzie, T. Matsuoka, P. Mikkelsen, B. Morton, T. Motokawa, D. Nichols, S. Palumbi, N. Suzuki, J. Taylor, P. Tyler, T. Uehara, J. Woodley, and C. Young. We would also like to thank Robin Gutell for providing us with the secondary structure of *Strongylocentrotus*. This study benefited from the SEQNET facilities and was directly supported by NERC grant GR3/7960.

Appendix: Echinoid larval characters used in this analysis, and the data matrix

1. Postoral arm rod fenestration: fenestrated along entire length (0); fenestrated only along distal portion (1); completely unfenestrated (2).
2. Postoral arm rod length: more than three times body length (0); less than two times body length (1).
3. Recurrent rod: present (0); absent (1).
4. Ventral transverse rod junction: rods do not meet at midline (0); rods meet in stage I larvae (1).
5. Body rod junction: rods do not meet at midline (0); rods meet in stage I larvae (1).
6. Posterodorsal rod junction: rods do not meet at midline (0); rods meet in stage I larvae (1).
7. Skeletal weight: gracile (0); robust (1).
8. Posterior connecting rod: absent (0); present (1).
9. Spines on body rod: absent (0); present (1).
10. Thorn on body rod: absent (0); present (1).
11. Postoral arm rod thickness: normal (0); exceptionally thick (1).
12. Anterolateral arm rod length: extends beyond oral hood (0); not extending beyond oral hood (1).
13. 'Extra' ventral transverse rod: absent (0); present (1).
14. Body rod: gently curved (0); straight (1).
15. Body rod termini: unbranched (0); branched (1).
16. Posterior meshwork in body rod: absent (0); present (1).
17. Stereom on body rod: absent prior to metamorphism (0); present at posterior terminus (1); present near junction with ventral transverse rod (2).
18. Orientation of recurrent rods: point posteriorly (0); point dorsally (1).
19. Body rod: less than two times the ventral transverse process length (0); more than two times the ventral transverse process length (1).
20. Terminal bulb on body rod: absent (0); present (1).
21. Structure of recurrent rods: single connecting rod (0); double connecting rod (1).
22. Proximal curvature of postoral arms: straight (0); gently curved (1).
23. Posterodorsal arms: present (0); absent (1).
24. Dorsal transverse process: present (0); absent (1).
25. Posterior extension on posterodorsal element: present (0); absent (1).
26. Posterodorsal arm rod fenestration: fenestrated along entire length (0); partial fenestration only (1); completely unfenestrated (2).
27. Posterodorsal arm: longer than body length (0); very much shorter than body length (1).
28. Posterodorsal arm shape: straight (0); curved (1).

29. Central process on dorsal arch: long (0); short (1).
30. Posterolateral processes on dorsal arch: present (0); absent (1).
31. Preoral arm rod shape: sigmoidally curved (0); straight (1).
32. Anterodorsal arms: absent (0); small flaps only (1); long (more than the diameter of the oral hood (2).
33. Preoral arm rods orientated: parallel (0); convergent (1).
34. Posterior element: present (0); absent (1).
35. Shape of posterior element: gently curved (0); strongly curved (1).
36. Posterolateral arm rods: absent or small (0); long (more than body length) and at right angles to the body axis (1); long (more than body length) and posteriorly swept (2).
37. Anterior central process on posterolateral rods: absent (0); present (1).
38. Weight of central portion of posterior element; rodlike (0); chunky (1).
39. Posterior 'arm': absent (0); present (1).
40. Central hole on posterior element: absent (0); present (1).
41. Decoration on posterolateral arm rods: termini bare (0); termini with fringe (1).
42. Number of termini on posterolateral rods: one (0); two (1).
43. Dorsal and ventral vibratile lobes: present (0); absent (1).
44. Posterior vibratile lobe or posterolateral arm: present (0); absent (1).
45. 'Extra' vibratile lobes: absent (0); present (1).
46. Shape of arm ectoderm in stage I larvae: appressed to arm rods (0); greatly inflated (1).
47. Posterior epaulettes: absent (0); present (1).
48. Dorsal and ventral epaulettes: absent (0); present (1).

Data matrix:

Cidaris	00000	00000	00000	00000	00000	00000	00000	00000	00001	000
Eucidaris	00000	00000	00000	00000	00000	10000	00000	01000	00001	000
Diadema	00011	11101	11100	00000	00121	21?11	10101	00100	00110	000
Echinolampas	01001	11110	00?00	00000	00001	00?11	1001?	?????	??110	000
Arachnoides	01011	11110	00100	10000	00000	00011	1001?	?????	??110	000
Echinocyamus	01011	11110	00100	10000	00000	00011	1001?	?????	??110	000
Echinodiscus	01011	11110	00100	10000	00000	00011	0001?	?????	??110	000
Encope	01011	11110	00100	10000	00000	00001	0001?	?????	??110	000
Brissus	10011	11110	00010	0?000	00000	10001	1?10?	00010	?0?00	000
Meoma	1?011	11110	00010	0?000	00000	?????	???0?	0????	???00	000
Echinocardium	10011	11110	00010	02000	00000	10001	12101	10010	00110	000
Spatangus	10011	11110	00010	02000	00000	10001	12101	10010	00110	000
Arbacia	10011	11100	00000	01000	00000	00111	01000	20101	00010	000
Temnopleurus	01000	01010	00001	00100	00000	00100	10100	00000	11000	101
Mespilia	01000	01010	00001	00100	00000	00100	10100	00000	11000	101
Salmacis	01000	01010	00001	00100	00000	00100	10100	00000	11000	101
Echinus	21000	00000	00000	00000	00010	00001	10000	00000	00010	111
Paracentrotus	21100	00000	00010	00?11	00010	10001	1001?	?????	??010	111
Psammechinus	21100	00000	00010	00?11	00010	10001	1001?	?????	??010	111
Strongylocentrot	21?00	00000	00010	00011	00010	10001	100?0	00000	000?0	111
Colobocentrotus	01011	00110	00000	00000	10000	00000	10100	01000	?0000	011
Stomopneustes	01011	1???0	00000	00000	0?000	00???	?0?00	0??0?	??000	0??
Glyptocidaris	01011	01010	00000	01000	0?000	00011	10000	00000	00000	011
Lytechinus	21011	11110	00000	00100	01000	10100	10000	00000	00000	011
Sphaerechinus	01011	01110	00000	00000	00000	00?00	10000	00000	00000	011
Tripneustes	01011	01110	00000	00000	00000	10100	10000	00000	10000	011

References

Archie, J. W. (1989). Homoplasy excess ratios: new indices for measuring levels of homoplasy in phylogenetic systematics, and a critique of the consistency index. *Syst. Zool.*, **38**, 253–69.

Brooks, D. R. and McLennan, D. A. (1991). *Phylogeny, ecology and behavior*. University of Chicago Press.

Coddington, J. A. (1988). Cladistic test of adaptational hypotheses. *Cladistics*, **4**, 3–22.

DeBeer, G. (1958). *Embryos and ancestors*. Clarendon, Oxford.

Deleporte, P. (1993). Characters, attributes and tests of evolutionary scenarios. *Cladistics*, **9**, 427–32.

De Queiroz, A. (1993). For consensus (sometimes). *Syst. Biol.*, **42**, 368–72.

Emlet, R. B. (1985). Crystal axes in Recent and fossil echinoids indicate trophic mode in larval development. *Science*, **230**, 937–40.

Emlet, R. B. (1989). Apical skeletons of sea urchins (Echinodermata: Echinoidea): two methods for inferring mode of larval development. *Paleobiology*, **15**, 223–54.

Eernisse, D. J. and Kluge, A. G. (1993). Taxonomic congruence versus total evidence, and amniote phylogeny inferred from fossils, molecules, and morphology. *Mol. Biol. Evol.*, **10**, 1170–95.

Fitch, W. M. and Ayala, F. J. (1994). The superoxide dismutase molecular clock revisited. *Proc. Natl Acad. Sci. USA*, **91**, 6802–7.

Gutell, R. R., Larsen, N., and Woese, C. R. (1994). Lessons from an evolving rRNA: 16S and 23S rRNA structures from a comparative perspective. *Microbiol. Rev.*, **58**, 10–26.

Jablonski, D. and Lutz, R. A. (1983). Larval ecology of marine benthic invertebrates: paleobiological implications. *Biol. Rev.*, **58**, 21–89.

Kluge, A. G. and Wolf, A. J. (1993). Cladistics: what's in a word. *Cladistics*, **9**, 183–200.

Larson, A. (1994). The comparison of morphological and molecular data in phylogenetic systematics. In *Molecular approaches to ecology and evolution*, (ed. B. Schierwater, B. Streit, G. P. Wagner, and R. De Salle), pp. 371–90. Birkhauser, Basel.

Levinton, J. S. (1988). *Genetics, paleontology, and macroevolution*. Cambridge University Press.

Li, W.-H. and Graur, D. (1991). *Fundamentals of molecular evolution*. Sinauer.

Littlewood, D. T. J. and Smith, A. B. (1995). A combined morphological and molecular phylogeny for sea urchins (Echinoidea: Echinodermata). *Phil. Trans. R. Soc.*, **B347**, 213–34.

Meier, R., Kores, P., and Darwin, S. (1991). Homoplasy slope ratio: a better measurement of observed homoplasy in cladistic analysis. *Syst. Zool.*, **40**, 74–88.

Olsen, G. J. and Woese, C. R. (1993). Ribosomal RNA: a key to phylogeny. *FASEB J.*, **7**, 113–23.

Runnegar, B. (1991). Nucleic acid and protein clocks. *Phil. Trans. R. Soc.*, **B333**, 391–7.

Sanderson, M. J. and Donoghue, M. J. (1989). Patterns of variation in levels of homoplasy. *Evolution*, **43**, 1781–95.

Smith, A. B. (1994). *Systematics and the fossil record; documenting evolutionary patterns*. Blackwell Scientific, Oxford.

Smith, A. B., Lafay, B., and Christen, R. (1992). Comparative variation of morphological and molecular evolution through time: 28S ribosomal RNA versus morphology in echinoids. *Phil. Trans. R. Soc.*, **B338**, 365–82.

Strathmann, R. R. (1988). Larvae, phylogeny, and von Baer's law. In *Echinoderm phylogeny and evolution*, (ed. C.R.C. Paul and A. B. Smith), pp. 53–67. Clarendon, Oxford.

Suter, S. J. (1994). Cladistic analysis of cassiduloid echinoids: trying to see the phylogeny for the trees. *Biol. J. Linn. Soc.*, **53**, 31–72.

Swofford, D. L. (1991). When are phylogenetic estimates from molecular and morphological data incongruent? In *Phylogenetic analysis of DNA sequences*, (ed. M. M. Miyamoto and J. Cracraft), pp. 295–333. Oxford University Press.

Swofford, D. L. (1993). *PAUP: Phylogenetic Analysis Using Parsimony*, version 3.1.1. (computer program and manual). Smithsonian Institution, Washington DC.

Vawter, L. and Brown, W. M. (1986). Nuclear and mitochondrial DNA comparisons reveal extreme rate variation in the molecular clock. *Science*, **234**, 194–6.

Vawter, L. and Brown, W. M. (1993). Rates and patterns of base change in the small subunit ribosomal RNA gene. *Genetics*, **134**, 597–608.

Wray, G. A. (1991). The evolution of developmental strategy in marine invertebrates. *TREE*, **6**, 45–50.

Wray, G. A. (1992). The evolution of larval morphology during the post-Paleozoic radiation of echinoids. *Paleobiology*, **18**, 258–87.

16

Molecular phylogenies and host–parasite cospeciation: gophers and lice as a model system

Roderic D. M. Page and Mark S. Hafner

16.1 Introduction

This chapter outlines a general theoretical and methodological framework for comparing phylogenies of hosts and parasites to address a broad variety of evolutionary questions that could not otherwise be investigated by independent study of either group. The rationale, advantages, and limitations of this approach (known generally as the comparative method) have been discussed elsewhere (Harvey and Pagel 1991), as has the long and illustrious history of host–parasite studies in general (Brooks and McLennan 1993). Here, we illustrate how the application of new and powerful molecular techniques to comparative studies of host–parasite phylogeny enables study of an entirely new domain of topics that could not be explored with non-molecular data.

Host-parasite systems are intrinsically interesting to evolutionary biologists because they signal a long and intimate association between two or more groups of organisms that are distantly related and quite dissimilar biologically. This long history of association often leads to reciprocal adaptations in the hosts and their parasites (classical coevolution or coadaptation) as well as contemporaneous cladogenic events in the two lineages (cospeciation). The phenomenon of cospeciation is of particular interest to comparative phylogeneticists because cospeciation events identify temporal links between the host and parasite phylogenies, and thus provide an internal time calibration for comparative studies of rates of evolution in the two groups. Evidence of cospeciation also can be used to test hypotheses of coadaptation in the hosts and parasites.

Although evidence of cospeciation in a host–parasite assemblage presents exciting opportunities for the study of evolution, the task of obtaining this evidence is fraught with theoretical and methodological challenges. The analysis involves three steps (tree building, tree comparison, and estimation of divergence), each of which is dependent on the prior step, and each of which requires a different set of experimental and analytical tools. To illustrate this three-tiered protocol for investigation of cospeciation, we use the example of pocket gophers and their chewing lice studied by Hafner *et al.* (1994).

16.2 Pocket gophers and their lice

The hosts in this example include several species of pocket gophers of the rodent family Geomyidae. Pocket gophers are fossorial and extremely asocial, and gopher species generally are allopatric. Nearly all species of pocket gophers are parasitized by chewing lice of the mallophagan family Trichodectidae. These lice are restricted to pocket gophers, and the entire life cycle of these wingless insects occurs on the host. Thus, the combined biological characteristics of pocket gophers and chewing lice (i.e. asocial hosts, well-dispersed host populations, and parasites with low vagility) suggest that the lice have few opportunities for colonization of new host species (Nadler and Hafner 1989; Nadler *et al.* 1990). This, in turn, has resulted in a high level of cospeciation in this host–parasite assemblage (Hafner and Nadler 1988).

Cospeciation in pocket gophers and their chewing lice has been investigated from a variety of perspectives, including morphology (Timm 1983), allozymes (for example Hafner and Nadler 1988; Demastes and Hafner 1993), and nucleotide sequences (Hafner *et al.* 1994). In each case, evidence of cospeciation in this assemblage has been so dramatic that the gopher–louse system has become literally a 'text-book example' of cospeciation (for example Noble *et al.* 1989; Esch and Fernández 1993; Ridley 1993). Most recently, Hafner *et al.* (1994) obtained DNA sequences (379 base pairs) from the same region of the cytochrome c oxidase subunit I (COI) gene from the mitochondria of 15 taxa of pocket gophers and 17 taxa of lice that parasitize these gophers. For details about the taxa, dataset, and analysis, the reader is referred to Hafner *et al.*'s paper. Herein, we focus on this study to illustrate our general method of investigation and to demonstrate the utility of molecular data for the study of cospeciation.

16.3 Reconstructing host and parasite phylogenies

Development of phylogenetic trees for the hosts and their parasites lays the foundation for subsequent tests of cospeciation. Because further analyses are dependent on the quality of these trees, the trees must be consistent, well-resolved, and independent (i.e. one phylogeny cannot be inferred from the other (Hafner and Nadler 1990)). Furthermore, if one intends to study comparative rates of molecular evolution, the trees must be based on genetic systems that are homologous in the hosts and parasites (such as the COI gene in gophers and lice compared by Hafner *et al.* (1994).

Systematists have developed a large number of methods for estimating phylogenies (Swofford and Olsen 1990; Hillis *et al.* 1993), each of which uses a different model of character evolution and potentially yields a different tree for the group studied. Because no single method of phylogenetic analysis is universally regarded as superior to others, it behoves the investigator to use multiple methods and to indicate how different analyses affect tree structure. Importantly, demonstration that the host and parasite phylogenies are reasonably robust to different methods of analysis will increase confidence in subsequent tests of cospeciation.

Hafner *et al.* (1994) used four tree-building methods to reconstruct gopher and louse relationships and showed that major portions of the phylogenies were insensitive to method of analysis. Nevertheless, the multiple analyses revealed phylogenetic uncertainty in certain regions of the host and parasite trees, which caused Hafner *et al.* to retain multiple trees (four host trees and five parasite trees) for subsequent tests of cospeciation. Because all systematic analyses will be hampered by some degree of phylogenetic uncertainty, and because the source of that uncertainty generally is unknown (limitations of the data, weakness of the analysis, or both), we recommend use of multiple host and parasite trees for tests of cospeciation in all but the most clear-cut cases. To simplify the following discussion, we will restrict our analysis to the gopher and louse phylogenies illustrated in Fig. 16.1 (data from Hafner *et al.* 1994).

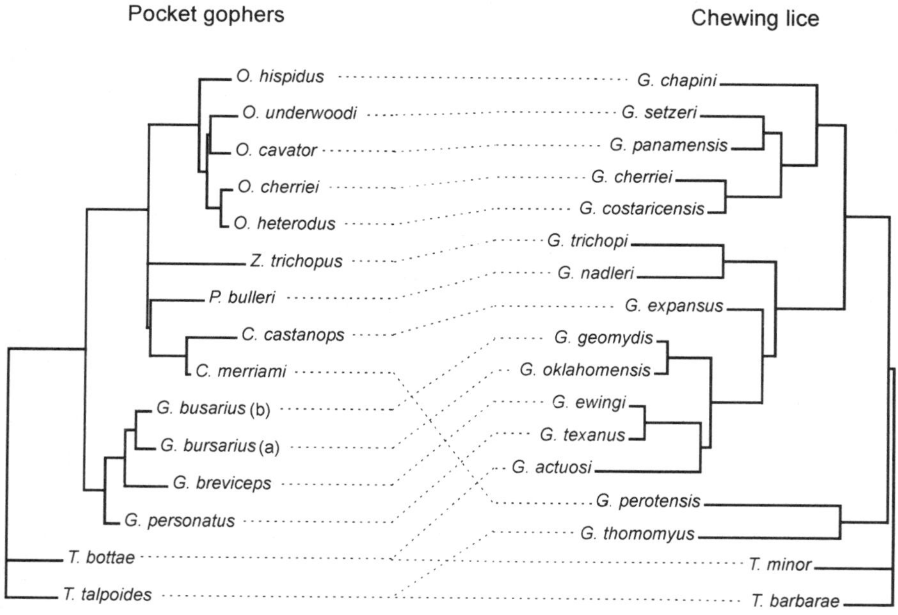

Fig. 16.1 Phylogenies for pocket gophers and their chewing lice based on nucleotide sequence data analysed by Hafner *et al.* (1994). Shown are composite trees based on multiple methods of phylogenetic analysis detailed by Hafner *et al.* (1994). Branch lengths are proportional to the expected numbers of substitutions at the third codon position in the COI gene estimated using Felsenstein's (1989) maximum likelihood algorithm (DNAML, with transition/transversion ratio of 4.0 for both clades). Coexisting hosts and parasites are connected by dotted lines. Pocket gopher genera are *Orthogeomys*, *Zygogeomys*, *Pappogeomys*, *Cratogeomys*, and *Thomomys*. *Geomys bursarius* is represented by two subspecies (a = *G. b. halli*; b = *G. b. majusculus*). Chewing louse genera are *Geomydoecus* and *Thomomydoecus*.

16.4 Reconstructing the history of a host–parasite association

The prerequisite for any comparison of host-parasite evolution is a reconstruction of the history of that association. Here, the challenge is to determine whether the degree of similarity observed between the host and parasite phylogenies exceeds the similarity we would expect to see by chance. At present there are two methods for obtaining such a reconstruction, Brooks' parsimony analysis (Brooks and McLennan 1991) and Page's (1990*a*, 1993*a*, 1994) component analysis. Brooks' parsimony analysis uses additive binary coding to represent parasite phylogenies, then optimizes the resulting codes on the host phylogeny. Page (1990*a*, 1994) has argued that this procedure can give spurious results, hence in this study we use the most recent refinement of component analysis (Page 1995).

Component analysis is a method developed originally to reconstruct biogeographic histories of taxa (Nelson and Platnick 1981), but its similarity to Goodman *et al.*'s (1979) procedure for comparing gene trees and species trees suggests that component analysis is sufficiently general to be applied to any historical association, including host–parasite systems (Page 1990*a*). To date, component analysis has been applied to the association of the tree genus *Nothofagus* and its fungal parasite *Cyttaria* (Page 1990*a*), pocket gophers and their lice (for example Page 1990*b*; Hafner *et al.* 1994), and seabirds and their lice (Paterson *et al.* 1993; Paterson 1994).

The analogy between comparing parasite and host phylogenies and comparing gene and organismal phylogenies is instructive. Parasitologists have generally assumed that unless host and parasite phylogenies are absolutely congruent, host switching has occurred (for example Brooks and McLennan 1991, p. 205). The complexity of the relationship between gene trees and species trees, even in the absence of horizontal transfer (such as introgression), suggests that this assumption may be unjustified. For example, if we view parasites as 'genes' of their hosts, passed from parent to offspring for multiple generations, we can imagine that the parasites might be subject to the same stochastic processes that affect genes in populations, such as loss through drift, retention of ancestral (plesiomorphic) character states, and lineage sorting (Avise *et al.* 1984). If so, it is likely that much of the incongruence between host and parasite trees may result simply from chance loss or retention of parasite lineages. Further, we might expect to see higher levels of incongruence between host and parasite phylogenies when younger lineages are studied (for example closely related species) simply because there has been insufficient time for lineage sorting of the parasites. In theory, chance extinction of the parasites should result eventually in reciprocal monophyly of parasite lineages on sister taxa of hosts (Demastes and Hafner 1993).

Has cospeciation occurred?

A simple test of the hypothesis of cospeciation is to ask whether the structure of the parasite tree is independent of that of its host. If so, we would expect the amount of cospeciation observed between the hosts and parasites (i.e. the number of cospeciation events in the two phylogenies) to be no greater than that expected between the host tree and random parasite trees (Page 1995). Applying this test to the phylogenies in Fig. 16.1, we reject the hypothesis that the louse phylogeny is independent of the gopher phylogeny ($p = 0.004$, computed using 1000 random trees). It is possible that recent host switching could produce spurious congruence between the host and parasite trees, especially if the parasites preferentially colonize hosts that are closely related. Similarly, incongruence between the host and parasite phylogenies could result from differential survival of multiple parasite lineages, rather than host switching (as discussed above) (see also Page 1993*b*). If genetic data are available for hosts and parasites, as in our gopher–louse example, information on amounts of genetic divergence (or relative coalescence times) can assist our efforts to discriminate between these possibilities (Page 1993*b*).

16.5 Studies of coadaptation and colonization in hosts and parasites

Component analysis (Page 1993*a*, 1995) identifies pairs of equivalent nodes in the host and parasite trees that reflect the same historical event. These equivalent nodes can be depicted visually by overlaying the parasite tree on the host tree, wherein each node of the parasite tree is adjacent to the corresponding node in the host tree (Fig. 16.2). Hypotheses of coadaptation in the hosts and parasites can be tested using these nodes. For example, Harvey and Keymer (1991) used simplified phylogenies of gophers and lice taken from Hafner and Nadler (1988) to show that evolution of body size in lice and their hosts is highly correlated. Numerous other morphological, physiological, and ecological attributes of the hosts and parasites can be compared using the cospeciation framework.

In a parasite clade that shows evidence for host switching, the investigator may wish to ask if there are geographic, morphological, or ecological correlates of host switching. Reconstruction of the biogeographic history of host-switching events may reveal whether colonization of new hosts is simply opportunistic (nearest neighbour), or whether parasites are tracking a particular resource in the host taxa that is not itself correlated with host phylogeny (such as quill size preferences shown by the feather mites of birds (Kethley and Johnston 1975). Knowledge of past host-switching events, coupled with genetic data for the hosts and parasites, permits detection of possible changes in rates of evolution in the hosts or parasites (or both) following colonization events.

A recent laboratory investigation of host switching (Reed 1994) illustrates how knowledge of phylogeny can illuminate studies of parasite transmission. In a series of controlled laboratory experiments, Reed (1994) transferred lice from their natural hosts to new pocket gopher hosts to determine whether the high level of host specificity observed in nature is the result of host-specific adaptations (wherein lice are uniquely adapted for life on a particular taxon of host), or

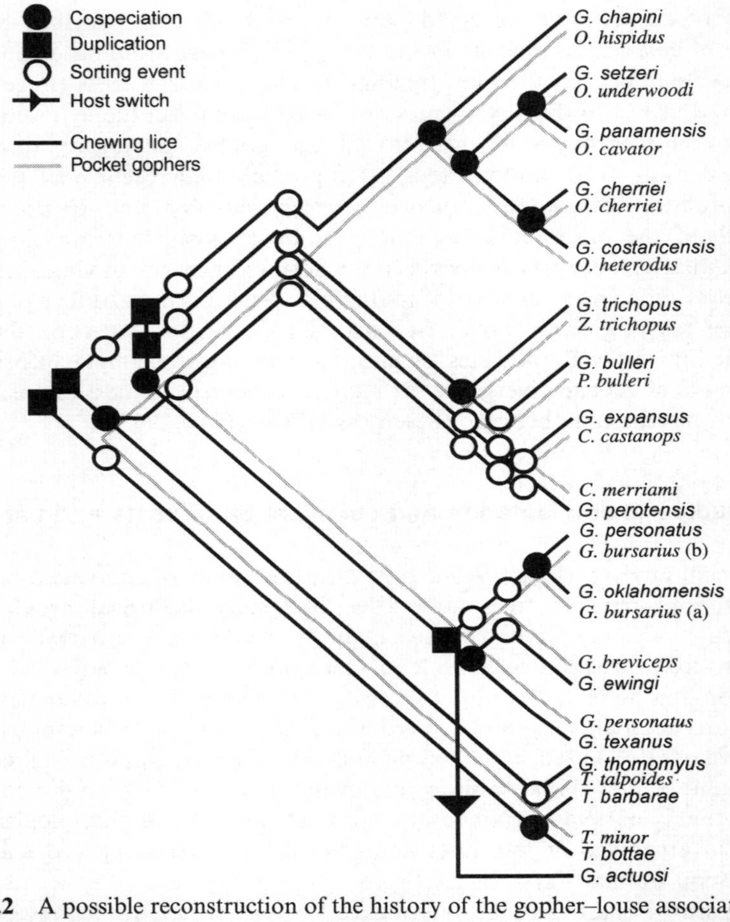

Fig. 16.2 A possible reconstruction of the history of the gopher–louse association that postulates 10 cospeciation events, five duplications (*in situ* speciation of the lice on the same host), 20 sorting events (instances where louse lineages have been lost or remain undetected), and a single host switch (by *Geomydoecus actuosi*).

is simply the result of the louse's lack of opportunity to colonize new hosts (resulting from the generally allopatric distribution of host taxa and low dispersal abilities of the parasites). To determine if there is a phylogenetic component to host specificity, Reed transferred lice between hosts of varying degrees of phylogenetic relatedness (Figure 16.3a). For example, the first set of experiments transferred lice between hosts belonging to different subspecies of *Thomomys bottae*. All such transfers were successful (i.e. the colonizing lice were able to survive and reproduce on the new host for approximately four louse generations). The second set of experiments transferred lice between hosts belonging to different species of *Thomomys* (*T. bottae* and *T. talpoides*). Three of four such transfers were successful. Finally, Reed made multiple attempts to transfer lice between hosts of different genera (*Thomomys* and *Cratogeomys*). Only 1 of 14 such attempts was successful (the louse *G. expansus*, whose natural host is *Cratogeomys*, was able to survive and reproduce on *Thomomys bottae*; Fig. 16.3b).

A

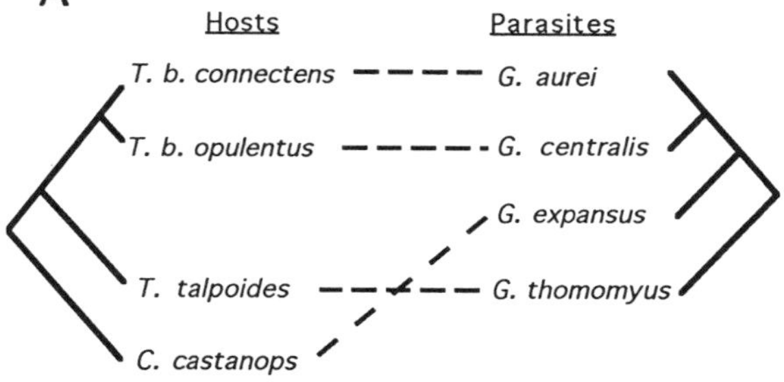

B

Recipient Host (columns); Source Host (rows)

Source Host	*T. bottae connectens*	*T. bottae opulentus*	*Thomomys talpoides*	*Cratogeomys castanops*
T. bottae connectens	yes (control)	yes	—	no
T. bottae opulentus	yes	—	yes	no
Thomomys talpoides	—	yes	yes (control)	no
Cratogeomys castanops	—	yes	—	—

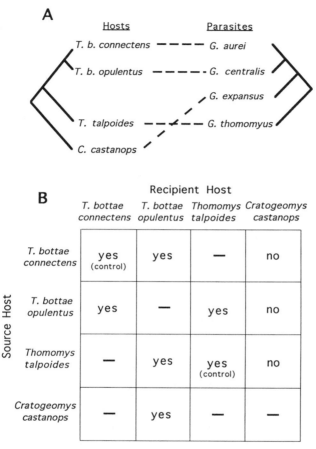

Fig. 16.3 Panel A illustrates relationships among pocket gophers and chewing lice studied by Reed (1994). The host cladogram is supported by morphological (Russell 1968) and molecular data (immunology and allozymes (Hafner 1982)). The parasite cladogram is based on morphological evidence (Page *et al.* 1995), and the structure of the tree is also consistent with allozymic data (Demastes 1990). Abbreviations as in Fig. 16.1. Broken lines connect parasites with their natural hosts. Panel B outlines the experimental design and results of the host-transfer experiments conducted by Reed (1994). Lice were transferred from the source host (the louse's natural host) to the recipient host to test the louse's ability to survive and reproduce on the new host. 'Yes' signifies that the introduced lice were able to survive and reproduce on the new host. 'No' indicates that the lice were unable to survive on the new host. Dashes indicate transfer experiments that were not attempted; these were considered redundant within the context of Reed's (1994) experimental design.

Reed's (1994) experiments suggest that a louse's ability to survive and reproduce on a new host decreases with increasing phylogenetic distance between the new host and the louse's natural host. The actual mechanism constraining successful colonization is unknown, but may be related to hair-shaft diameter in the host (Reed 1994). Regardless of the causal mechanism involved, these results suggest that the high level of host specificity observed between gophers and lice in nature is the combined result of the louse's general lack of opportunity to colonize new hosts (because of the typically dispersed distribution of host taxa), reinforced by some degree of host-specific adaptation by the lice.

Reed (1994) used knowledge of the phylogenetic relationships among the gophers and lice he studied to suggest that the louse *G. expansus* is a relatively recent colonist of the host *C. castanops* (Fig. 16.3a). However, Reed's hypothesis seems less tenable when the taxa he studied are viewed in a broader phylogenetic context. For example, morphological evidence (Page *et al.* 1995) indicates that the lice *G. aurei* and *G. centralis* (Fig. 16.3a) are closely related to *G. actuosi* (Fig. 16.1). *G. actuosi*, in turn, appears to be a relatively recent colonist of *Thomomys bottae* (Fig. 16.2). Thus, it is more parsimonious to hypothesize that the common ancestor of the louse taxa *G. actuosi*, *G. aurei*, and *G. centralis* switched to the new host (*T. bottae*) prior to speciation of these louse taxa. The alternative explanation requires two host-switching events (*G. expansus* to *castanops* and *G. aurei* to *T. bottae*). This example also illustrates how limited taxon sampling can influence the historical reconstruction of coevolutionary events.

As mentioned earlier, Reed (1994) showed that the louse *G. expansus* (whose natural host is *Cratogeomys*) is able to survive and reproduce on a new host belonging to a different genus (*Thomomys*). In contrast, none of the lice naturally found on *Thomomys* was able to survive on *Cratogeomys* in Reed's experiments (Fig. 16.3b). These results suggest that the louse habitat provided by *Cratogeomys* somehow lacks a resource that is critical to the survival of lice that normally occur on *Thomomys* hosts. Considering that gophers of the genus *Cratogeomys* are considerably larger in body size than those of the genus *Thomomys*, it is possible that their hair shafts are also larger in diameter and, perhaps, are too large for grasping and feeding by lice that normally occur on *Thomomys*. A relationship between body size and hair shaft diameter in pocket gophers has yet to be demonstrated. However, if such a relationship exists, it may help to explain the significant positive relationship observed between gopher body size and louse body size by Harvey and Keymer (1991).

16.6 Comparisons of genetic divergence in hosts and parasites

There are many ways to convert molecular data (including data from allozymes, restriction-fragment patterns, and protein and DNA sequences) into estimates of genetic divergence (Swofford and Olsen 1990). Each method has inherent advantages and limitations, and each involves assumptions about the nature of evolutionary change at the molecular level. Recent comparative studies of genetic differentiation in hosts and parasites have used either pairwise estimates of genetic distance (for example Hafner and Nadler 1990; Page 1990*a*) or estimates of length of homologous branches in the host and parasite trees (for example Hafner *et al.* 1994). The former method, although easy to apply, has fundamental statistical limitations because of the non-independence of pairwise measurements. The latter method (branch length comparisons) generally avoids the problem of statistical dependence, but requires an often complex model of evolution to apportion change on to branches. As we will illustrate later, different models often yield different estimates of branch lengths, which may result in different interpretations of relative rates of change in the hosts and their parasites. Until knowledge of molecular evolution advances to the point where generally accepted models are available, the researcher should be explicit about the model selected and should be aware of the implications of that model.

Molecular clocks and relative timing of cospeciation events

Once estimates of branch lengths are calculated, lengths of equivalent branches in the host and parasite trees can be compared. Although the comparison may seem straightforward, interpretation of differences in branch length may be confounded by multiple factors. For example, host branches may be consistently longer than parasite branches because the hosts are evolving more rapidly, or because the hosts consistently diverged prior to their parasites, or both. Thus, meaningful interpretation of this comparison requires knowledge of relative rates of change in the hosts and parasites, which assumes that genetic change in each group is roughly clocklike. Our reliance on molecular clocks for this part of the analysis requires that we test for the existence of a clock, rather than simply assume one exists. Various tests are available for this purpose (for example Muse and Weir 1992; Goldman 1993; Adell and Dopazo 1994).

Hafner and Nadler (1990) proposed a framework for comparing host and parasite divergence, given molecular clocks (which may tick at different rates) in both groups. Fitting a line to a plot of parasite divergence against host divergence (Fig. 16.4) allows us to describe simultaneously two aspects of host–parasite divergence. The slope of the line (Fig. 16.4a) is an estimate of the relative rate of genetic change in the two groups. The *y*-intercept of the line (Fig. 16.4b) measures genetic divergence in the parasites at the time of host speciation. For example, an intercept of zero indicates synchronous cospeciation, wherein hosts and parasites speciate simultaneously. A negative intercept suggests delayed cospeciation, in which case the parasites tend to speciate consistently after their hosts. Finally, a positive intercept signals preemptive cospeciation, in which case the parasites diverge prior to their hosts.

Returning to the analogy with gene trees, if the bivariate plots shown in Fig. 16.4 were instead plotting sequence divergence for a given gene against time of taxon divergence, then a positive intercept would reflect the average sequence divergence that exists among populations of a species (Lynch and Jarrell 1993). By analogy, a positive intercept in the comparison of host–parasite sequence divergence would reflect average differentiation among parasite populations of a single host species. Thus, we might expect the intercept to be positive in situations where parasite populations living on different hosts of the same species are genetically divergent. It is perhaps significant, therefore, that louse populations living on different individual hosts at a single locality show moderately high levels of genetic divergence (Nadler *et al.* 1990). Whether or not this population-level differentiation has long-term evolutionary consequences has yet to be explored.

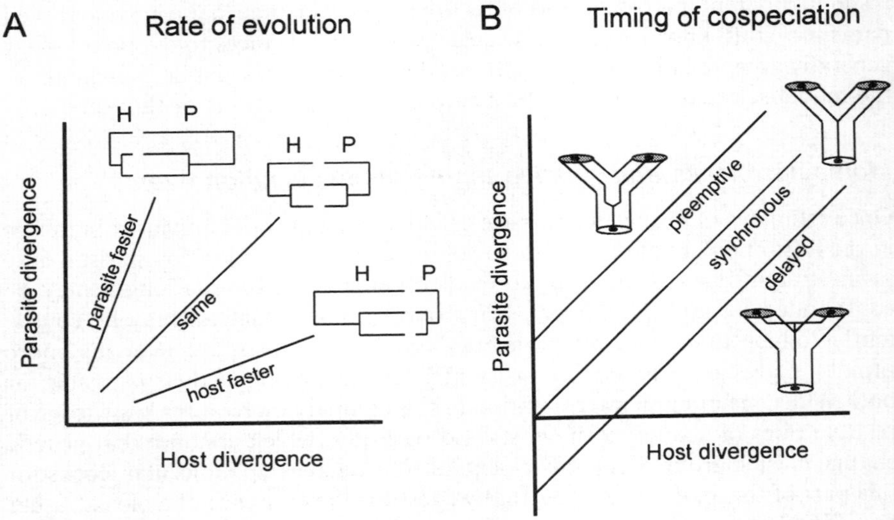

Fig. 16.4 Bivariate plots of the relationship between parasite divergence and host divergence. The slope of the relationship (panel A) indicates relative rates of evolution in the two clades. The trees (inset in panel A) are drawn with branch lengths proportional to amount of genetic change in the hosts (H) and parasites (P). The *y*-intercept (panel B) indicates the relative timing of speciation events. The inset figures in panel B illustrate relative timing of speciation events in the hosts (outer portion of figure) and their parasites (thin line within each figure). (Modified from Hafner and Nadler (1990, figure 2).)

Estimating branch lengths

There are many advantages to using DNA sequence data in studies of host–parasite cospeciation. Among these is the fact that the characters being compared have a known genetic basis. In contrast, morphological characters may be polygenic or lack a genetic basis altogether. With DNA sequence data we also are able to compare homologous sequences in the hosts and parasites, thereby avoiding the problem of comparing non-homologous morphological characters, or allozyme characters of dubious homology. Finally, DNA sequence data are relatively easy to generate and potentially provide a large number of characters for high-resolution analyses.

If we consider the gopher–louse data, the majority of substitutions in the gopher and louse COI sequences are synonymous (silent) substitutions at the third codon position (Hafner *et al.* 1994). In fact, third-position substitutions are so numerous that almost any pairwise comparison will suffer from the effects of multiple substitutions at the same nucleotide position. Unless corrected for, this substitutional saturation will lead to an underestimate of the genetic distance between taxa (or underestimates of branch lengths), which is why multiple methods have been developed to compensate for the effects of saturation (Tajima and Nei 1984). We should also note that if the substitution process differs at different sites along the sequence (for example first, second, and third codon positions), then the utility of a single overall measure of sequence divergence is dubious (Irwin *et al.* 1991).

Although it is widely acknowledged that estimates of DNA sequence divergence should be adjusted for the effects of saturation, there is no general consensus as to how this should be done. For example, Hafner *et al.* (1994) attempted to correct for transition bias in the gopher and louse COI data by using the largest observed pairwise transition bias in a maximum likelihood phylogeny reconstruction. They reasoned that this value, which is usually measured between the most recently diverged taxa, is least likely to be affected by saturation and, therefore, is the most reasonable estimate of the actual transition bias for this gene region. In contrast, Page (in preparation) recommends use of the transition bias estimate that maximizes the likelihood of the phylogeny. Use of these different correction factors can have profound influence on estimates of branch length. For example, Hafner *et al.*'s analysis suggests lice are evolving 10 to 11 times more rapidly than their hosts at selectively neutral sites. In contrast, Page's analysis suggests that lice are evolving only twice as fast as gophers. Research into the effects of transitional saturation (and evolutionary models in general) is now moving at a rapid pace (Goldman 1993; Yang 1994), and we expect that some degree of consensus will be reached in the near future.

Phylogenetic sampling

Another correlate of accuracy of branch length estimation is phylogenetic sampling. The denser the sampling of lineages, the greater the chances of detecting evolutionary change (Langley and Fitch 1974; Moore *et al.* 1976; Fitch and Bruschi 1987; Fitch and Beintema 1990). In the gopher–louse study (Hafner *et al.* 1994), the 17 louse species examined tend to represent single exemplars from larger clades containing a total of 122 recognized species (Page *et al.* 1995). Pocket gophers are also taxonomically diverse (approximately 40 species and 450 subspecies), and relatively few taxa have been examined from a molecular perspective. Ideally, future studies will involve exhaustive sampling of gopher and louse clades so that different lineages within each group can be compared to determine if there are lineage-specific molecular clocks. The DNA data from Hafner *et al.* (1994) suggest that there may be lineage-specific rate differences, although these deviations may result from sampling error (Page, in preparation).

Stochasticity

The DNA sequences analysed by Hafner *et al.* (1994) represent relatively short regions (379 base pairs) of a single mitochondrial gene (COI). As such, extrapolation from these data to the entire COI gene, or to the entire mitochondrial genome, are tenuous. In addition, random events, such as lineage sorting (Avise *et al.* 1984) may have resulted in a mitochondrial genealogy ('gene tree') that is quite different from the nuclear genealogy ('species tree'). Thus, it is important for researchers studying organellar genomes to compare their phylogenies with those based on nuclear-encoded characters (for example nuclear sequences, morphology, or allozymes). So far, the nuclear and mitochondrial phylogenies for gophers and lice are in close agreement (Hafner and Nadler 1988). However, we expect to see increased discordance between mitochondrial and nuclear genealogies as we explore cospeciation on a finer scale (for example within species). For example, Patton and Smith (1994) have shown that the mitochondrial and allozyme trees for pocket gophers of the genus *Thomomys* are incongruent. If chewing lice are transmitted primarily from mother to offspring in *Thomomys* (as are mitochondrial haplotypes), then we predict that the phylogeny of lice from *Thomomys* will be more similar to the host mitochondrial tree than nuclear tree (Nadler *et al.* 1990). We are currently testing this hypothesis.

Because of the relatively small number of nucleotides sampled in the gopher-louse study, maximum likelihood confidence limits on the estimates of branch lengths are quite broad, hampering efforts to compare host and parasite evolution. Although sampling error (both genome sampling and taxon sampling) certainly contributes to this decreased resolution, it is also likely that stochasticity of the substitution process and clade-specific variation in rates of substitution decrease our ability to see clear, assemblage-wide trends. Where general trends are evident (for example the gopher-louse rate difference reported by Hafner *et al.* (1994), they are not particularly strong. Clearly, more sequence data and increased taxonomic sampling are needed to increase our confidence in these preliminary findings.

16.7 Conclusions and perspectives

Gillespie (1991, p. 139) distinguishes between two uses of molecular clocks: as a source of data on times of divergence between lineages (coalescence times), or as tests of hypotheses about molecular evolution. Similarly, we can use measures of molecular divergence to test our reconstructions of host and parasite phylogenies (and to calculate time since divergence of cospeciating clades), or we can endeavour to probe in detail the mechanics of molecular evolution and evolutionary rates in the hosts and parasites. The latter approach has more general appeal because it has the potential to yield findings that transcend the particular host–parasite system studied. For example, discovery of rate correlates or other evolutionary patterns shared between distantly related hosts and parasites (for example mammals and insects in the Hafner *et al.* study) may signal underlying evolutionary processes that have a high degree of universality. In this regard, T. Spradling (in MSH's laboratory) is currently sequencing the COI gene of whipworms (endoparasitic nematodes) that parasitize pocket gophers. If cospeciation is evident in all three symbionts (gophers, lice, and whipworms), this framework can be used to test for rate correlates or other evolutionary patterns that show even greater universality.

Future studies comparing population structure of hosts and their parasites will reveal if the structuring of a parasite population on an individual host (and founder events as new hosts are colonized) tend to accelerate long-term parasite evolution relative to that of their hosts (Nadler *et al.* 1990). To be convincing, such a test will have to demonstrate that short-term population-level phenomena (such as decreased heterozygosity and polymorphism in the parasites) have persistent and long-term phylogenetic consequences. Similarly, studies of the molecular genetics of parasites at zones of hybridization between host taxa can yield important information about the history of the zone (for example Patton *et al.* 1984; Nadler *et al.* 1990) or about modes of parasite transmission (J. Demastes, in preparation). If genetic introgression is present in both groups, then rate and pattern of introgression can be compared to reveal common demographic patterns. In other cases, parasites can be treated as 'genes' of their hosts to serve as an independent measure of extent of host introgression (Bohlin and Zimmerman 1982; Patton *et al.* 1984).

Although, at present, there are few published studies of cospeciation explored from a molecular perspective, we anticipate rapid growth in this research area as molecular tools become more widely available and the advantages of this approach better known. Unfortunately, many host–parasite systems will show little or no evidence of cospeciation (for example Baverstock *et al.* 1985), which will preclude comparative studies of higher-order phenomena, such as evolutionary rates. However, in those systems with appreciable cospeciation, the researcher will have the unparalleled opportunity to compare evolution in the same gene(s), over the same period of time, in distantly related organisms. Within this framework, the potential is great for discovery of large-scale evolutionary patterns that apply to diverse groups of organisms.

Acknowledgements

We thank Paul Harvey, Dale Clayton, James Demastes, Theresa Spradling, and Xuhua Xia for helpful comments on this manuscript. This research was supported in part by a National Science Foundation grant to MSH.

References

Adell, J. C. and Dopazo, J. (1994). Monte Carlo simulation in phylogenies: an application to test the constancy of evolutionary rates. *J. Mol. Evol.*, **38**, 305–9.

Avise, J. C., Neigel, J. E., and Arnold, J. (1984). Demographic influences on mitochondrial DNA lineage survivorship in animal populations. *J. Mol. Evol.*, **20**, 99–105.

Baverstock, P. R., Adams, M., and Beveridge, I. (1985). Biochemical differentiation in bile duct cestodes and their marsupial hosts. *Mol. Biol. Evol.*, **2**, 321–37.

Bohlin, R. G. and Zimmerman, E. G. (1982). Genic differentiation of two chromosomal races of the *Geomys bursarius* complex. *J. Mamm.*, **63**, 218–28.

Brooks, D. R. and McLennan, D. A. (1991). *Phylogeny, ecology, and behavior*. University of Chicago Press.

Brooks, D. R. and McLennan, D. A. (1993). *Parascript: parasites and the language of evolution*. Smithsonian Institution Press, Washington, DC.

Demastes, J. W. (1990). Host–parasite coevolutionary relationships in two assemblages of pocket gophers and chewing lice. M.S. Thesis. Louisiana State University, Baton Rouge.

Demastes, J. W. and Hafner, M. S. (1993). Cospeciation of pocket gophers (*Geomys*) and their chewing lice (*Geomydoecus*). *J. Mamm.*, **74**, 521–30.

Esch, G. W. and Fernández, J. C. (1993). *A functional biology of parasitism: ecological and evolutionary implications*. Chapman and Hall, London.

Felsenstein, J. (1989). PHYLIP-Phylogenetic inference package, version 3.2. *Cladistics*, **5**, 164–6.

Fitch, W. M. and Beintema, J. J. (1990). Correcting parsimomious trees for unseen nucleotide substitutions: the effects of dense branching as exemplified by ribonuclease. *Mol. Biol. Evol.*, **7**, 438–43.

Fitch, W. M. and Brushci, M. (1987). The evolution of prokaryotic ferredoxins—with a general method correcting for unobserved substitutions in less branched lineages. *Mol. Biol. Evol.*, **4**, 381–94.

Gillespie, J. H. (1991). *The causes of molecular evolution*. Oxford University Press, New York.

Goldman, N. (1993). Statistical tests of models of DNA substitution. *J. Mol. Evol.*, **36**, 182–98.

Goodman, M., Czelusniak, J., Moore, G. W., Romero-Herrera, E., and Matsuda, G. (1979). Fitting the gene lineage into its species lineage, a parsimony strategy illustrated by cladograms constructed from globin sequences. *Syst. Zool.*, **28**, 132–63.

Hafner, M. S. (1982). A biochemical investigation of geomyoid systematics (Mammalia: Rodentia). *Z. zool. Syst. Evolut.-Forsch.*, **20**, 118–30.

Hafner, M. S. and Nadler, S. A. (1988). Phylogenetic trees support the coevolution of parasites and their hosts. *Nature*, **332**, 258–9.

Hafner, M. S. and Nadler, S. A. (1990). Cospeciation in host-parasite assemblages comparative analysis of rates of evolution and timing of cospeciation events. *Syst. Zool.*, **39**, 192–204.

Hafner, M. S., Sudman, P. D., Villablanca, F. X., Spradling, T. A., Demastes, J. W. and Nadler, S. A. (1994). Disparate rates of molecular evolution in cospeciating hosts and parasites. *Science*, **265**, 1087–90.

Harvey, P. H. and Keymer, A. E. (1991). Comparing life history using phylogenies. *Phil. Trans. R. Soc.*, **B332**, 31–9.

Harvey, P. H. and Pagel, M. D. (1991). *The comparative method in evolutionary biology*. Oxford University Press.

Hillis, D. M., Allard, M. W., and Miyamoto, M. M. (1993). Analysis of DNA sequence data: phylogenetic inference. *Methods Enzymol.*, **224**, 456–87.

Irwin, D. M., Kocher, T. D., and Wilson, A. C. (1991). Evolution of the cytochrome b gene in mammals. *J. Mol. Evol.*, **32**, 128–44.

Kethley, J. B. and Johnston, D. E. (1975). Resource tracking patterns in bird and mammal ectoparasites. *Misc. Publ. Entomol. Soc. Am.*, **9**, 231–6.

Langley, C. H. and Fitch, W. M. (1974). An examination of the constancy of the rate of molecular evolution. *J. Mol. Evol.*, **3**, 161–77.

Lynch, M. and Jarrell, P. E. (1993). A method for calibrating molecular clocks and its application to animal mitochondrial DNA. *Genetics*, **135**, 1197–208.

Moore, G. W., Goodman, M., Callahan, C., Holmquist, R., and Moise, H. (1976). Stochastic versus augmented maximum parsimony method for estimating superimposed mutations in the divergent evolution of protein sequences. Methods tested on cytochrome c amino acid sequences. *J. Mol. Biol.*, **105**, 15–37.

Muse, S. V. and Weir, B. S. (1992). Testing for equality of evolutionary rates. *Genetics*, **132**, 269–76.

Nadler, S. A. and Hafner, M. S. (1989). Genetic differentiation in sympatric species of chewing lice (Mallophaga: Trichodectidae). *Ann. Entomol. Soc. Am.*, **82**, 109–13.

Nadler, S. A., Hafner, M. S., Hafner, J. C., and Hafner, D. J. (1990). Genetic differentiation among chewing louse populations (Mallophaga: Trichodectidae) in a pocket gopher contact zone (Rodentia: Geomyidae). *Evolution*, **44**, 942–51.

Nelson, G. and Platnick, N. I. (1981). *Systematics and biogeography: cladistics and vicariance*. Columbia University Press, New York.

Noble, E. R., Noble, G. A., Schad, G. A., and MacInnes, A. J. (1989). *Parasitology. The biology of animal parasites*, 6th edn. Lea and Febiger, Philadelphia.

Page, R. D. M. (1990*a*). Component analysis: a valiant failure? *Cladistics*, **6**, 119–36.

Page, R. D. M. (1990*b*). Temporal congruence and cladistic analysis of biogeography and cospeciation. *Syst. Zool.*, **39**, 205–26.

Page, R. D. M. (1993*a*). *COMPONENT 2.0. Tree comparison software for use with Microsoft® Windows™*. The Natural History Museum, London.

Page, R. D. M. (1993*b*). Genes, organisms, and areas: the problem of multiple lineages. *Syst. Biol.*, **42**, 77–84.

Page, R. D. M. (1994). Maps between trees and cladistic analysis of historical associations among genes, organisms, and areas. *Syst. Biol.*, **43**, 58–77.

Page, R. D. M. (1995). Parallel phylogenies: reconstructing the history of host–parasite associations. *Cladistics*, **10**, 155–73.

Page, R. D. M., Price, R. D., and Hellenthal, R. A. (1995). Phylogeny of *Geomydoecus* and *Thomomydoecus* pocket gopher lice (Phthiraptera: Trichodectidae) inferred from cladistic analysis of adult and first-instar morphology. *Syst. Entomol.*, **20**, 129–43.

Paterson, A. M. (1994). Coevolution of seabirds and feather lice: A phylogenetic analysis of cospeciation using behavioural, molecular and morphological characters. Ph.D. Dissertation. University of Otago, Dunedin.

Paterson, A. M., Gray, R. D., and Wallis, G. P. (1993). Parasites, petrels and penguins: Does louse presence reflect seabird phylogeny? *Int. J. Parasitol.*, **23**, 515–26.

Patton, J. L. and Smith, M. F. (1994). Paraphyly, polyphyly, and the nature of species boundaries in pocket gophers (genus *Thomomys*). *Syst. Biol.*, **43**, 11–26.

Patton, J. L., Smith, M. F., Price, R. D., and Hellenthal, R. A. (1984). Genetics of hybridization between the pocket gophers *Thomomys bottae* and *Thomomys townsendii* in northeastern California. *Great Basin Naturalist*, **44**, 431–40.

Reed, D. L. (1994). Possible mechanism for cospeciation in pocket gophers and lice, and evidence of resource partitioning among coexisting genera of lice. M.S. Thesis. Louisiana State University, Baton Rouge.

Ridley, M. (1993). *Evolution.* Blackwell Scientific, Boston, MA.

Russell, R. J. (1968). Evolution and classification of the pocket gophers of the subfamily Geomyinae. *Univ. Kansas Publ. Mus. Nat. Hist.*, **16**, 473–579.

Smith, M. F., Thomas, W. K., and Patton, J. L. (1992). Mitochondrial DNA-like sequence in the nuclear genome of an akodontine rodent. *Mol. Biol. Evol.*, **9**, 204–15.

Swofford, D. L. and Olsen, G. J. (1990). Phylogeny reconstruction. In *Molecular systematics*, (ed. D. M. Hillis and C. Moritz), pp. 411–501. Sinauer, Sunderland, MA.

Tajima, R. and Nei, M. (1984). Estimation of evolutionary distance between nucleotide sequences. *Mol. Biol. Evol.*, **1**, 269–85.

Timm, R. M. (1983). Farenholz's Rule and resource tracking: A study of host-parasite coevolution. In *Coevolution*, (ed. M. H. Nitecki), pp. 225–66. University of Chicago Press.

Yang, Z. (1994). Statistical properties of the maximum likelihood method of phylogenetic estimation and comparison with distance matrix methods. *Syst. Biol.*, **43**, 329–42.

Character evolution

17

A microevolutionary link between phylogenies and comparative data

Emília P. Martins and Thomas F. Hansen

17.1 Introduction

Usually, interspecific or 'comparative' data are expected to be similar to one another because the species from which the data are measured have been evolving together for some period of evolutionary time (for example Felsenstein 1985; Harvey and Pagel 1991). We term this similarity due to common ancestry the 'phylogenetic correlation', referring also to an expected relationship between interspecific trait variation and the phylogeny along which the measured species evolved. In this chapter we discuss how the extent and pattern of phylogenetic correlation depends not only on the phylogeny of the species but also on the microevolutionary processes underlying genetic and phenotypic change. As microevolution can differ among characters or among clades of organisms, explicit determination of the microevolutionary processes underlying the assumptions and statistical models of any phylogenetic method can be one of the most difficult, yet important, aspects of conducting interspecific analyses. An awareness of these underlying processes can also provide us with an exciting new set of opportunities for the use of phylogenies in inferring microevolutionary process from comparative data.

Traditionally, systematists have used phylogenetic correlation to infer the branching patterns of speciation (i.e. the phylogeny) underlying extant organisms (see Felsenstein 1988*a*; Swofford and Olsen 1990 for reviews). Several phenotypic traits are chosen which fit the assumptions of some numerical algorithm (for example, traits that are believed to have evolved neutrally or parsimoniously), and the relationships between interspecific measurements of these traits are used to infer the historical relationships between the measured species. Throughout this procedure, the emphasis is on the evolution of species (or other taxonomic group).

In contrast, the 'comparative method' is a family of techniques in which interspecific measurements are used to infer something about the biology of particular traits (for reviews see Harvey and Pagel 1991; Miles and Dunham 1993; Maddison 1994; Martins and Hansen 1996). One branch of comparative methods views phylogenetic correlation primarily as a statistical nuisance which may obscure our ability to detect adaptation. Independence of data points is one of the primary assumptions of most parametric statistical procedures, and when ordinary statistics are used to analyse phylogenetically correlated data, this assumption is regularly violated. From this perspective, several methods have been proposed to partition variation in a set of interspecific data into a phylogenetic component (describing the relationship between trait variation and the phylogeny) and a current adaptation component (describing the response to natural selection in recent times). These methods strive to 'correct' for the problem of phylogenetic correlation by filtering it out of a set of comparative data using implicit or explicit models of phenotypic evolution and a phylogeny. Unlike phylogeny reconstruction, in the statistical correction of comparative data the emphasis is on the organismal traits rather than on the species which exhibit those traits. Traits are chosen because of some particular hypothesis rather than because they fit a set of predefined assumptions, and generally, the phylogeny and models of phenotypic evolution are assumed to be known.

Another branch of comparative methods views adaptation as an integral component of long-term evolutionary history. Natural selection, random genetic drift, and other forces work together throughout evolutionary time, rather than simply at the end points of a phylogeny. Thus, adaptation is one of the many factors working together to create phylogenetic correlation in a set of comparative data. The comparative methods developed from this perspective use the expected and observed patterns of phylogenetic correlation to infer the microevolutionary processes underlying phenotypic change. To do this, we specify a particular model of phenotypic evolution (for example random genetic drift, stabilizing selection with random genetic drift) and then estimate specific parameters from that model (for example the strength of selection) using the interspecific data and a known phylogeny. Using this general approach, we can now infer the ancestral states of phenotypes, the strength and type of microevolutionary forces acting on characters, the relationships between evolutionary changes in two or more traits, and the degree of phylogenetic 'inertia' in a character.

Thus, systematists and comparative biologists use measurements of extant species to infer the historical relationships among species (i.e. reconstruct phylogenies), to correct statistical problems in the analysis of comparative data, and to infer the detailed evolutionary history of particular characters. Each of these situations requires the use of three entities: (1) the phenotypes or genotypes of extant species, (2) a phylogeny describing the evolutionary history of the species that were measured, and (3) the microevolutionary processes underlying character evolution along the phylogeny and leading to the species phenotypes observed today. Given data or assumptions regarding any two of these three entities, we can infer some of the properties of the third. For example, systematists regularly use measurements of the phenotypes and/or genotypes of extant species (#1) and assumptions regarding the underlying microevolutionary processes (#3) to infer the phylogeny of a group of species

(#2). Comparative biologists, on the other hand, use sets of comparative data (#1) and a phylogeny (#2) to infer something about the underlying evolutionary process (#3), including the relationship between evolutionary changes in two traits.

In this chapter we discuss how a microevolutionary perspective can be used to link these three entities, and thereby clarify the issues underlying all three. Using the general framework proposed in Hansen and Martins (1996), we discuss what can and cannot be inferred from comparative data and a phylogeny, and how that framework can be used to evaluate and compare proposed methods. In many cases, statistical techniques to estimate the desired parameters have not yet been developed, and/or phylogenetic information may not be adequate to answer the questions. At the risk of being overly optimistic, this chapter strives only to discuss what *may* be possible given appropriate information.

17.2 Phylogenetic correlation

Phylogenetic correlation refers to the similarity between species expected due to their shared evolutionary histories. Here we are concerned primarily with the predictive information we can obtain for one species through measures of other species in the same clade. Hence, we define 'phylogenetic correlation' as the expected similarity between measurements of different taxa due to the fact that those taxa once evolved together as a single common ancestor. These taxa can be extant species, hypothetical common ancestors of groups of species, or extinct taxa represented in the fossil record. Thus, phylogenetic correlation also describes the expected relationship between measurements of an ancestral taxon and its descendants or between two ancestral taxa on a single phylogeny. In more general terms, phylogenetic correlation refers to the relationship between interspecific trait variation and the phylogeny along which the species evolved. Comparative data may also exhibit some dependence between species phenotypes (i.e. relationships between trait variation and the phylogeny) due to convergent evolution in similar environments or character displacement in different environments. Following Hansen and Martins (1996), when these relationships are not mediated through a common ancestor, we do not consider them to be part of the phylogenetic correlation.

Evolutionary theory predicts that all species share at least one common ancestor. Any set of comparative data will thus have some phylogenetic correlation due to the species' shared evolution as the common ancestor of the entire clade. In most cases, this ancient phylogenetic correlation is not relevant to the questions of interest, and we restrict our discussion of phylogenetic correlation to the pairwise relationships among interspecific measures given the phenotype of their single common ancestor at the root of the phylogeny. Thus, for example, data measured from a group of species that have exhibited a perfect 'star' radiation, in which all species radiated essentially instantaneously from a single common ancestor, are *not* phylogenetically correlated, as we have defined it here.

Although phylogenetic correlation (as we have defined it) arises only through the similarities between ancestors and their descendants, this similarity can have many sources. Usually, the similarity to the ancestral state dissipates over time as mutations appear and species phenotypes respond to natural selection. Therefore we expect species that are more distantly related to be less correlated with one another. Naively, we would also expect traits that change rapidly on a macroevolutionary time-scale to show less phylogenetic correlation. This is true if rapidly evolving traits are somewhat constrained in the possible states they can attain. If not, rapid evolution generates more variation overall so that there is more variation available to be attributed to common ancestry (phylogenetic correlation) as well as to more recent evolutionary events. If phenotypic evolution is totally unconstrained, as in some of the neutral drift models we shall consider, there is no relationship between evolutionary rate and phylogenetic correlation. Many behavioural, morphological, and life-history traits show high levels of within-species genetic variation and are capable of rapid, fluent adaptation to new conditions. Such traits would tend to be much more influenced by very recent selective history than ancestral condition and are therefore not expected to show much phylogenetic correlation directly. They may, however, be phylogenetically correlated due to selective or developmental constraints that are themselves inherited from common ancestors. Selective constraints may be inherited when descendants seek out and live in niches and environments similar to those of their ancestors and developmental constraints may be inherited through deeply integrated developmental pathways that bias the variation available to selection on the trait of interest.

Similarity can be measured in a number of different ways. We can use the pair-wise relationships between species to describe the phylogenetic correlations summarized by a set of interspecific data. For any particular model of the microevolutionary process, covariances or correlations can be used to describe the expected relationship between measures of each pair of species. The set of all possible relationships can be described as an $N \times N$ matrix, where N is the number of species in the clade. We will, as in Martins (1995), refer to this matrix of phylogenetic covariances as the expected relationship matrix (**ERM**). In an **ERM**, the diagonal elements describe the expected variance of each species phenotype, whereas the off-diagonal elements are the expected covariances between each pair of species phenotypes.

The above rough classification of the causes of phylogenetic correlation is not sufficiently precise to be useful in making specific predictions or describing particular patterns. To do this, we need explicit formulations of the mechanisms of microevolutionary change, and a detailed description of how those mechanisms lead to the macroevolutionary patterns observed in a set of comparative data. These are described in detail in Hansen and Martins (1996), and more generally below.

17.3 A microevolutionary framework

During each generation, forces such as selection, mutation, random genetic drift, inheritance, and environmental fluctuations combine to produce evolutionary changes in the mean phenotype of a population. Because the microevolutionary forces acting in each generation have both deterministic and random elements, we can model the mean phenotype of a continuous character or the species-typical state of a categorical character as a random variable, X, and the series of phenotypic states existing through time, t, as a stochastic process, $X(t)$. If we assume that within-species variability and the general form of the microevolutionary changes in phenotype are constants, then the phenotype at any generation, $X(t)$, is a function only of the phenotype during the previous generation, $X(t-1)$.

We can use a general model of this type to describe the evolutionary process that unfolds along the branches of a phylogenetic tree, eventually resulting in a set of comparative data. The process begins with a single ancestor, and ends in the phenotypes of each of the extant species at the tips of the phylogeny. We consider evolution along each branch of the phylogeny as a separate stochastic process with these separate processes joining at nodal points of the phylogeny with the final value of one process being the initial value of the following process. Because our aim is to study the phylogenetic source of correlation, we disregard all sources of correlation not mediated through common ancestry by assuming that the character evolves independently along each of the descendant branches resulting from a split. Thus, the evolutionary process starts at the root of the tree with some initial value $X_0(t_0)$, evolves along the stem according to the process $X_0(t)$ and reaches some value $X_0(t_1)$ at the first branch point. There, two new independent processes $X_1(t)$ and $X_2(t)$ are started along the descendant branches 1 and 2, both with initial condition $X_0(t_1)$. The process develops along the tree in this way until the end points, the extant species, are reached. Comparative or interspecific data are a snapshot of this process at a single point in time. The species phenotypes will be phylogenetically correlated with one another due to their shared evolutionary history along the phylogeny.

Because the stretches of time between nodes on a phylogeny usually encompass thousands or even millions of generations, we will limit our discussion to continuous time models of the evolutionary process. Hence, the general framework of comparative evolution proposed in Hansen and Martins (1996) views comparative data as the result of a continuous time stochastic process, $X(t)$, describing the phenotypes of species as they evolve along a phylogenetic tree. The framework unites a set of comparative data, the phylogeny, and the underlying microevolutionary processes.

In Hansen and Martins (1996), we used this framework to derive statements about the statistical properties of a set of comparative data given a particular phylogeny and model of the microevolutionary process. Our primary goal was to determine how the pattern of phylogenetic correlation present in a set of comparative data depends on the microevolutionary processes underlying phenotypic change. In very general terms, we showed that under the framework described above, the phylogenetic covariance or relationship between the phenotypes of two species is equal to the covariance between the expected phenotypes of the two species given the phenotype of their most recent common ancestor. More formally, the covariance between species phenotypes equals the covariance between the regression of each species on the common ancestor ($\text{Cov}[X_i, X_j] = \text{Cov}[E(X_i|X_a), E(X_j|X_a)]$, where X_i and X_j are the two species phenotypes and X_a is the phenotype of the common ancestor). Because these regressions can usually be computed directly from the parameters of a specified microevolutionary model (for example random genetic drift alone, weak stabilizing selection with random genetic drift), this result reduces the problem of calculating the phylogenetic correlation across the entire phylogeny to simpler calculations involving the microevolutionary process along each branch of the phylogeny. The expected phylogenetic correlation between each pair of species can thus be computed with relatively little complication, and the resulting matrix of phylogenetic correlations (the **ERM** above) can be used to develop new phylogeny reconstruction and comparative methods, or to explore existing methods from a microevolutionary perspective.

17.4 The pattern of phylogenetic correlation expected under different microevolutionary processes

Different patterns of phylogenetic correlation are expected for different microevolutionary models. Phenotypic evolution under stabilizing selection can lead to comparative data with very different patterns of phylogenetic correlation than will phenotypic evolution under random genetic drift. Felsenstein (1988*b*) and Hansen and Martins (1996) have discussed the expected correlation pattern from a range of simple microevolutionary models reflecting many of the processes known to constitute evolution. We review a few of these below.

For continuous traits, the most common explicit microevolutionary assumption made by comparative and phylogenetic methods is that of evolution by random genetic drift alone or by random genetic drift and mutation together. The most common way of modelling phenotypic evolution of this sort is to use Brownian motion, a mathematical stochastic process in which evolutionary changes in a character are (1) independent of the previous state of the character, (2) normally distributed with zero mean (such that overall there is no net change in the phenotype), and (3) have a variance proportional to the length of time over which the change takes place (such that the divergence between two species or populations increases at a constant rate with time). Brownian motion was originally proposed by Edwards and Cavalli-Sforza (1964) as a model of neutral phenotypic evolution and has been used by many population and quantitative geneticists thereafter (for reviews see Felsenstein 1988*a*, *b*; Lynch 1993).

Under a Brownian motion model, the change in a species phenotype during t generations is normally distributed with zero mean and variance equal to $\sigma^2 t$, where σ^2 is the variance of evolutionary changes occurring at each generation (assumed to be constant over the entire phylogeny). In models of neutral evolution via random genetic drift, σ^2 is usually interpreted as the genetic variance in the trait. In models which include the effects of mutation as well as random genetic drift, σ^2 is interpreted as half the mutational variance arising each generation. Using the framework proposed in Hansen and Martins (1996), for a Brownian motion, the regression of a species' phenotype on the state of its ancestor is equal to the ancestral phenotype, $E[X_i|X_a]=X_a$. Thus, the covariance between any two species phenotypes reduces to the variance of the phenotype of their most recent common ancestor ($\text{Cov}[X_i, X_j]=\text{Cov}(E[X_i|X_a],E[X_j|X_a])=\text{Cov}[X_a, X_a]=\text{Var}[X_a]$). The variance of the ancestral phenotype is $\sigma^2 t_a$, where t_a is the time from the ancestor to the root of the tree (i.e. the time the two species evolved together as one). Thus, the **ERM** for a set of phenotypes that evolved via a Brownian motion model of phenotypic evolution consists of a matrix in which each entry is of the form $\sigma^2 t_a$, where t_a is the time between the common ancestor of each pair of species and the root of the tree. Thus, the phylogenetic correlation (described by the **ERM**) decreases linearly with the time that the species have been separated.

Genetic drift and mutation are unlikely to be the sole mechanisms for the evolution of continuous characters, both because they are weak forces that can be modified by even extremely weak selection and because empirically measured rates of evolution and levels of interspecific variation are often much lower than what is expected from drift alone (for example Lynch 1990; Cheetham *et al.* 1994). Still, Brownian motion can also be used to model several other evolutionary scenarios. For example, Brownian motion has been used to describe phenotypic evolution under directional selection, when the direction of selection fluctuates up and down at random (Felsenstein 1988*b*). Similarly, we can use Brownian motion to describe phenotypic evolution under strong stabilizing selection in which the species phenotypes are closely tracking optima that fluctuate as if by Brownian motion. (Brownian motion is not a particularly good model of weak stabilizing selection when the species phenotypes do not track the optimal phenotype closely.) More surprisingly, a large class of models developed to describe phenotypic evolution occurring in discrete, punctuated bursts yield the same patterns of phylogenetic correlation (but not usually the normal distribution) expected from Brownian motion (Hansen and Martins 1996). Thus, phylogenetic reconstruction or comparative methods based on the pattern of phylogenetic correlation produced via Brownian motion (for example Felsenstein 1973, 1985) have some robustness to different microevolutionary scenarios. However, the interpretation of results from these methods depends critically on the details of the microevolutionary process being considered. We discuss these issues in technical detail in Hansen and Martins (1996).

Other microevolutionary processes produce different expected patterns of phylogenetic correlation. For example, the correlation between species evolving under weak stabilizing selection as well as random genetic drift (for example Lande 1976, 1979) decays exponentially with increasing time as opposed to the linear decay expected from Brownian motion. This means that species far apart in the phylogeny will be relatively much less similar if stabilizing selection is present than if drift or directional selection are the sole microevolutionary forces. Selection acting to keep the phenotype near an optimum essentially erases the historical record stored in the phylogenetic correlation. Similar effects can be expected from any microevolutionary process which includes a constraint on phenotypic divergence, as well as from certain models of neutral evolution (for example the mutation-drift model of Cockerham and Tachida (1987)).

Although categorical characters are usually treated in terms of their probabilities of being in different states rather than in terms of the correlations between species phenotypes, we can still think of the dependency between species phenotypes (i.e. phylogenetic correlation) as being mediated through an expected similarity between the species and their most recent common ancestor. As with continuous characters, different types of microevolutionary processes will lead to different patterns of phylogenetic dependences. For example, a common model of the evolution of a binary (0/1, yes/no) characters is that there is a constant probability of changing from state A to state B over each interval of time, and a possibly different probability of reverting from state B to state A. Under such a model, phylogenetic correlation is lost exponentially with increasing time between species in much the same way as under stabilizing selection. Unlike Brownian motion, in both these models, characters are constrained from diverging freely.

In the next sections we consider some methods of phylogeny reconstruction and comparison in the framework just described, and discuss how this framework can be used to understand the microevolutionary assumptions underlying each method, and to develop new methods with different assumptions.

17.5 Phylogeny reconstruction

Imagine that we have actually measured six species in an existing clade, and that we would like to infer the phylogeny underlying the evolution of their phenotypes. We can estimate the phylogenetic correlation among species empirically, for example by calculating the phenetic distances between all species in the clade. Given a known model of the microevolutionary process we can also calculate the **ERM** predicted for any phylogeny containing these six species. If we calculate the **ERM** predicted for many different phylogenies, then we might define the 'best' phylogeny as one which gives us the closest match between empirical measurements and theoretical expectations (**ERM**).

In very broad terms, most phylogeny reconstruction methods can be represented as a similar comparison of observed patterns of phylogenetic correlation and the **ERM**. Different methods use different criteria for determining whether the observed and expected patterns are similar. For example, the maximum likelihood approach proposed by Felsenstein (1973) begins by assuming that the characters evolve as if by a Brownian motion model of phenotypic evolution. Given a particular phylogeny, we can describe the expected pattern of phylogenetic correlation for a trait undergoing Brownian motion evolution along that phylogeny as an **ERM**. As mentioned above, the elements of this **ERM** take the form $\sigma^2[t_a]$ where t_a is the time that each pair of species evolved together from the root of the tree to the most recent common ancestor of the pair. We can compare this **ERM** with the observed data, and find the probability of the observed species data given the **ERM**. Felsenstein uses this probability to find an estimate of the best phylogeny using maximum likelihood techniques. The maximum likelihood phylogeny is the tree that gives the highest probability of the observed data. Despite formidable computational and conceptual difficulties, it is often possible to find that phylogeny by trying out several possible phylogenies and calculating the probability of the measured data given each one.

Consider now what would happen if the microevolutionary model used to develop a maximum likelihood phylogeny reconstruction algorithm had been one incorporating the effects of weak stabilizing selection in addition to random genetic drift. As described above, the pattern of phylogenetic correlation as described by the **ERM** looks quite different in this situation. Instead of containing elements that are directly proportional to t_a, the **ERM** now consists of elements $(\sigma^2/2\alpha)\exp(-\alpha t_{ij})$, where t_{ij} is the time separating the two species and α refers to the magnitude of a restraining force pushing the phenotype towards an optimum. The phylogenetic correlation between any pair of species will depend on the magnitude of the restraining force (α). Thus, the likelihood of a phylogeny for a given set of data under this model also depends on the restraining force (α) and will be different from the likelihood under Felsenstein's Brownian motion model. Because we need to estimate an extra parameter, the computational difficulties are hugely magnified.

Parsimony methods implicitly assume a microevolutionary model in which evolutionary changes are rare. Parsimony approaches choose the phylogeny that minimizes the number of evolutionary changes that must have occurred. Although an **ERM** is always implicit in these approaches, we can work backward within a microevolutionary framework to determine the details of the **ERM** underlying the assumption of parsimony. Felsenstein (1981) showed that the basic parsimony assumption is equivalent to maximum likelihood based on a microevolutionary process in which the rate of phenotypic evolution is constant, very small, equal, and independent for all characters. Weighted parsimony and compatibility methods assume a process in which the rates of evolution can be somewhat higher, and need not be the same for all characters.

We have not discussed the many practical difficulties involved in reconstructing phylogenies. For example, there are 945 possible phylogenies for six species and the number of possible trees increases rapidly with the number of species

(Felsenstein 1978). Therefore, it is not usually possible to try all of the possible phylogenies, and numerous algorithms for searching among possible phylogenies have been developed. It is also not clear which measure of 'match' between observed data and **ERM** is the most appropriate. Computer simulation studies suggest that most of the existing approaches give reasonable answers (for example Kuhner and Felsenstein 1994), and still other methods may be possible. Finally, and probably most importantly, there is no agreement in systematics about whether a single model is sufficient in most situations and, if such a model exists, which one it might be. The Brownian motion model of molecular evolution used by Felsenstein (1973, 1981) in his maximum likelihood method, the parsimony model of phenotypic evolution developed by Farris (1970) and other authors, and the 'distance method' models all have proponents and detractors (for reviews see Felsenstein 1988*a*; Swofford and Olson 1990). Our general framework allows for the comparison of different methods of phylogeny reconstruction in similar ways. Given that we try to reconstruct phylogenetic information using vastly different types of data (molecular and morphological, functional and non-functional, slowly evolving and rapidly evolving, categorical and continuous, variable and non-variable), that the selective, genetic, and demographic parameters of the species involved may differ tremendously, it may be more reasonable to consider different phylogenetic reconstruction methods for different situations. By understanding the microevolutionary assumptions underlying the various methods, we can make informed choices among them.

17.6 'Correcting' for phylogeny

Most standard parametric statistics (for example regression, ANOVA) make three assumptions when estimating parameters and conducting hypothesis tests. These are that the error terms in general linear models underlying the statistical procedure: (1) are statistically independent of one another, (2) have equal variances, and (3) are normally distributed. From an evolutionary perspective, these assumptions can be translated into assumptions regarding the observed species data, the underlying phylogeny and the nature of the evolutionary process. If the comparative data are statistically independent, they will be uncorrelated with each other. An **ERM** describing this situation would have off-diagonal elements equal to zero. Similarly, an **ERM** describing the situation in which the error term is homoscedastic would have all diagonal elements equal to one another. The two assumptions together are equivalent to requiring that the observed species data have a variance-covariance matrix of the form $\sigma^2 \boldsymbol{I}$, where σ^2 is the variance of the species data and $\boldsymbol{I}$ is the identity matrix (with 1s along the diagonal, and 0s elsewhere). In evolutionary terms, only comparative data measured from species related to one another by star phylogenies, or with characters evolving under certain microevolutionary models (for example strong stabilizing selection) will exhibit patterns of phylogenetic correlation of this type. The final requirement is that the error terms (and thus, usually, the measured data) be normally distributed. Again, only certain microevolutionary processes will lead to normally distributed species phenotypes with patterns of phylogenetic correlation of the above type. These assumptions are clearly not fulfilled for many comparative datasets.

Different comparative methods propose different ways of ensuring that these assumptions are met. Most can be viewed as a transformation of the raw species data into statistics which meet these assumptions. Many comparative methods use a known phylogeny and an assumed model of phenotypic change to transform the observed species data and measures of their phylogenetic correlation into the form described above. One unavoidable problem with any of these methods is that the general pattern of expected phylogenetic correlations (the **ERM**) must be known before the methods can be applied. We can only know the **ERM** if we also have some information about the phylogeny and the microevolutionary processes underlying phenotypic evolution along that phylogeny. Thus, all comparative methods make microevolutionary assumptions of one sort or another. They simply differ in the quantity and quality of those assumptions, and in whether those assumptions are made implicitly or explicitly.

For example, Felsenstein's (1985) method uses an assumption that the characters have been evolving as if by Brownian motion to generate his 'standardized, independent contrasts'. Felsenstein's procedure might be described as using a model $y' = \boldsymbol{C}y$ to transform raw species data (y) into a set of contrasts (y') which have 0 covariance and the same variance by using the information contained in the **ERM** for that phylogeny and the Brownian motion model of change (combined in a complex way to form the transformation matrix, C). Under Brownian motion, the species mean phenotypes will be normally distributed. Thus, if the Brownian motion model is appropriate for the measured data, Felsenstein's contrasts are guaranteed to fit the three primary assumptions of most parametric statistics.

The result of a Felsensteinian analysis can also be directly interpreted in evolutionary terms. For example, a correlation coefficient describing the relationship between Felsenstein contrasts in two characters is an estimate of the relationship between evolutionary changes in the two traits averaged across the entire phylogeny. When the underlying Brownian motion model is interpreted more narrowly as describing the evolution of two traits due only to random genetic drift, the estimated correlation coefficient can also be used as an estimate of the genetic correlation between the two traits. Alternatively, if the Brownian motion model is interpreted instead as a description of phenotypic evolution under strong stabilizing selection around a fluctuating optimum, the correlation between Felsenstein contrasts is an estimate of the correlation between evolutionary changes in the optima of the two traits. Thus, a precisely formulated microevolutionary model allows for a much more explicit evolutionary interpretation of the results.

A correlation between Felsenstein contrasts is *not* an estimate of the relationship between the two traits which is free of phylogenetic correlation. Phylogenetic correlation is explicitly built into the Brownian motion model underlying Felsenstein's (1985) method. Other methods were designed to find and remove the phylogenetic correlation in a set of comparative data before calculating further statistics. For example, the spatial autoregressive methods proposed by Cheverud *et al.* (1985) and Gittleman and Kot (1990) partition variation in the comparative data into a phylogenetic component ($\rho \boldsymbol{Wy}$, where ρ is an estimated autocorrelation coefficient, $\boldsymbol{W}$ is a symmetric matrix that describes the phylogenetic relationship or similarity between each pair of species, and $\boldsymbol{y}$ is a vector of the measured data) and a non-phylogenetic residual component ($\varepsilon = \boldsymbol{y} - \rho \boldsymbol{Wy}$). These residual components can then be used in further statistical analyses. The spatial autoregressive method can be seen as a way of transforming the raw data ($\boldsymbol{y}$) into statistics ($\boldsymbol{y}' = \varepsilon = \boldsymbol{y} - \rho \boldsymbol{Wy}$) which are free of phylogenetic correlations. The use of statistical parameters such as ρ gives the method some flexibility to find the best statistical fit to the observed data.

If we consider the spatial autoregressive method from the evolutionary framework described above, we immediately realize that there is an expected pattern of phylogenetic correlation (an **ERM**) underlying the transformation of the raw comparative data into the desired form, as there is for all statistical models of comparative data. To find this **ERM**, we start by assuming that the transformation works and that the residual component of the model (ε) is normally distributed with variance–covariance matrix, $\sigma^2 \boldsymbol{I}$, as assumed by standard statistical procedures. Since the residual components are a complex function of the observed data, we can do some algebra and show that this assumption will only be true if the data are normally distributed with variance–covariance matrix equal to $\sigma^2(\boldsymbol{I} - \rho \boldsymbol{W})^{-1}(\boldsymbol{I} - \rho \boldsymbol{W})^{-1}$, a fairly complex function of the $\boldsymbol{W}$ matrix. For any microevolutionary model with corresponding **ERM** we can choose the $\boldsymbol{W}$ matrix (through the formula $\boldsymbol{W} = \rho^{-1}(\boldsymbol{I} - \mathbf{ERM}^{-1/2})$ such that the method finds and removes the phylogenetic correlation. However, the simple $\boldsymbol{W}$ matrices recommended by Cheverud *et al.* (1985) and Gittleman and Kot (1990) are not compatible with any of the microevolutionary processes considered in Hansen and Martins (1996) (or any other microevolutionary models of which we are aware).

17.7 Inferring evolutionary parameters

With the availability of so many approaches to the problem of incorporating phylogenetic information into comparative analyses, researchers have also begun to explore a number of other interesting evolutionary questions that can be answered by combining comparative data with phylogenies. For example, comparative analyses can be used to estimate the degree of 'phylogenetic effect' in a character, the magnitude of the relationship between evolutionary changes in two characters, or the rate of phenotypic evolution of different characters. They can be used to infer the ancestral states of characters, and to test whether selection has been acting on a character or group of organisms, and to estimate the strength of that selection. Although statistical methods to conduct these analyses have not all been fully developed, new methods are being proposed every year, and we concentrate on discussion of which techniques are theoretically possible, rather than which have already been implemented.

Answering the above questions requires reference to an explicit microevolutionary framework and a known phylogeny. The general microevolutionary process outlined above (and described at length in Hansen and Martins (1996)) provides that framework. In essence, given a known phylogeny and a model of the phenotypic evolutionary process, we can find the expected phylogenetic correlation structure (**ERM**) which corresponds to those assumptions. Given an explicit ERM, we can compare the underlying model of the microevolutionary process to the actual patterns observed in a set of comparative data, and use the relationships between these to answer the above questions.

For example, as pointed out originally by Cheverud *et al.* (1985), a measure of the statistical fit of a phylogenetic transformation model to a set of interspecific data can be used as a reasonable estimate of the degree of phylogenetic 'effect' or 'inertia' inherent in a character. Similarly, Lynch (1991) developed an estimator of 'phylogenetic heritability' which is analogous to the concept of heritability for family trees in quantitative genetics. With any method based on standard regression techniques (e.g., Cheverud et al. 1985, Grafen 1989, Lynch 1991), a coefficient of determination (r^2) can be used as a reasonable estimate of the fit of the model to the data. This statistic summarizes the correspondence between the patterns found in the measured comparative data and the statistical or evolutionary model, and can thus serve as a reasonable estimate of the 'phylogenetic effect' of a character. In theory, other measures of similarity can be developed for other methods such as Felsenstein's maximum likelihood approaches and parsimony.

Another important question is whether a character has been evolving purely by random genetic drift or whether selection has played an important role. Martins (1994) developed a generalized least squares (GLS) procedure to estimate the rate of phenotypic evolution which can also be used as a test of whether the Brownian motion model of phenotypic evolution via random genetic drift provides an adequate fit to a set of comparative data, or whether a model of weak stabilizing selection would be more appropriate. This method consists of calculating the divergence between pairs of species on a phylogeny and relating this divergence to the time that the species have been evolving independently of one another. Under a Brownian motion model of evolution, the relationship between divergence and time is expected to be linear with pairs of species that have been separated for long periods of time also exhibiting greater phenotypic differentiation (this is the **ERM**). Under a model of stabilizing selection, the relationship between divergence and time is expected to be exponential rather than linear. Given a known phylogeny, either or both models can be fitted to a set of comparative data, and the fit of the models to the data can be assessed using GLS regression procedures. The relative appropriateness of the two competing models can also be compared using likelihood, and estimates of the internal parameters of the model (for example the rate of phenotypic evolution, the strength of stabilizing selection) can be obtained. Again, in theory, similar methods could be developed to compare the fit of any microevolutionary model for which an **ERM** can be defined.

An **ERM** can also be used to develop estimators for ancestral states of a character. Such ancestral states are needed in a number of comparative methods, yet their estimation theory is largely nonexistent (but see Maddison 1991). Given the phylogenetic correlation structure described by an **ERM**, standard GLS methods can be used to find optimal estimators and their standard errors.

17.8 Discussion

Evolutionary biology is by nature an inferential science. We do empirical studies to monitor current evolutionary forces in the field, and then infer that the same forces were prevalent in earlier times. Artificial selection experiments show us how microevolutionary forces can act on existing phenotypes, and suggest what the result would be if those forces continued over geological time. Even paleontologists are often forced to infer the former existence and phenotypes of species from the tiniest traces of evidence in the fossil record. Rarely does evolution occur sufficiently quickly or within our view, so that we can observe and measure it.

One of the few types of information that we have about how evolution actually occurred can be measured as the phenotypic and genetic diversity of existing species. By assuming that extant species are the result of long-term evolution along phylogenetic trees, with shared ancestors reaching back to the very beginning of life, we can work backward and infer the history of those species and their phenotypes. We can observe existing species to measure their phenotypes, their properties of inheritance, and the types of environment in which they live. Using this information and some assumptions about the type and magnitude of microevolutionary processes (for example selection, drift) that were active in the past, we can develop hypotheses about the patterns of species diversification and division that led to the patterns of phenotypic and genetic similarity observed in extant species. Similarly, if we assume that certain microevolutionary processes were acting, we can work backward and infer the ancestral states of particular characters, how quickly they evolved, and whether or not evolutionary changes in those characters have been subjected to various types of constraints through evolutionary time. Although there are many problems that still need to be addressed regarding inferences made inappropriately from methods that have already been developed (for example Leroi *et al.* 1994) and although some authors (for example Harvey and Pagel 1991) may despair at ever having sufficiently good phylogenies or methods that are sufficiently free of assumptions, the phylogenetic comparative method remains one of the most powerful techniques in modern evolutionary biology.

In this chapter we have shown how phylogeny reconstruction, removal of phylogenetic effects and estimation of microevolutionary parameters from comparative data all centre around assumptions regarding the type of phylogenetic correlation found in the data. We reviewed a previous paper (Hansen and Martins 1995) in which we show how this phylogenetic correlation may take many forms, each of which depends strongly on the mode of evolution of the characters in question. This earlier study also shows how the phylogenetic correlation of a set of comparative data expected for a particular microevolutionary model can be computed and thus how the microevolutionary model, the phylogeny, and the comparative data are interrelated. Most of all, we hope that we have helped to eliminate the common belief that certain comparative procedures and methods are free of evolutionary assumptions. All methods of phylogeny reconstruction and comparative analysis make some microevolutionary assumptions. Thus, use of explicit assumptions should be considered a strength rather than a weakness of different methods. Furthermore, a microevolutionary perspective gives us a broader perspective on evolutionary pattern and process.

Acknowledgements

We would like to thank Paul Harvey and Joe Felsenstein for the many discussions which led to the ideas in this paper. This work was supported by a grant from the National Science Foundation (#DEB-9406964) to EPM.

References

Cheetham, A. H., Jackson, J. B. C., and Hayek, L.-A. C. (1994). Quantitative genetics of Bryozoan phenotypic evolution. II. Analysis of selection and random change in fossil species using reconstructed genetic parameters. *Evolution*, **48**, 360–75.

Cheverud, J. M., Dow, M. M., and Leutenegger, W. (1985). The quantitative assessment of phylogenetic constraints in comparative analyses: sexual dimorphism in body weights among primates. *Evolution*, **39**, 1335–51.

Cockerham, C. C. and Tachida, H. (1987). Evolution and maintenance of quantitative genetic variation by mutation. *Proc. Natl. Acad. Sci. USA*, **84**, 6205–9.

Edwards, A. F. W. and Cavalli-Sforza, L. L. (1964). Reconstruction of evolutionary trees. In *Phenetic and phylogenetic classification* (ed. W. H. Heywood and J. McNeill), pp. 67–76. London. Systematics Association Publication Number 6.

Farris, J. S. (1970). Methods for computing Wagner trees. *Syst. Zool.*, **19**, 83–92.

Felsenstein, J. (1973). Maximum-likelihood estimation of evolutionary trees from continuous characters. *Am. J Hum. Genet.*, **25**, 471–2.

Felsenstein, J. (1978). The number of evolutionary trees. *Syst. Zool.* **27**, 27–33.

Felsenstein, J. (1981). Evolutionary trees from gene frequencies and quantitative characters: finding maximum likelihood estimates. *Evolution*, **35**, 1229–42.

Felsenstein, J. (1985). Phylogenies and the comparative method. *Am. Nat.*, **125**, 1–15.

Felsenstein, J. (1988*a*). Phylogenies from molecular sequences: Inferences and reliability. *Ann. Rev. Genet.*, **22**, 521–65.

Felsenstein, J. (1988*b*). Phylogenies and quantitative characters. *Ann. Rev. Ecol. Syst.*, **19**, 445–71.

Gittleman, J. L. and Kot, M. (1990). Adaptation: statistics and a null model for estimating phylogenetic effects. *Syst. Zool.*, **39**, 227–41.

Grafen, A. (1989). The phylogenetic regression. *Phil. Trans. R. Soc.*, **B326**, 119–57.

Hansen, T. F. and Martins, E. P. (1996). Translating between microevolutionary process and macroevolutionary patterns: a general model of the correlation structure of interspecific data. *Evolution*. (In press.)

Harvey, P. H. and Pagel, M. D. (1991). *The comparative method in evolutionary biology*. Oxford University Press.

Kuhner, M. K. and Felsenstein, J. (1994). A simulation comparison of phylogeny algorithms under equal and unequal evolutionary rates. *Mol. Bio. Evol.*, **11**, 459–68.

Lande, R. (1976). Natural selection and random genetic drift in phenotypic evolution. *Evolution*, **30**, 314–34.

Lande, R. (1979). Quantitative genetic analysis of multivariate evolution, applied to brain: body size allometry. *Evolution*, **33**, 402–16

Leroi, A. M., Rose, M. R., and Lauder, G. V. (1994). What does the comparative method reveal about adaptation? *Am. Nat.*, **143**, 381–402.

Lynch, M. (1990). The rate of morphological evolution in mammals from the standpoint of the neutral expectation. *Am. Nat.*, **136**, 724–41.

Lynch, M. (1991). Methods for the analysis of comparative data in evolutionary ecology. *Evolution*, **45**, 1065–80.

Lynch, M. (1993). Neutral models of phenotypic evolution. In *Ecological genetics*, (ed. L. Real), pp. 86–108. Princeton University Press.

Maddison, D. R. (1994). Phylogenetic methods for inferring the evolutionary history and processes of change in discretely valued characters. *Ann. Rev. Entomol.*, **39**, 267–92.

Maddison, W. (1991). Squared-change parsimony reconstructions of ancestral states for continuous-valued characters on a phylogenetic tree. *Syst. Zool.*, **40**, 304–14.

Martins, E. P. (1994). Estimating rates of character change from comparative data. *Am. Nat.*, **14**, 193–209.

Martins, E. P. (1995). Phylogenies and comparative data, a microevolutionary perspective. *Phil. Trans. R. Soc.*, **B349**, 85–91.

Martins, E. P. and Hansen, T. F. (1995). Phylogenetic comparative methods: the statistical analysis of interspecific data. In *Phylogenies and the comparative method in animal behavior*, (ed. E.P. Martins). Oxford University Press. (In press.)

Miles, D. and Dunham, A. (1993). Historical perspectives in ecology and evolutionary biology: the use of phylogenetic comparative analysis. *Ann. Rev. Ecol. Syst.*, **24**, 587–619.

Swofford, D.L. and Olsen, G. J. (1990). Phylogeny reconstruction. In *Molecular systematics.*, (ed. D. M Hillis and C. Moritz). Sinauer, Sunderland, MA.

18

Comparative tests of evolutionary lability and rates using molecular phylogenies

John L. Gittleman, C. Gregory Anderson, Mark Kot, and Hang-Kwang Luh

18.1 Introduction

Molecular phylogenies provide an estimate of the evolutionary history of extant taxa. If, for example, it is possible to reconstruct hierarchical changes in a clade within a given time frame, then evolutionary rates and lability can be estimated (Harvey *et al.* 1991, 1994). This is an important step in our understanding of the evolutionary process because previously we relied on measures taken from the fossil record thus restricting study to well-preserved traits. Via molecular phylogenies, it is now possible to trace evolutionary patterns in behavioural, ecological, or other poorly preserved traits. Here we shall test the idea that behavioural and ecological characters evolve at faster rates than morphological traits. Conceptually, our work parallels cladistic analyses of the origin, transitions, and historical (homology) constraints of behaviour and ecology (see for example Lauder 1986; Coddington 1988; de Queiroz and Wimberger 1993; Greene 1994; for review see Brooks and McLennan 1991). However, in contrast to that body of work, we examine continuous traits using statistical comparative methods and molecular phylogenies for measuring taxonomic divergence and rate. Using quantitative techniques may permit a broader comparative sweep (both in traits and taxa) for analysing evolutionary trends and, more importantly, may permit full use of the more precise nature (branch lengths and divergence times) of molecular phylogenies. We emphasize, however, that the present work is inherently descriptive, and not of a causal form.

18.2 Evolutionary lability

Much conjecture relates to the idea that different kinds of traits are relatively more labile than others through evolutionary time (Gause 1947; Schmalhausen 1949; Bradshaw 1965; Mayr 1974; Arnold 1992). Generally, it is thought that behavioural traits are more labile than morphological or physiological traits. West-Eberhard (1987) suggests that four mechanisms may causally allow for greater lability in a trait: genetic (allelic-switch) polymorphism, disruptive selection, pleiotropic effects or origin via contextual shift (preadaptation).

To address the problem of relative lability in phenotypic traits, we describe a method for examining the degree to which different kinds of traits change through phylogenetic time. In contrast to many analogous questions pertaining to 'phenotypic plasticity' (see Scheiner 1993), we are concerned with phylogenetic history and its relationship with trait variation (see also Gittleman *et al.* 1996).

18.3 Evolutionary rates

Because the issue of 'evolutionary rate' is so large (see for example reviews in Simpson 1953; Hecht 1965; Gingerich 1983, 1993; Fenster and Sorhannus 1991), we focus our attention on: (1) differential rate change among traits; (2) methods for estimating evolutionary rate; and (3) reconstructing phylogenies to measure rate change.

Differential rate change among traits

Many authors have suggested that rates of evolution are at least partly determined by overall behavioural complexity or that behaviour acts as a 'pacemaker' of evolution (Schmalhausen 1949; Simpson 1958; Mayr 1963). Typically, it is argued, behavioural plasticity allows individuals or populations to enter new niches and, over time, reveals that behaviour must evolve at a more rapid pace (Bateson 1988). Although well-studied examples of behavioural evolution are available at the population level (see Endler 1986; Endler and McLellan 1988; Scheiner 1993), cross-species studies are restricted to cladistic analyses which show transformational change among discrete traits (see for example Greene and Burghardt 1978; Lauder 1986; McLennan 1991; for review see Brooks and McLennan 1991). These studies do not directly assess rate in a quantitative sense, in contrast to approaches in macroevolution and paleontology (see below) which do measure degrees of phenotypic evolution along branching (time) intervals.

Measuring evolutionary rate

Gingerich (1993) recently reviewed and criticized three methods for calculating evolutionary rates:

1. Haldane (1949) proposed a unit of change, the 'darwin', which he defined as evolutionary change in the population mean by a factor of e (the base of natural logarithms) per million years. A rate of one darwin corresponds to a difference of one logarithmic unit per million years.
2. Haldane further suggested that the phenotypic standard deviation is an alternative to measuring 'e', pointing out that population variation is the

target of selection and thus a more appropriate measure; Gingerich notes that rates of evolution calculated in standard deviations per million years may be referred to as the 'simpson'.

3. Haldane also suggested that evolutionary rate be measured in generations (one standard deviation per generation) rather than years because this would incorporate the reproductive cycling underlying evolutionary change; Gingerich calls this unit of measure the 'haldane'.

There are various strengths and weaknesses of each method depending on the traits in question, quality of data, and what inferences are being made. The present study, using molecular phylogenies, relies on extant taxa which do not contain standard measures of time-scale or generation time; variances and standard deviations of morphological and behavioural traits are not always known, thus precluding the use of simpsons or haldanes. Therefore, given constraints imposed by the comparative data, darwins will be used in the present work. The calculation of evolutionary rate in darwins provides a useful and analytically appropriate estimate for 'descriptive comparisons' (Gingerich 1993, p. 458). As an aside, even though we use the mathematical calculation of, and term, darwin, a new term should be coined that is more in line with measuring rate of molecular evolution (perhaps the 'wilson').

Phylogenetic information

Essential to measuring evolutionary rates is that time, in terms of ancestral-descendent (i.e. phylogenetic) relationships, is either known or inferred. As with most problems of evolutionary history, issues of rate change are hampered by the fact that we can rarely observe evolution at the species level. Except for unusually complete fossil records, our primary evidence for evolutionary change is through inferences about sequences of taxonomic change based on end-point analysis. To study times of change and, consequently, rate patterns it is necessary to reconstruct past events, as in standard phylogenetic reconstruction. The present work accepts that we are at a stage in which 'working phylogenies' are realistic estimates of 'real phylogenies' (Harvey *et al.* 1991; see also Purvis *et al.* 1994). That is, phylogenies should provide information about topological changes among taxa: points of divergence (nodes) and branch lengths (time) in phylogenies can be used to estimate rates of trait evolution. For this information, we do not argue that molecular information is inherently superior to morphological information. Rather, phylogenetic information must be selected for the specific questions at hand (Hillis 1987; Donoghue and Sanderson 1992). Because we are interested in new uses of molecular phylogenies, specifically in terms of detecting relative lability and rates of evolution among phenotypic traits, we want to employ phylogenies which are independent of the traits being studied. We therefore exclusively use molecular phylogenies. Here, we restrict our analysis to various mammalian taxa because we are familiar with them.

18.4 Methods

To examine relative lability and rate of evolution in different kinds of traits, we use: (1) molecular phylogenies; (2) quantitative species values of morphological, life-history (ecological), and behavioural traits; and (3) comparative statistical methods. All of this information is drawn from the literature. We present weaknesses and restrictive assumptions in each element of the study as they become apparent and might influence our conclusions.

Molecular phylogenies

We found 97 molecular phylogenies of mammalian taxa published since 1978. Of these, only 8 were useable (see Table 18.1). The other 89 were discarded for the following reasons: inappropriate genetic samples (total 31: 12 included albumin/immunological samples; 19 were based on karyotypic G- and C-bands) for detecting changes over relatively short time-scales as well as the phenetic nature of these studies; no time-scale and/or divergence information (12), as necessary for analysing evolutionary lability and rates; including only higher taxa (i.e. families and above), thus not amenable to species-level study; those phylogenies at the species level, 17 were of prohibitively small sample sizes (< 10 species) and 16 included species for which little, if any, behavioural, life-history, or morphological information is available.

Table 18.1 Molecular phylogenies of mammals used in comparative analyses

Taxon	#Species	Genetic data	Phylogenetic method	Source
Primates	26	103 beta-type sequences	Parsimony	Goodman *et al.* (1982)
Ceboidea	19	27 e-globin gene sequences	Parsimony	Schneider *et al.* (1993)
Carnivora	39	DNA hybridization	Distance	Wayne *et al.* (1989)
Canidae	28	DNA hybridization	Distance	Wayne *et al.* (in press)
Bovidae	27	40 allozyme loci	Distance	Georgiadis *et al.* (1990)
Cervidae	11	mtDNA restriction sites	Parsimony	Cronin (1991)
Cetacea	13	Cytochrome b (mtDNA)	Parsimony	Arnason and Gullberg (1994)
Arvocolinae	15	Protein electrophoresis	Distance	Chaline and Graf (1988)

There is no consensus as to which phylogenetic reconstruction methods are best for producing phylogenies from molecular data (Felsenstein 1988; Moritz and Hillis 1990; Hillis *et al.* 1994). Precedence was given to phylogenetic estimates based on DNA sequences rather than protein-based methods because of their inherently greater accuracy (see Moritz and Hillis 1990). The selected molecular phylogenies are generally appropriate for the present study because phylogenetic inferences for mammals involve ranges of 0–5 or 5–50 million year divergence times (see Hillis and Moritz 1990). Nevertheless, given the disparate nature of the genetic material and reconstruction methods used in the phylogenies of this study, as indeed in most molecular phylogenies (Felsenstein 1988), we anticipate that error is introduced when comparing differential trait evolution from different kinds of molecular information. Certainly, as in other comparative studies using molecular phylogenies (Harvey *et al.* 1991, 1994), our results will be wrong if the timing and branching patterns are wrong.

None of the published sources for the eight identified phylogenies presented the original distance matrices for generating the phylogenies. Thus, to estimate distances and points of divergence, we measured in millimetres all pairwise distances between species in a published phylogeny.

Trait data

For species represented in the molecular phylogenies, trait values were derived from published literature for the following: brain weight—average brain weight (g) of adult male and adult female calculated from volumetric measures of skull braincases; body weight—average body weight (kg or g in the case of Arvicolinae) of adult male and adult female; gestation length—average time from conception to birth (days), minus any possible period of delayed implantation (as for example in some mustelids and ursids); birth weight—average weight (g) of young at birth; home-range size—average total area (km^2) used by an individual (or group in social species) during normal activities; group size—average number of individuals which regularly associate together and share a common home range.

All analyses were performed on logarithmically transformed data because cross-species values of the examined trait variables approximate a lognormal distribution; logarithmic transformations are necessary to reduce skew. For the rates analysis, standardization procedures followed Gingerich (1983, 1993) in which logarithmic trait values were used to calculate darwins and then darwins along with time intervals were similarly transformed; this helps reduce differences in measurement of the traits and in the time-scales among taxa.

The original data files, including both the molecular distance matrices and the species trait values, are available from the authors.

Analyses: definitions and procedures

We define *evolutionary lability* in phylogenetic terms: a trait has greater lability if it is not correlated or has a low correlation with phylogenetic distance, measured from node to node at speciation events. Our definition also resembles the concept as discussed by de Queiroz and Wimberger (1993). We recognize, however, that lability itself is subject to selection (Williams 1966) and therefore we do not consider it to be a 'non-genetic' phenomenon.

To measure relative phenotypic change with phylogenetic distance, we borrow concepts and methods from the extensive literature on spatial autocorrelation (Cliff and Ord 1973, 1981; Upton and Fingleton 1985; see also

Legendre (1993) for review of ecological applications). Moran's (1950) I statistic is used to examine the relationship between phylogenetic distance and phenotypic variation. We denote the observed phenotypic trait for species i by y_i and write $\bar{y}$ for the average of y_i over the n species of the dataset. Using this notation, Moran's coefficient is given as

$$I=\frac{n}{S_0}\,\frac{\sum_{i=1}^{n}\sum_{j=1}^{n} w_{ij}(y_i-\bar{y})(y_j-\bar{y})}{\sum_{i=1}^{n}(y_i-\bar{y})^2}$$

where

$$S_0=\sum_{i=1}^{n}\sum_{j=1}^{n} w_{ij}.$$

Moran's I is, in essence, an (estimated) autocorrelation coefficient: the numerator is a measure of covariance among the y_i and the denominator is a measure of variance. At the heart of this statistic lies the weighting matrix W, $W=[w_{ij}]$. In searching out phylogenetic autocorrelation at some level, I will compare the phenotypic trait of a species with a weighted average of the trait over a set of neighbours. The ijth element of the W matrix, wij, is the weight assigned to the jth species in computing the weighted average for species i. w_{ii} is always set to 0 (to circumvent the fact that a species is its own best predictor), and because we average over neighbours, we will, for non-zero rows, transform the non-zero rows of the matrix so that

$$\sum_{j=1}^{n} w_{ij}=1,\ i=1,\ 2,\ldots,n.$$

The w_{ij} are determined by the correlation we are trying to ascertain. With specified (molecular) branch lengths, as in the present study, we average over all species within some finite interval of distance, typically set in accord with the frequency distribution of branch length intervals (see below). We use W, in effect, to flag species that are to be averaged.

Because we refer to I as an autocorrelation, it is tempting to assume that $-1 \leqslant I \leqslant 1$, with values ± 1 indicating a strong phylogenetic correlation. However, for this to be true, we must first standardize values of I by its maximum possible value. To interpret the values of I, we also tabulate associated standard deviates, z, in which z-scores greater than 1.96 may be used to reject the null hypothesis of no phylogenetic correlation at the 0.05 level.

To assess relative lability among traits, we employ a 'phylogenetic correlogram' (see Gittleman and Kot 1990). This is simply a graph showing how

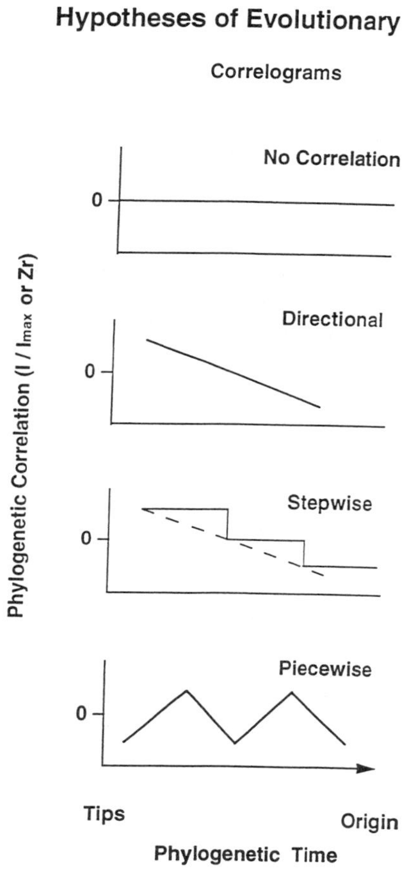

Fig. 18.1 Hypothetical patterns of phylogenetic change in quantitative traits. Patterns are detected using a spatial autocorrelation statistic, Moran's I (see text for details). Evolutionary lability of traits can be measured by relative differences in observed Is (or associated z transformations) between traits, with no phylogenetic pattern expressed by zero and strong phylogenetic pattern represented by ± 1.

autocorrelation (observed Is or associated zs) varies with phylogenetic distance, as hypothesized in Fig. 18.1. With detailed, complete phylogenetic information (i.e. branching patterns), we first calculate a set of cut-offs. The cut-offs are formed by analysing frequency distributions of the distances observed for each phylogeny and then selecting those distances with the highest frequency ('D'); this *maximizes* the chance of finding trait change at a given distance in a phylogeny (i.e. phylogenetic correlation). Weighting matrices are then defined by applying the Moran's I as follows:

$w_{ij} = 1$ if the distance of species i and j lies between a fixed pair of consecutive cut-offs,
$= 0$, otherwise.

We then use the correlogram to show where phylogenetic correlation occurs with divergence and branching patterns in the phylogeny.

From these correlograms, we can detect four hypothetical patterns (Fig. 18.1):

1. No relationship with phylogeny, as possibly observed under strong stabilizing selection.
2. A specific autoregressive ('directional') form with correlation falling off with phyletic distance, a pattern possibly observed from a simple Brownian motion random walk (i.e. pure random genetic drift).
3. A 'stepwise' function with phylogenetic distance, representing stasis followed by change between nodes (note: this is not, in reality, an alternative hypothesis to the second hypothesis because the overall pattern of change is in proportion to phyletic distance).
4. A 'piecewise' function with phylogenetic distance, representing linear change of significant positive and negative (phylogenetic) correlation and, importantly, greater phenotypic plasticity than in the 'directional' or 'stepwise' patterns; this pattern might reflect changes in evolutionary rate (for example, punctuational change) between speciation events.

Our aim is to examine these patterns among traits, with the prediction that behavioural traits (home-range size; population group size) will show either no phylogenetic pattern or less phylogenetic correlation than morphological and possibly life-history traits.

To examine *evolutionary rate* in different phenotypic traits (behaviour, life histories, morphology), we measure 'darwins' for each trait. Similar to Haldane's (1949) definition and Gingerich's (1983, 1993) usage, darwins are calculated as

$$r = \frac{\ln(x_2) - \ln(x_1)}{\Delta t}$$

where r is the rate of change (in darwins), x_1 is the initial dimension of a character, x_2 is the final dimension of a character, and $\Delta t = t_1 - t_2$ is the amount of time elapsed between the ages of x_1 and x_2. Because rates of evolution change with temporal scale (i.e. evolutionary rate appears inversely proportional to time interval (Gingerich 1983, 1993)), log (darwins) will be plotted against log (time intervals) as measured from phylogenetic distances and calibrated from a molecular clock specific for each taxonomic group. Such plots will produce estimates of evolutionary rate across taxa that can be statistically represented by intercepts and slopes (Gingerich 1983, 1993). These statistical measures reveal different kinds of rate information. Observed intercepts represent an estimate of change per $10^0 = 1$ time unit (i.e. per million years in this study) among traits. Slopes reflect the extent to which rates of trait evolution occur in a random walk, directional, or stabilizing (stasis) manner (see Gingerich 1993). During a random walk, changes in one direction rapidly cancel changes in the other direction: net evolutionary rate rapidly falls off with time interval and the slope on a log–log plot is negative. In a directional process, there is little or no cancellation and rates stay fairly constant, this leads to a log–log plot of shallow slope. We are aware of the problem that statistical intercepts and slopes for representing rate change are potentially influenced by different units of measurement among variables. For example, rates of change for areas and volumes cannot be compared directly with rates for linear measurements. The double logarithmic plot helps ameliorate this problem. Because time calibration for molecular phylogenies can only be made within each group of independently derived organisms (Hillis and Moritz 1990), patterns of rate are detected for each taxon and then compared.

Measurement of evolutionary rate in 'darwins' may well be a biased estimate of rate due to statistical non-independence in the data and lack of an explicit null model of phenotypic evolution. Following Lynch's (1990) metric for measuring evolutionary rate under neutral selection, Martins (1994) proposed a statistical approach for estimating evolutionary rate under neutral evolution or stabilizing selection while also controlling for statistical non-independence of trait values. At present, we are applying this method for estimating evolutionary rate (Gittleman *et al.*, in preparation).

18.5 Results and discussion

In the following sections, empirical results and discussion are combined for each problem. We do this because, analytically, evolutionary lability and rates should be addressed separately. Throughout, we emphasize comparisons of traits rather than comparisons between taxa. This reflects both the nature of the questions asked as well as a limitation in using, at present, molecular data: molecular phylogenies across independent monophyletic groups cannot be collated because of differences in rates of molecular evolution, types of characters and methods of analysis, and the complexity in inferring 'homology' in molecular systematics (see Moritz and Hillis 1990; Hillis 1994).

Evolutionary lability

General trends regarding evolutionary lability are plotted in Figs. 18.2a and 18.2b. Note that the scale is not constant across correlograms: observed z-scores and time intervals for different taxa are not drawn on the same scale so that actual comparative patterns are more easily observed. Moreover, comparisons of the correlograms among taxa are qualitative and do not necessarily indicate that observed differences are statistically significant.

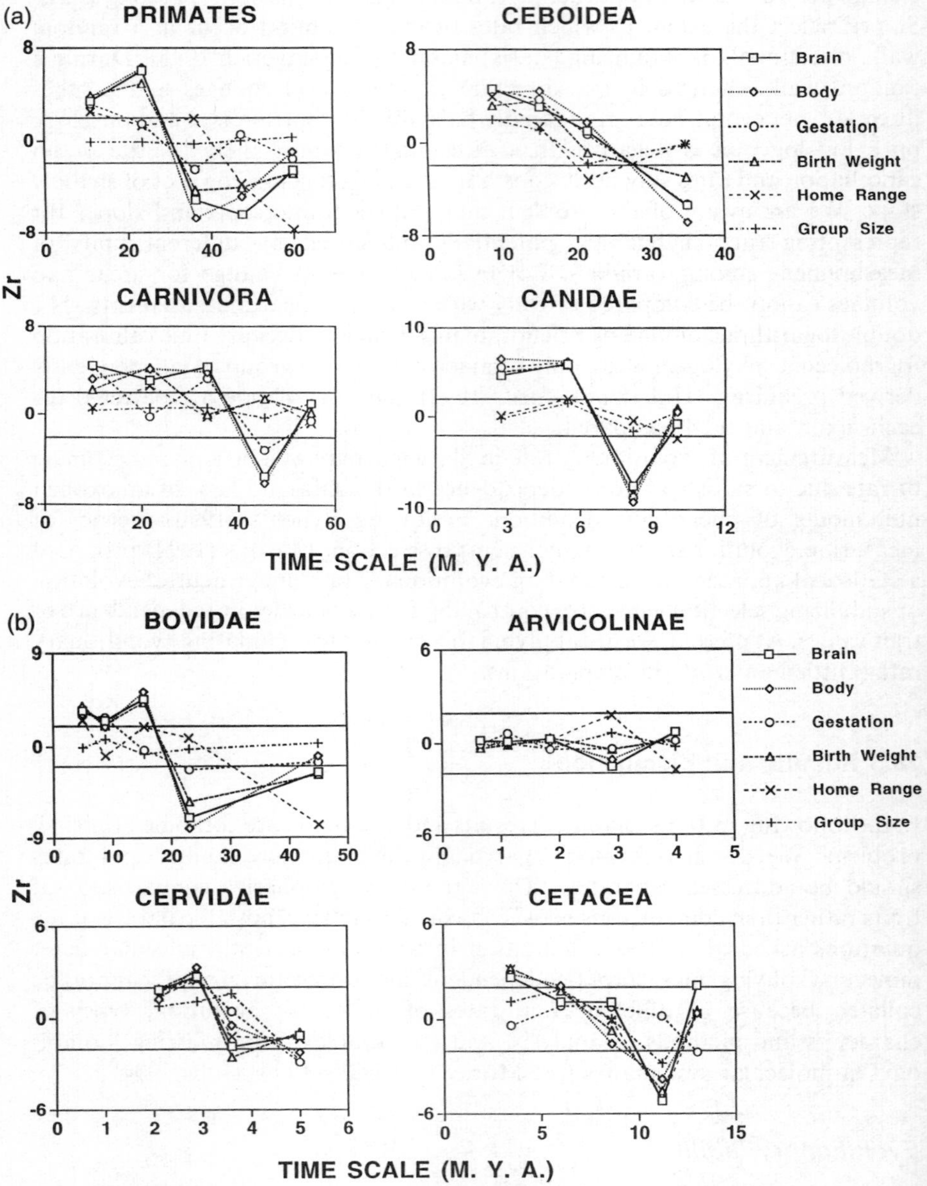

Fig. 18.2 Observed patterns of *z*-scores (from Moran's *I*) for the quantitative traits of: brain weight, body weight, gestation length, birth weight, home-range size, and group size (see text for definitions). Phylogenetic patterns of each trait are shown across various mammal taxa including: (a) Primates, Carnivora, Ceboidea, Canidae, and (b) Bovidae, Cervidae, Arvicolinae, and Cetacea. Intervals of time-scale and *z*-scores are not standardized in order to accurately measure relative correlation among traits for each taxon.

Do different types of traits change differently through time? The answer is 'yes'. As expected, behavioural traits, especially group size, show no relationship with phylogeny at any interval of time. This generally indicates that there is little if any behavioural change with cladogenesis. This is an important finding for two reasons. First, even in relatively old taxonomic groups (bovids, carnivores) there is little evidence of phylogenetic relations at different points (nodes) of speciation, at least in the variable examined. Second, a trait like group size which has important ecological functions and relations with morphological traits, may reveal an evolutionary pattern quite independent of other traits even across vast (phylogenetic) time-scales.

In terms of characterizing trait change in relation to our hypothesized patterns (see Fig. 18.1), all of the traits generally follow a 'directional' or 'piecewise' pattern, essentially revealing an autoregressive model. Closely related taxa, especially at the species level, reveal positive and significant correlation with a precipitous drop-off at greater phylogenetic distances. This general pattern emerges independent of different types of traits, molecular phylogenies, time-scale and, except for Arvicolinae, sample size.

Interesting variation occurs, however, in relative correlation (z-scores) among traits and in considering the time-scale in which significant negative correlation occurs. The observed correlations for life-history traits are generally lower than for morphological traits; this holds at most time intervals across the bovids, cetacea, primates, carnivores, ceboids, and canids (see Figs. 18.2a and 18.2b). Such a result is consistent with the idea that life-history traits are relatively more flexible than many morphological traits (Gittleman 1993). Following significant positive correlation at more recent time periods, significant negative correlation is observed at later times, specifically points of time occurring at two (cervids, primates, canids) or three (ceboids, carnivores, bovids, cetacea) speciation events. Although this result could simply reflect tree topology, a possibility that warrants further study, this pattern may suggest some parallelism at lower taxonomic levels due to genetic and developmental similarities and, as consequence, similar phenotypic responses to selection. Sanderson and Donoghue (1989) argue differently, showing that consistency indices (CIs) for a wide variety of characters across plants and animals are not affected by taxonomic rank. Although it is difficult to draw direct comparisons between these studies because of different comparative methods, taxa, and characters, there is the possibility that the lack of change in the CIs in Sanderson and Donoghue's study relates to sample size; in one illustration, there appears to be a decline in CIs from lower to higher taxa and then an increase in CIs at the highest rank, but this is not statistically signficant perhaps because of small samples.

Last, it is observed that in all cases but two, phylogenetic correlations generally converge around a value of 0 at the initial (basal) ancestor; this is due to there being little observed variation, positive or negative, at the split from the initial ancestor at the oldest time interval. The interesting exceptions are in most traits across the Ceboidea and home range size across the Bovidae. In the ceboids the molecular tree is quite unbalanced between two major clades: one group, including the subfamily Callitrichinae and the genera *Cebus*, *Saimiri*, and *Aotus*, has different distance intervals and topologies from another group including the tribe Pitheciini, the subfamily Atelinae, and the genus *Callicebus* (Schneider *et al.* 1993). In the bovids, there is a significant negative correlation with home-range size. This may be because the oldest phylogenetic node is associated with a comparatively large value of home-range size in the (presumed, or actually an outgroup) ancestral species of the bovids, the giraffe. Both of these patterns underscore the need to investigate the effects of tree topology on observed phylogenetic correlation, both throughout the entire tree and at different time intervals.

Evolutionary rate

We estimate evolutionary rate, as measured in plots of 'darwins' against time intervals, among traits for each taxonomic group examined. Results are discussed in terms of observed rates over a million years of evolution (i.e. observed intercepts) and the directedness of evolution (i.e. observed slopes). We urge caution when interpreting the results given here because of potential problems in different measures of evolutionary rate, possible non-independence of the data, lack of a null model for trait evolution, and, perhaps most importantly, the literal usage of divergence time in the molecular phylogenies. Yet we are encouraged by these results because they show that macroevolutionary measures of evolutionary rate (see Gingerich 1983, 1993) can be co-opted for comparing rates of differential trait evolution using molecular phylogenies.

Initially, to get an overall impression of the kind of patterns that emerge, we present a plot of evolutionary rate on time across the primates for one morphological trait (body weight; Fig. 18.3a) and one behavioural trait (group size; Fig. 18.3b). Four patterns are evident in this one example. First, both variables show the expected inverse relationship between evolutionary rate and time-scale, though the values of slope are shallow. Second, the observed slopes are more negative in group size than in body weight, which suggests that evolutionary change is more of a random walk in the behavioural trait and relatively more directional in the morphological trait (see Gingerich 1993). Third, the rate of change over a million years (i.e. the intercept) is greater in group size than body weight, which suggests that the instantaneous evolutionary rate is greater in the behavioural variable. Fourth, even though group size appears to have a higher instantaneous rate, the net rate over long intervals may be lower due to the lower level of directionality. Due to problems of sample size and standardizing time-scale, it is difficult to directly compare these results with studies of other taxonomic groups and traits or even to make taxonomic comparisons within this study. Indeed, all of the values of slope observed in this study are within a standard deviation of -0.50, which is expected for a random walk model of evolutionary rate based on experimental and fossil data (Gingerich 1993).

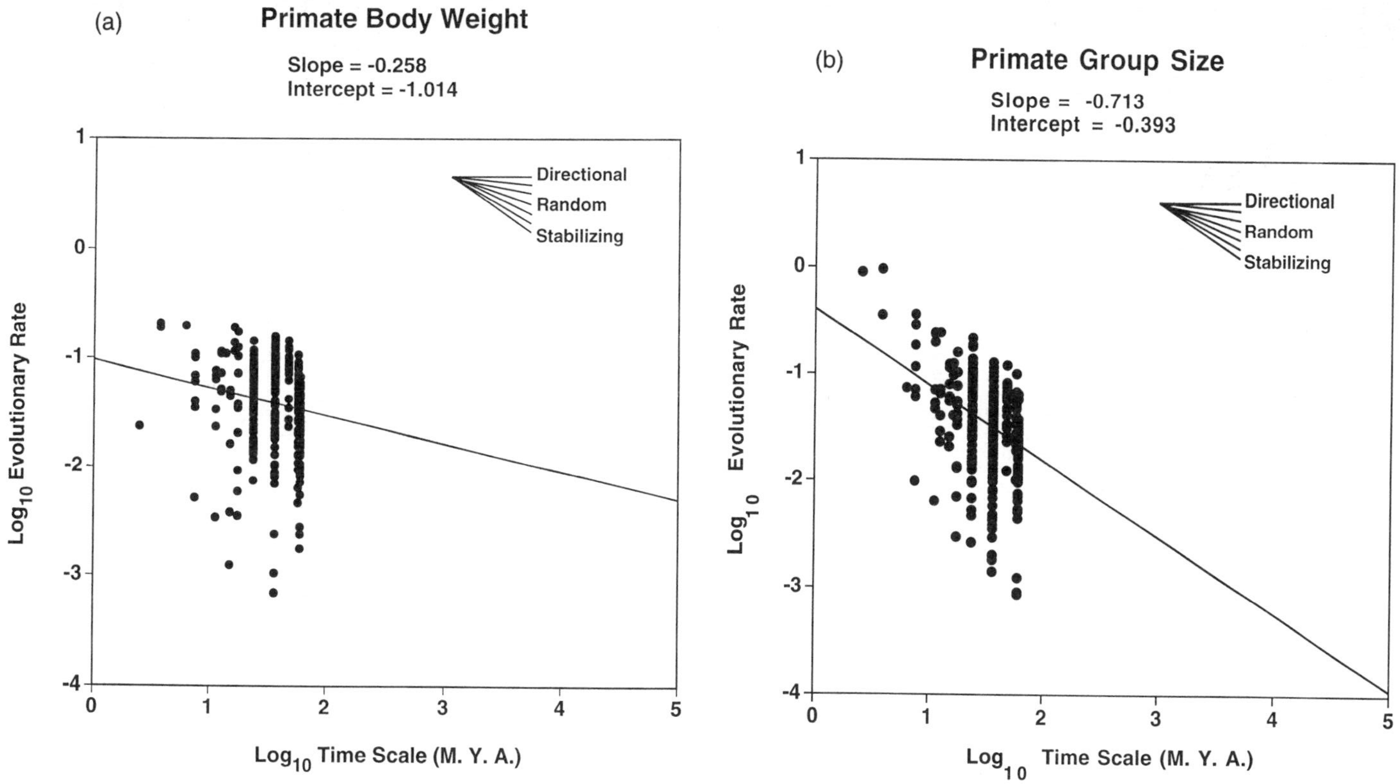

Fig. 18.3 Log evolutionary rate ('darwins') versus log time interval (million years ago) observed across Primates for (a) body weight and (b) group size. Values of slope represent directedness of evolution and values of intercept represent rate over a million years.

It is instructive, however, to compare observed slopes and intercepts of rate change. Comparisons of the directedness of evolutionary change can be made by plotting values of slope for each trait across taxa, as shown in Fig. 18.4. Observed slopes are significantly different between traits ($F_{4,70} = 6.91$, $P < 0.0001$), with values for the two behavioural variables (home-range size and group size) steeper than the morphological and life-history variables. This indicates that, within taxa, directedness of evolution in the behavioural traits may be different, as expected, and may change more randomly over time than the other traits. The slopes of the morphological and life-history traits are almost all negative; however, in comparison to the behavioural traits, it is interesting that they approximate more of a directional evolutionary process, as expressed by a horizontal distribution of average slope zero (0.0) on a rate versus interval plot (see Gingerich 1993).

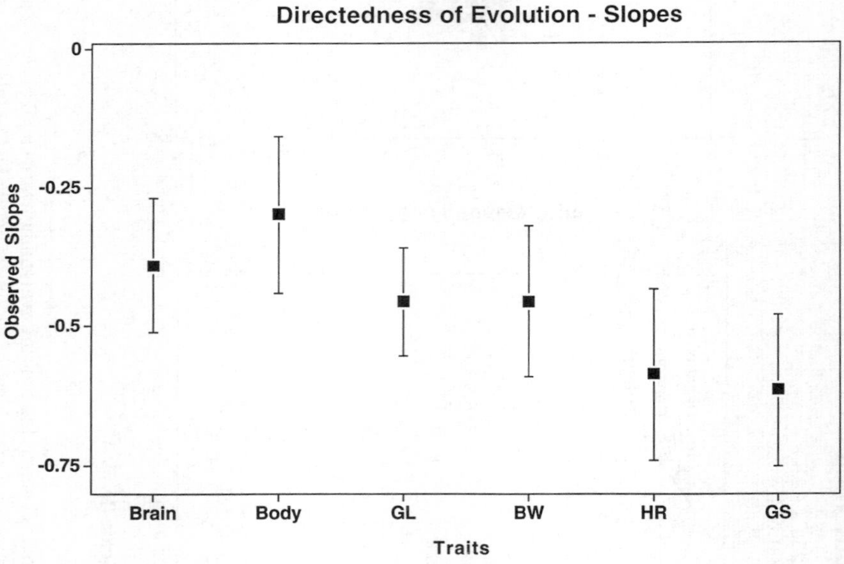

Fig. 18.4 Combined values of slope across all mammalian taxa observed for each quantitative trait. Traits are denoted by: brain—brain weight; body—body weight; GL—gestation length; BW—birth weight; HR—home-range size; GS—group size. Means and standard errors are included for each trait.

Similarly, observed intercepts (see Fig. 18.5) are significantly different between traits ($F_{4,70} = 2.29$, $P = 0.0467$), with intercepts for the two behavioural traits consistently higher than for the other traits. This reflects that evolutionary rates for short intervals of time are higher in the behavioural traits, as expected. As with the slopes, the intercepts for the two behavioural traits generally cluster together and the four other traits cluster together with the exception of gestation length. Gestation has an unusually low intercept relative to the other traits, which suggests a slower evolutionary rate. This may result from gestation length indirectly changing with female size (allometry) whereas selection directly operates on the other traits. Similar arguments have been suggested for various life-history traits which are composites of other morphological traits (Riska 1989).

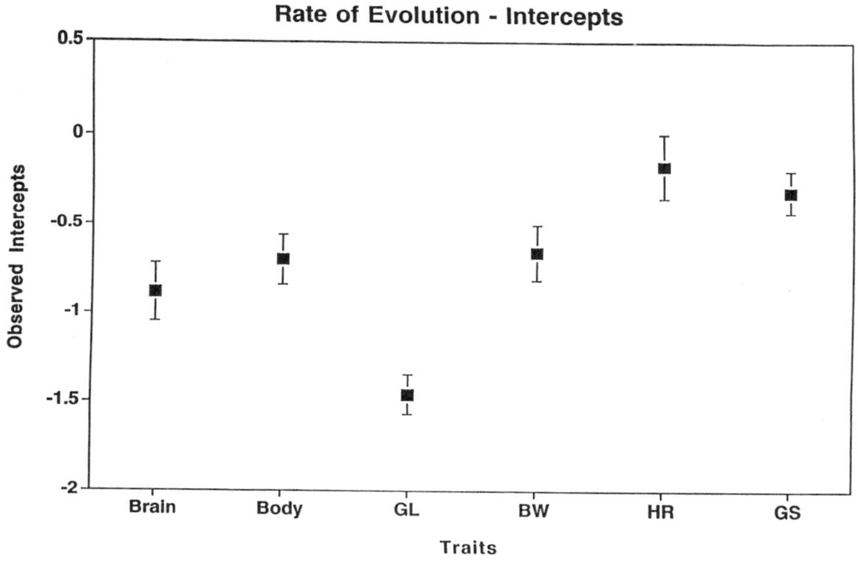

Fig. 18.5 Combined values of intercept across all mammalian taxa observed for each quantitative trait. Traits are denoted as in Fig. 18.4. Means and standard errors are included for each trait.

18.6 Using molecular phylogenies for comparative tests

We end with some general thoughts about applications and future uses of molecular phylogenies in comparative biology. Unifying molecular phylogenies and comparative data is currently frustrating. We were startled that, despite mammals receiving a disproportionate amount of attention in phylogenetic studies relative to other taxa (Sanderson *et al.* 1993), so few molecular phylogenies were useable for this comparative study. The main reason for this is that most molecular phylogenies of the mammals are at higher levels or within small clades (Novacek 1992; Honeycutt and Adkins 1993); a prime example is the most intensively studied of all mammals, the primates, in which molecular study primarily comprises familial groups (Miyamoto and Goodman 1990) or extant hominids (*Gorilla*, *Homo*, *Pan* (Patterson *et al.* 1993)). Clearly, analytical methods for combining phylogenies from different sources will aid in providing more information (see for example Purvis 1995). Nevertheless, there is evidently a clear lack of phylogenies for taxa which are the focus of study in animal behaviour, ecology, and general natural history; this omission will continue to impede further work in analysing the phylogeny and lability of trait evolution.

The patterns detected in this study are clearly determined by the quality of the molecular phylogenies; quite literally, 'garbage in, garbage out'! Although we assume the phylogenies to be accurate, we realize that tree topologies are significantly affected by the rate of genomic evolution which in turn varies between genes and between taxa. More careful analysis is required to evaluate the quality of the molecular information used in our analysis. As others have noted, it is unlikely that one gene will elucidate phylogenetic relationships across taxa (Hillis and Moritz 1990; Graybeal 1994). Thus, for comparative purposes, future work will have to match specific genes with particular types of divergences such as with long time frames or speciose regions of a tree.

Last, the markedly different patterns of evolutionary lability and rates among morphological, life-history, and behavioural traits begs the question about what factors are causally interrelated; that is, does behavioural evolution drive evolutionary change? Given that the comparative databases and precise molecular phylogenies are now in hand, there is a need for further work on multivariate comparative methods.

Acknowledgements

We thank Robert Wayne for discussion on the availability and use of mammalian molecular phylogenies and Philip Gingerich, Paul Harvey, Emília Martins, Trevor Price, Harry Greene, Gareth Russell, and Gordon Burghardt for helpful comments on the manuscript. Financial support was received from the Department of Zoology, Science Alliance and Dean of Liberal Arts, all of The University of Tennessee (JLG, CGA), and the National Science Foundation (H-KL, MK).

References

Arnason, U. and Gullberg, A. (1994). Relationship of baleen whales established by cytochrome b gene sequence comparison. *Nature*, **367**, 726–8.

Arnold, S. J. (1992). Constraints on phenotypic evolution. *Am. Nat.*, **140**, S85–S107.

Bateson, P. (1988). The active role of behaviour in evolution. In *Evolutionary processes and metaphors*, (ed. M.-W. Ho and S. W. Fox), pp. 191–207. Wiley, London.

Bradshaw, A. D. (1965). Evolutionary significance of phenotypic plasticity in plants. *Adv. Genet.*, **13**, 115–55.

Brooks, D. R. and McLennan, D. A. (1991). *Phylogeny, ecology, and behavior*. University of Chicago Press.

Chaline, J. and Graf, J.-D. (1988). Phylogeny of the Arvicolidae (Rodentia): biochemical and paleontological evidence. *J. Mamm.*, **69**, 22–33.

Cliff, A. D. and Ord, J. K. (1973). *Spatial autocorrelation*. Pion, London.

Cliff, A. D. and Ord, J. K. (1981). *Spatial processes: models and applications*. Pion, London.

Coddington, J. A. (1988). Cladistic tests of adaptational hypotheses. *Cladistics*, **4**, 3–22.

Cronin, M. A. (1991). Mitochondrial-DNA phylogeny of deer (Cervidae). *J. Mamm.*, **72**, 533–66

De Queiroz, A. and Wimberger, P. H. (1993). The usefulness of behavior for phylogeny estimation: levels of homoplasy in behavioral and morphological characters. *Evolution*, **47**, 46–60.

Donoghue, M. J. and Sanderson, M. J. (1992). The suitability of molecular and morphological evidence in reconstructing plant phylogeny. In *Molecular systematics of plants*, (ed. P. S. Soltis, D. E. Soltis, and J. J. Doyle), pp. 340–68. Chapman and Hall, New York.

Endler, J. A. (1986). *Natural selection in the wild*. Princeton University Press.

Endler, J. A. and McLellan, T. (1988). The processes of evolution: toward a newer synthesis. *Ann. Rev. Ecol. Syst.*, **19**, 395–421.

Felsenstein, J. (1988). Phylogenies from molecular sequences: inference and reliability. *Ann. Rev. Genet.*, **22**, 521–65.

Fenster, E. J. and Sorhannus, U. (1991). On the measurement of morphological rates of evolution. A review. *Evol. Biol.*, **25**, 375–410.

Gause, G. F. (1947). Problems of evolution. *Trans. Conn. Acad. Sci.*, **37**, 17–68.

Georgiadis, N. J., Kat, P. W., Oketch, H., and Patton, J. (1990). Allozyme divergence within the Bovidae. *Evolution*, **44**, 2135–49.

Gingerich, P. D. (1983). Rates of evolution: effects of time and temporal scaling. *Science*, **222**, 159–61.

Gingerich, P. D. (1993). Quantification and comparison of evolutionary rates. *Am. J. Sci.*, **293A**, 453–78.

Gittleman, J. L. (1993). Carnivore life histories: a re-analysis in the light of new models. In *Mammals as predators*, (ed. by N. Dunstone and M. L. Gorman), pp. 65–86. Oxford University Press.

Gittleman, J. L. and Kot, M. (1990). Adaptation: statistics and a null model for estimating phylogenetic effects. *Syst. Zool.*, **39**, 227–41.

Gittleman, J. L., Anderson, C. G., Kot, M., and Luh, H.-K. (1996). Comparing behavioral and morphological evolution: using molecular phylogenies to measure phylogeny, plasticity and rates. In *Phylogenies and the comparative method in animal behavior*, (ed. E. P. Martins), pp. 167–206. Oxford University Press.

Goodman, M., Romero-Herrara, A. E., Dene, H., Czelusniak, J., and Tashian, R. E. (1982). Amino acid sequence evidence on the phylogeny of primates and other eutherians. In *Macromolecular sequence in systematic and evolutionary biology*, (ed. M. Goodman), pp. 115–91. Plenum, New York.

Graybeal, A. (1994). Evaluating the phylogenetic utility of genes: a search for genes informative about deep divergences among vertebrates. *Syst. Biol.*, **43**, 174–93.

Greene, H. W. (1994). Homology and behavioral repertoires. In *Homology: the hierarchical basis of comparative biology*, (ed. B. K. Hall), pp. 369–91. Academic, San Diego.

Greene, H. W. and Burghardt, G. M. (1978). Behavior and phylogeny: constriction in ancient and modern snakes. *Science*, **200**, 74–7.

Haldane, J. B. S. (1949). Suggestions as to quantitative measurement of rates of evolution. *Evolution*, **3**, 51–6.

Harvey, P. H., May, R. M., and Nee, S. (1994). Phylogenies without fossils: estimating lineage birth and death rates. *Evolution*, **48**, 523–29.

Harvey, P. H., Nee, S., Mooers, A., and Partridge, L. (1991). These hierarchical views of life: phylogenies and metapopulations. In *Genes in ecology*, (ed. R. J. Berry and T. J. Crawford), pp. 123–37. Blackwell Scientific, Oxford.

Hecht, M. K. (1965). The role of natural selection and evolutionary rates in the origin of higher levels of organization. *Syst. Zool.*, **14**, 301–17.

Hillis, D. M. (1987). Molecular versus morphological approaches to systematics. *Ann. Rev. Ecol. Syst.*, **18**, 23–42.

Hillis, D. M. (1994). Homology in molecular biology. In *Homology: the hierarchical basis of comparative biology*, (ed. B. K. Hall), pp. 339–68. Academic, San Diego.

Hillis, D. M. and C. Moritz. (1990). An overview of applications of molecular systematics. In *Molecular systematics*, (ed. D. M. Hillis and C. Moritz), pp. 502–15. Sinauer, Sunderland, MA.

Hillis, D. M., Helsenbeck, J. P., and Cunningham, C. W. (1994). Application and accuracy of molecular phylogenies. *Science*, **264**, 671–7.

Honeycutt, R. L. and Adkins, R. M. (1993). Higher level systematics of eutherian mammals: an assessment of molecular characters and phylogenetic hypotheses. *Ann. Rev. Ecol. Syst.*, **24**, 279–305.

Lauder, G. V. (1986). Homology, analogy, and the evolution of behavior. In *Evolution of animal behavior*, (ed. M. H. Nitecki and J. A. Kitchell), pp. 9–40. Oxford University Press, New York.

Legendre, P. (1993). Spatial autocorrelation: trouble or new paradigm? *Ecology*, **74**, 1659–73.

Lynch, M. (1990). The rate of morphological evolution in mammals from the standpoint of the neutral expectation. *Am. Nat.*, **136**, 727–41.

McLennan, D. A. (1991). Integrating phylogeny and experimental ethology: from pattern to process. *Evolution*, **45**, 1773–89.

Martins, E. P. (1994). Estimating the rate of phenotypic evolution from comparative data. *Am. Nat.*, **144**, 193–209.

Mayr, E. (1963). *Animal species and evolution*. Harvard University Press, Cambridge, MA.

Mayr, E. (1974). Behavior programs and evolutionary strategies. *Am. Sci.*, **62**, 650–9.

Miyamoto, M. M. and Goodman, M. (1990). DNA systematics and evolution of primates. *Ann. Rev. Ecol. Syst.*, **21**, 197–220.

Moran, P. A. P. (1950). Notes on continuous stochastic phenomena. *Biometrika*, **37**, 17–23.

Moritz, C. and Hillis, D. M. (1990). Molecular systematics: context and controversies. In *Molecular systematics*, (ed. D. M. Hillis and C. Moritz), pp. 1–10. Sinauer, Sunderland, MA.

Novacek, M. J. (1992). Mammalian phylogeny: shaking the tree. *Nature*, **356**, 121–5.

Patterson, C., Williams, D. M., and Humphries, C. J. (1993). Congruence between molecular and morphological phylogenies. *Ann. Rev. Ecol. Syst.*, **24**, 153–88.

Purvis, A. (1995). A composite estimate of primate phylogeny. *Phil. Trans. R. Soc.*, **348**, 405–21.

Purvis, A., Gittleman, J. L., and Luh, H.-K. (1994). Truth or consequences: effects of phylogenetic accuracy on two comparative methods. *J. Theor. Biol.*, **167**, 293–300.

Riska, B. (1989). Composite traits, selection response, and evolution. *Evolution*, **43**, 1172–91.

Sanderson, M. J. and Donoghue, M. J. (1989). Patterns of variation in levels of homoplasy. *Evolution*, **43**, 1781–95.

Sanderson, M. J., Baldwin, B. G., Bharathan, G., Campbell, C. S., von Dohlen, C., Ferguson, D., Porter, J. M., Wojciechowski, M. F., and Donoghue, M. J. (1993). The growth of phylogenetic information and the need for a phylogenetic data base. *Syst Biol.*, **42**, 562–8.

Scheiner, S. M. (1993). Genetics and evolution of phenotypic plasticity. *Ann. Rev. Ecol. Syst.*, **24**, 35–68.

Schmalhausen, I. I. (1949). *Factors of evolution: the theory of stabilizing selection.* Blakiston, Philadelphia.

Schneider, H., Schneider, M. P. C., Sampaio, I., Harada, M. L., Stanhope, M., Czelusniak, J., and Goodman, M. (1993). Molecular phylogeny of the New World monkeys (Platyrrhini, Primates). *Mol. Phylogenet. Evol.*, **2**, 225–42.

Simpson, G. G. (1953). *The major features of evolution.* Columbia University Press, New York.

Simpson, G. G. (1958). Behavior and evolution. In *Behavior and evolution*, (ed. A. Roe and G. G. Simpson), pp. 507–35. Yale University Press, New Haven, CT.

Upton, G. J. G. and Fingleton, G. (1985). *Spatial data analysis by example.* Wiley, Chichester.

Wayne, R. K., Benveniste, R. E., Janczewski, D. N., and O'Brien, S. J. (1989). Molecular and biochemical evolution of the Carnivora. In *Carnivore behavior, ecology, and evolution*, (ed. J. L. Gittleman), pp. 465–94. Cornell University Press, Ithaca, NY.

Wayne, R. K., Geffen, E., Girman, D. J., and Marshall, C. (In press.) Molecular systematics of the Canidae. *Syst. Biol.*

West-Eberhard, M. J. (1987). Flexible strategy and social evolution. In *Animal societies: theories and facts*, (ed. Y. Ito, J. L. Brown, and J. Kikkawa), pp. 35–51. Japan Scientific Societies Press, Tokyo.

Williams, G. C. (1966). *Adaptation and natural selection.* Princeton University Press.

19

Community evolution in Greater Antillean *Anolis* lizards: phylogenetic patterns and experimental tests

Jonathan B. Losos

19.1 Introduction

The use of phylogenies in all manner of comparative studies has increased remarkably in recent years. Five years ago, few would have predicted the integral role that phylogenetic approaches play today in fields as disparate as epidemiology, population genetics, and macroevolution, as this volume attests. The goal of this chapter is two-fold: first, to illustrate the importance of a phylogenetic approach to another field, community ecology, and, second, to discuss an important new use to which phylogenies have been put recently: rather than being used to test previously derived hypotheses, phylogenies are now being used to derive hypotheses which can be tested subsequently using data from extant taxa.

Phylogenetic approaches to community ecology

Two traditions exist in the study of species diversity. Biogeographers and paleontologists have long been cognizant of historical influences on the structure of communities. By contrast, ecologists have been more concerned with the role that present-day processes play in shaping community composition. Of course, rather than being antagonistic, these approaches are complementary and should be integrated (for example MacArthur 1972; Ricklefs 1987). A phylogenetic context provides the appropriate setting for such integration (for example Gorman 1992).

Hypotheses derived from phylogenies

Phylogenetic comparative methods were developed to permit appropriate and statistically valid means of testing hypotheses (Gittleman 1982; Ridley 1983; Felsenstein 1985). A large body of literature has developed to elaborate on these methods (reviewed in Maddison and Maddison 1992; Miles and Dunham 1993;

Losos and Miles 1994), but these methods all have one feature in common: they are methods for using phylogenies to evaluate previously established hypotheses. By contrast, several workers have recently used phylogenies as the *source* of hypotheses, which can then be tested in the laboratory or in the field (Lauder 1989; Futuyma and McCafferty 1990; McLennan 1991).

In this chapter, I will use my studies of Caribbean *Anolis* communities to illustrate, first, the important role that phylogenetic approaches can play in studies of the causes of similarities and differences between communities and, second, the manner in which phylogenetic studies can suggest testable hypotheses. To accomplish these goals, I will review previously published work and provide preliminary conclusions from unpublished and on-going studies.

19.2 Background on Greater Antillean *Anolis* communities

Anolis is one of the most speciose vertebrate genera, with well in excess of 300 described species. Approximately 150 of these species occur on Caribbean islands, with the majority being found on the Greater Antilles (Cuba, Hispaniola, Jamaica, and Puerto Rico). Two patterns characterize the anole communities on these islands. First, on any of the islands, species within a community utilize different microhabitats. These ecological types, termed 'ecomorphs', display morphological and behavioural adaptations to their different microhabitats. For example, 'trunk–ground' anoles (ecomorphs are named for the microhabitats they most frequently utilize) have long legs and run and jump frequently on broad surfaces, whereas 'twig' anoles have short legs and move slowly on narrow surfaces.

Most remarkably, when one compares communities across islands, one finds essentially the same set of ecomorphs represented on each island. Despite uncertainty about higher-level anole phylogeny (Guyer and Savage 1986, 1992; Cannatella and de Queiroz 1989; Williams 1989; Hass *et al.* 1993), we can conclude confidently that each of the ecomorphs has evolved independently on different islands a minimum of three times (the extent to which the Cuban and Hispaniolan radiations are independent of each other is not clear). Of the six ecomorph types, four occur on all islands (trunk–ground, trunk–crown, crown-giant, and twig), one occurs on all islands except Jamaica (grass–bush ecomorph), and one occurs only on Cuba and Hispaniola (trunk ecomorph).

To determine whether the ecomorph concept is valid, I measured the morphology and ecology of adult males of a large number of species. Examination of the position of the species in a multidimensional morphological space indicates that species group by ecomorph type, rather than by phylogeny (Fig. 19.1). Similarly, the ecomorphs differ in habitat use, although there is some overlap between the crown-giant, trunk, and trunk–crown ecomorphs (Fig. 19.2). Thus, the ecomorph concept is valid; the same ecomorph types have evolved convergently multiple times.

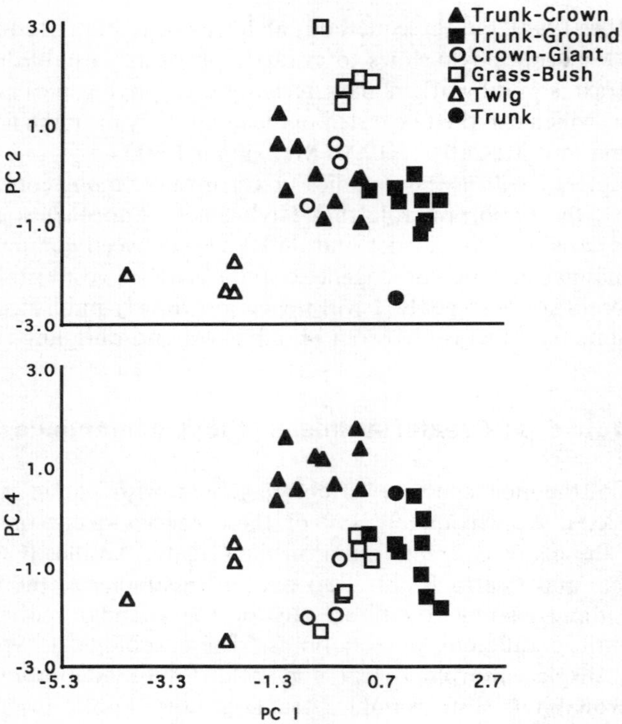

Fig. 19.1 Position of species in a morphological space determined by a principal components analysis. Effects of size were removed from the variables; a variable representing overall size was also included in the analysis. PC 3 is not shown because it loads only for size and serves only to distinguish the crown-giants from the other ecomorphs. Species include at least one representative of each ecomorph type on each of the Greater Antilles, except that data are not available for several Cuban ecomorphs (from Losos and de Queiroz, submitted).

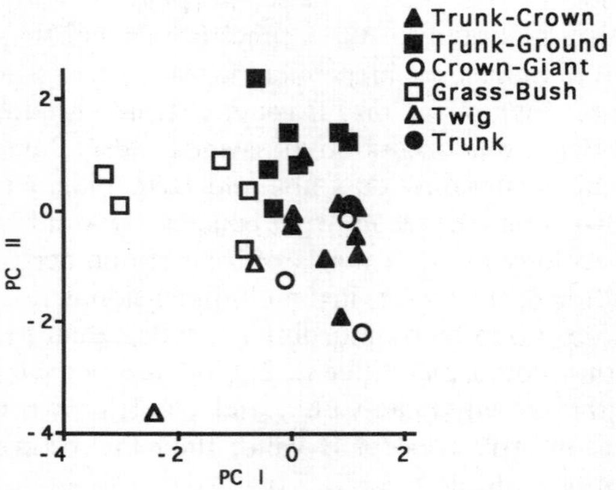

Fig. 19.2 Habitat use by the ecomorphs (from Losos and de Queiroz, submitted).

19.3 Evolution of community structure

Two models—the character displacement and microhabitat specialization hypotheses—have been put forward to explain patterns of community evolution in Greater Antillean anoles (Williams 1972; Losos 1992). Both models invoke interspecific competition as the driving force behind community evolution. The implication of interspecific competition is reasonable because a large body of evidence affirms its importance in extant anole communities. This evidence comes from a variety of sources, including behavioural studies, geographic and temporal comparisons, analyses of the effects of introduced species on native species, and experimental studies (reviewed in Losos 1994).

In a study well ahead of its time, Williams (1972) mapped ecological and morphological characters onto a phylogeny of Puerto Rican *Anolis*. Based on this exercise, Williams concluded that early diversification within the Puerto Rican radiation was a result of two instances of character displacement in body size producing a community composed of three species divergent in size but all living high in the tree. Subsequent evolutionary events then led to species moving out of the crown and occupying other habitats.

Re-examination of this analysis using parsimony techniques to reconstruct the evolution of body size indicates that Williams' analysis was correct: the first two divergence events in Puerto Rico were accompanied by substantial divergence in body size (Losos 1992; Fig. 19.3). However, since Williams' analysis, a phylogeny for the anoles of Jamaica has become available (Hedges and Burnell 1990; Losos 1992). Reconstruction of character evolution on this island provides no evidence of character displacement in the early stages of the Jamaican radiation (Fig. 19.3).

An alternative hypothesis is that the addition of new species to an expanding anole community forces the species to alter their habitat use to minimize competitive pressures. Substantial evidence exists for such habitat resource partitioning among extant anole species (Losos 1994). Over evolutionary time, one would expect these species to evolve adaptations to their altered habitat use.

Note that this hypothesis does not specify how speciation occurs in anoles. Rather, it only requires that once speciation has occurred, the result is two ecologically similar species that compete for resources when they come into sympatry. Little is known about how speciation in anoles occurs, although most discussion has invoked scenarios of allopatric speciation (Losos 1994).

To test the hypothesis that the addition of species to a community leads to shifts in habitat use, I employed parsimony to reconstruct the evolution of the ecomorphs in the Jamaican and Puerto Rican anole radiations (Losos 1992; Fig. 19.4). I based this analysis on morphology, but because morphology and ecology are intimately linked in anoles (Moermond 1979; Pounds 1988; Losos 1990*a*) this analysis also indicates the evolution of habitat use.

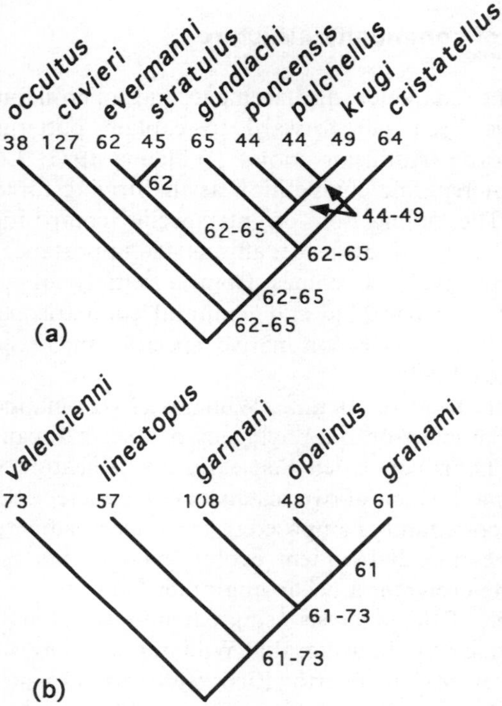

Fig. 19.3 Size evolution (as represented by snout–vent length) in (a) Puerto Rico and (b) Jamaica. Parsimony was used to reconstruct the size of ancestral taxa. In some cases there was a range of equally parsimonious possibilities for an ancestral taxon; this ambiguity does not affect interpretation of general patterns of size evolution. The first two divergence events in Puerto Rico produced communities with small and medium (two species) and small, medium, and large (three species), in agreement with the character displacement hypothesis. By contrast, substantial change in size did not occur in Jamaica in early divergence events. From Losos (1992) with permission.

In Jamaica, the analysis indicates that the initial divergence event led to a species clearly identifiable as a twig anole and another that occurs in a central position in morphological space that does not correspond to the location of any of the ecomorph types. Due to its central position, I refer to this taxon as a generalist (Fig. 19.5). The next divergence event produces a three-species community comprising a twig anole, a trunk–ground anole, and a species occurring in the crown of the tree, but not clearly identifiable as either a trunk–crown or a crown-giant anole. Finally, the four-ecomorph community contains the four ecomorphs present in Jamaica today.

In Puerto Rico, community evolution has followed an almost identical trajectory. The first divergence event again produces a twig anole and a generalist. The three-ecomorph stage comprises a twig anole, a trunk–ground anole, and again a species in the canopy, only in Puerto Rico this taxon is clearly

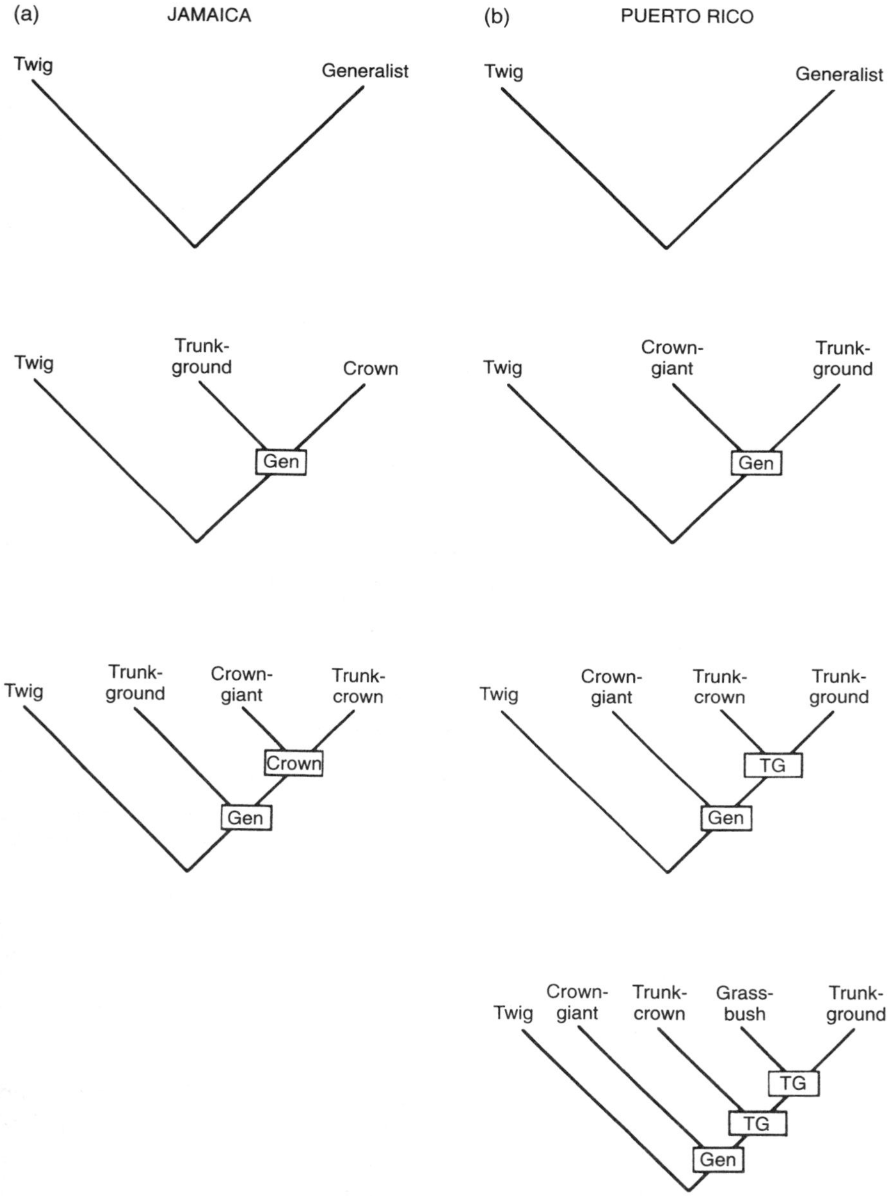

Fig. 19.4 Ecomorph evolution in Jamaica (a) and Puerto Rico (b). Parsimony was used to reconstruct the ecomorph type of ancestral data. From Losos (1992) with permission.

identifiable as a crown-giant. The four-ecomorph community is identical to that in Jamaica, and the fifth ecomorph to evolve in Puerto Rico is the one absent in Jamaica, the grass–bush anole.

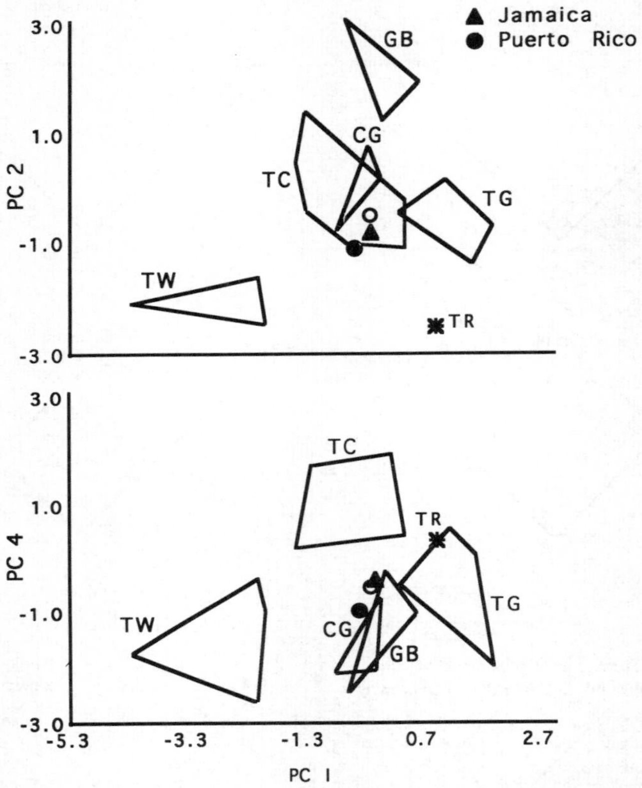

Fig. 19.5 Position of hypothetical ancestral taxa in morphological space. The outlines refer to the position of the ecomorphs in Fig. 19.1. The full symbols are the inferred morphological position of the generalized ancestral taxa at the two-species stage of community evolution (Fig. 19.4). Open circles represent the hypothetical position of a generalist species, which was determined by taking the mean position of the centroids for each of the ecomorphs. The hypothetical generalist and the inferred ancestral taxa are similar on PC 3 (which is not shown) and all are greatly different from crown-giant anoles on this axis. Consequently, the two inferred ancestral taxa are more similar to the generalist than they are to any of the ecomorphs (Losos and de Queiroz, submitted).

Thus, this analysis indicates that anole communities in Puerto Rico and Jamaica have evolved by sequential occupation of new habitats, as predicted by the habitat specialization hypothesis. Further, not only are communities convergent in structure today, but they have achieved this convergence by evolving through a nearly identical sequence of intermediate communities. This is a finding that would not necessarily be predicted by community ecological theory and that could not be made except in the context of an historical approach to community structure. Further, this similarity in community evolution trajectory suggests the hypothesis that there is a deterministic pathway by which anole communities evolve. This hypothesis, if correct, holds interesting implications for adaptive radiation and the anole adaptive landscape (Losos 1992).

One might question to what extent these results might be an artifactual consequence of the use of parsimony to reconstruct the sequence of evolution of ecomorphs. Certainly, the reconstruction of ancestral states by parsimony is prone to error, particularly when the assumptions underlying parsimony (for example rates of change being relatively slow and constant among lineages) are not met (see Maddison and Maddison 1992). Consequently, I would not be unduly confident that the reconstructed ecomorph type of any particular ancestor is correct. However, the most important finding of this study is not the ecomorph state of any particular ancestor, but, instead, the conclusion that the sequence of ecomorph evolution has been virtually identical on the two islands. Although uncertainty may exist in parsimony reconstructions, there is no reason to suspect that parsimony would artifactually reconstruct the same sequence of ecomorph evolution in two independent radiations. Indeed, a randomization analysis indicated that reconstructing two sequences of community evolution as similar as those reconstructed for Puerto Rico and Jamaica is significantly unlikely to have resulted by chance (Losos 1992).

The historical approach provides another insight into anole communities that would not be apparent without recourse to a phylogeny. A number of hypotheses based on current conditions could be proffered to explain why Jamaica lacks the grass–bush ecomorph found on the other three islands. For example, perhaps Jamaica does not have appropriate habitats (unlikely because grass is, in fact, readily available in Jamaica) or perhaps these habitats have been usurped by other taxa. However, the historical perspective indicates that the wrong question is being asked (Williams 1972; Losos 1992). The grass–bush anole is the fifth ecomorph to evolve in the ecomorph sequence, but Jamaica has only advanced to the four-ecomorph stage. Hence, the appropriate question is not 'Why is there no grass–bush anole in Jamaica?', but, rather, 'Why has Jamaica only progressed to the four-ecomorph stage?'. Rather than focusing on the grass–bush anole *per se*, this approach would concentrate on why Jamaica has not developed as many ecomorphs as the other islands. Possibilities include the age of Jamaica (and how long it has been above seawater) and whether Jamaican anoles are more tightly packed in ecological space than other Greater Antillean anole communities.

The hypotheses developed in this analysis are testable in two ways. First, the evolution of community structure has been examined in only two of the four Greater Antillean islands. As our understanding of anole phylogeny progresses, we will be able to examine whether communities have evolved in the same sequence on all islands. In particular, we would expect the grass–bush anole to evolve fifth on all islands and the trunk ecomorph, absent from Jamaica and Puerto Rico, to be the last to evolve on the other two islands. Second, we can test predictions stemming from this analysis about the role of interspecific competition in driving anole adaptive radiation.

19.4 Using phylogenetic approaches to derive testable hypotheses

The phylogenetic analysis suggests that the driving force behind the anole adaptive radiation is interspecific competition, which forces species to use new habitats to which they subsequently adapt. This model of community evolution suggests two predictions that can be tested among extant anole populations:

1. The presence of competing anole species should lead to shifts in habitat use.
2. Anole populations using new habitats should adapt morphologically.

Several colleagues and I have conducted experimental studies to address these questions. I will begin with the second prediction, for which the experiments are complete, and then discuss preliminary results from an on-going experiment that examines both predictions.

Populations using new habitats should adapt morphologically

In the late 1970s and early 1980s, Thomas Schoener introduced small propagules of *A. sagrei* to tiny islands ('rocks') in the vicinity of Staniel Cay, Exuma Cays, Bahamas (Schoener and Schoener 1983). These islands do not naturally support anoles; assuming that the islands were thus too small to maintain viable populations, Schoener conducted the introductions to study the process of extinction. In fact, most of the populations have survived and several have exploded in population size (Schoener and Schoener 1983). In retrospect, occasional hurricanes, rather than small island size, probably account for the lack of lizards on these islands (Schoener and Schoener 1983).

The scrubby small-diameter vegetation covering the experimental islands differs considerably from the large trees found at the locality of the source population on Staniel Cay. Further, the experimental islands differ among themselves in vegetational characteristics. Based on the relationship between limb length and mean perch diameter among Greater Antillean anoles (Losos 1990*a*; Losos *et al.* 1994), we made two predictions: first, that the experimental populations should have shorter limbs than the source population and, second, that among the experimental islands, a relationship should exist between limb length and perch diameter.

Examination of the populations in 1991 supported both predictions (J. Losos, K. Warheit, and T. Schoener, unpublished). The experimental populations have evolved non-randomly with regard to limb length; 12 of 14 experimental populations have shorter limbs than the source population. Further, among populations, a significant positive relationship exists between limb length and perch diameter. At present, laboratory rearing experiments are under way to determine whether such results could be the result of phenotypic plasticity; preliminary results are equivocal.

Thus, these results indicate that over a period of approximately 20 years (40 lizard generations at most), these populations have evolved adaptively in response to their new environmental setting. Although statistically significant, these morphological differences are relatively minor in absolute terms. However, comparisons of Bahamian populations of *A. sagrei* and *A. carolinensis*, which have been diverging for periods from 10 000 to hundreds of thousands of years, reveal the same trend in both species, only at a greater scale (Losos *et al.* 1995). This congruent relationship between limb length and perch diameter, manifested at different scales in species diverging for millions of years, populations for tens or hundreds of thousands of years, and experimental populations for twenty years, provides strong evidence for the adaptive nature of the relationship.

The presence of competing species leads to shifts in habitat use and subsequent adaptation

The experiment just discussed indicates that anole populations will adapt, and rapidly, to new environmental circumstances. However, the study does not indicate that interspecific competition can drive this evolution by causing shifts in habitat use. Habitat shifts in response to the presence of competitors have been noted repeatedly for anoles (reviewed in Losos 1994), including Bahamian populations of *A. sagrei* and *A. carolinensis*. None the less, few studies have documented that these shifts result in adaptive evolution (but see Lister 1976).

To examine this hypothesis, David Spiller and I have introduced populations of *A. sagrei* and *A. carolinensis* to small islands in the Exuma Cays, Bahamas. Some islands only received one species, whereas both species were placed on other islands. One year after the introduction, both species had achieved higher population densities on islands on which they were allopatric compared with islands on which they co-occurred. These results indicate the existence of interspecific interactions, but do not reveal the mechanism; possibilities include exploitative competition, interference competition, and intra-guild predation. We will continue monitoring these populations to determine whether interspecific effects are apparent in subsequent years and whether the populations show evolutionary responses to these effects.

Is evolutionary specialization a one-way street?

The phylogenetic analysis of community evolution suggests another testable hypothesis. If specialization to specific microhabitats is driven by increases in the number of sympatric congeners, as the phylogenetic analysis suggests, then if these competitors are removed, one might expect a species to reverse evolutionary direction and become less specialized. A number of small islands in the northern Caribbean contain one or two species that have clearly descended from a trunk–ground or trunk–crown anole. These species provide an appropriate test of the hypothesis that in the absence of competitors, specialized species should evolve a more generalized condition. The null hypothesis would be that these species have remained specialized and are still identifiable as a member of their ancestral ecomorph type.

An initial test of this hypothesis focused on the anoles of the Bahamas, which exhibit islands containing one to four species of anoles (Losos *et al.* 1995). We focused on *A. carolinensis* and *A. sagrei*, which both occur allopatrically on some islands. Of the eight populations of *A. carolinensis* and 13 of *A. sagrei*, none provided evidence of evolutionary generalization; all were clearly identifiable as trunk-crown (*A. carolinensis*) or trunk–ground (*A.* sagrei) anoles.

A subsequent analysis (Losos and de Queiroz, submitted) examined 10 species, five descended from trunk-crown anoles and five from trunk–ground anoles, on islands elsewhere in the Caribbean (Cayman Islands, Desecheo, Inagua, Mona, Navassa, St Croix, and Swan Island). Whereas most of the trunk–crown anoles had diverged relatively little from their ancestral mould, several of the trunk–ground anoles had diverged substantially. Consequently, these data suggest that some members of some ecomorph types are more capable than others of reversing evolutionary direction and becoming less specialized when environmental conditions change. Thus, we concluded that evolutionary specialization is a two-way thoroughfare for some ecomorphs and lineages, but only a one-way avenue for others.

Anolis radiation in the Lesser Antilles

The Lesser Antilles are considerably smaller than the Greater Antilles and are occupied only by one to two species of *Anolis*. Williams (1972) suggested that the simple faunas of the Lesser Antilles might be comparable to early stages in the Greater Antillean radiation. If this were the case, then Lesser Antillean anoles should be similar to inferred early stages in Greater Antillean community evolution.

Examination of Lesser Antillean anoles does not offer much support to this hypothesis (Losos and de Queiroz, submitted). Most Lesser Antillean species, including all species on one-species islands (the species for whom selective pressure to become a generalist might be strongest), are identifiable as trunk–crown ecomorphs. The phylogenetic analysis, however, suggests that generalists occurred in the early stages of Greater Antillean community evolution and that trunk–crown ecomorphs did not evolve until the four-ecomorph stage.

Hence, we conclude that the Lesser Antilles are not analogous to early stages in the evolution of Greater Antillean communities, as Roughgarden and Pacala (1989) have argued. Why community evolution should differ between the Greater and Lesser Antilles is not immediately obvious. Possible explanations include differences in the non-anole faunas of the two islands (the Lesser Antilles are relatively depauperate faunistically) or differences in the initial starting conditions of the respective radiations. A better phylogeny for *Anolis* will allow evaluation of this latter hypothesis.

Another possibility is that the phylogenetic analysis has reconstructed incorrectly the early stages of Greater Antillean community evolution. This explanation is unlikely because the most basal lineage in both Jamaica and Puerto Rico is a twig anole, but nothing at all similar to a twig anole occurs in the Lesser Antilles. Evolution in the Greater Antilles would have to be both remarkably non-parsimonious and parallel if the early stages of the Greater Antillean radiation were actually similar to communities seen today in the Lesser Antilles.

19.5 Conclusions

For years, many ecologists ignored history in their studies of community structure. Implicitly, such studies assumed that present-day processes are sufficiently powerful to lead to the same result regardless of historically determined initial conditions. Recent years, however, have seen a widespread appreciation of the importance of historical effects; consequently, studies of community organization are increasingly incorporating phylogenetic information (for example Losos 1990*b*, 1992; Richman and Price 1992; Gorman 1992; Grandcolas 1993).

None the less, a distinction must be made between which questions can be addressed by historical studies and which cannot. In particular, historical studies investigate patterns of evolutionary change through time, but cannot directly examine whether a particular process operated historically. Of course, one can make predictions about what patterns should be observed if a process has been important over time and then compare such predictions with observed patterns. But such studies can only demonstrate consistency of results with predictions, rather than directly demonstrating the operation of a particular process over evolutionary time.

The most effective way to test whether a particular process has been important in community evolution is to show that: (1) the process is important in extant communities; (2) over microevolutionary time, the process leads to evolutionary change in the predicted direction; and (3) historical analyses are consistent with the process being the driving force behind evolutionary change (Losos 1994). I suggest that these conditions have been met for the hypothesis that interspecific competition is the driving force behind adaptive radiation in Caribbean *Anolis* lizards.

More generally, these studies highlight the important role that phylogenetic approaches can play in the study of ecological phenomena. Such approaches are important not only for testing hypotheses, but also as the source of hypotheses that are ecologically testable. Indeed, without considering historical information, ecologists risk not only getting the wrong answers, but not even asking the correct questions.

Acknowledgements

Thanks to D. Irschick and J. Bergelson for helpful comments and advice, to P. Harvey for inviting me to participate in the Royal Society symposium, and to the National Science Foundation for support (DEB 9318642).

References

Cannatella, D. C. and de Queiroz, K. (1989). Phylogenetic systematics of the anoles: is a new taxonomy warranted? *Syst. Zool.*, **38**, 57–68.

Felsenstein, J. (1985). Phylogenies and the comparative method. *Am. Nat.*, **125**, 1–15.

Futuyma, D. J. and McCafferty, S. S. (1990). Phylogeny and the evolution of host plant associations in the leaf beetle genus *Ophraella* (Coleoptera, Chrysomelidae). *Evolution*, **44**, 1885–913.

Gittleman, J. L. (1982). The phylogeny of parental care in fishes. *Animal Behav.*, **29**, 936–41.

Gorman, O. (1992). Evolutionary ecology and historical ecology: assembly, structure, and organization of stream fish communities. In *Systematics, historical ecology, and North American freshwater fishes*, (ed, R. L. Mayden), pp. 659–88. Stanford University Press.

Grandcolas, P. (1993). The origin of biological diversity in a tropical cockroach lineage: a phylogenetic analysis of habitat choice and biome occupancy. *Acta Oecol.*, **14**, 259–70.

Guyer, C. and Savage, J. M. (1986). Cladistic relationships among anoles. *Syst. Biol.*, **35**, 509–31.

Guyer, C. and Savage, J. M. (1992). Anole systematics revisited. *Syst. Biol.*, **41**, 89–110.

Hass, C. A., Hedges S. B., and Maxson L. R. (1993). Molecular insights into the relationships and biogeography of West Indian anoline lizards. *Biochem. Syst. Ecol.*, **21**, 97–114.

Hedges, S. B. and Burnell, K. L. (1990). The Jamaican radiation of *Anolis* (Sauria: Iguanidae): an analysis of relationships and biogeography using sequential electrophoresis. *Carib. J. Sci.*, **26**, 31–44.

Lauder, G. V. (1989). Caudal fin locomotion in ray-finned fishes: historical and functional analyses. *Am. Zool.*, **29**, 85–102.

Lister, B. C. (1976). The nature of niche expansion in West Indian *Anolis* lizards II. Evolutionary components. *Evolution*, **30**, 677–92.

Losos, J. B. (1990*a*). Ecomorphology, performance capability, and scaling of West Indian *Anolis* lizards: an evolutionary analysis. *Ecol. Monogr.*, **60**, 369–88.

Losos, J. B. (1990*b*). A phylogenetic analysis of character displacement in Caribbean *Anolis* lizards. *Evolution*, **44**, 558–69.

Losos, J. B. (1992). The evolution of convergent structure in Caribbean *Anolis* communities. *Syst. Biol.*, **41**, 403–20.

Losos, J. B. (1994). Integrative approaches to evolutionary ecology: *Anolis* lizards as model systems. *Ann. Rev. Ecol. Syst.*, **25**, 467–93.

Losos, J. B. and Miles, D. B. (1994). Adaptation, constraint, and the comparative method: phylogenetic issues and methods. In *Ecological morphology: integrative organismal biology*, (ed. P. C. Wainwright and S. M. Reilly), pp. 60–98. University of Chicago Press.

Losos, J. B., Irschick, D. J., and Schoener T. W. (1994). Adaptation and constraint in the evolution of specialization of Bahamian *Anolis* lizards. *Evolution*, **48**, 1786–98.

MacArthur, R. (1972). *Geographical ecology*. Princeton University Press.

McLennan, D. A. (1991). Integrating phylogeny and experimental ethology: from pattern to process. *Evolution*, **45**, 1773–89.

Maddison, W. and Maddison, D. (1992). *MacClade version 3: analysis of phylogeny and character evolution*. Sinauer, Sunderland, MA.

Miles, D. B. and Dunham, A. E. (1993). Historical perspectives in ecology and evolutionary biology: the use of phylogenetic comparative analyses. *Ann. Rev. Ecol. Syst.*, **24**, 587–619.

Moermond, T. C. (1979). Habitat constraints on the behavior, morphology, and community structure of *Anolis* lizards. *Ecology*, **60**, 152–64.

Pounds, J. A. (1988). Ecomorphology, locomotion, and microhabitat structure: patterns in a tropical mainland *Anolis* community. *Ecol. Monogr.*, **58**, 299–320.

Richman, A. D. and Price, T. (1992). Evolution of ecological differences in the old world leaf warblers. *Nature*, **355**, 817–21.

Ricklefs, R. E. (1987). Community diversity: relative roles of local and regional processes. *Science*, **235**, 167–71.

Ridley, M. (1983). *The explanation of organic diversity: the comparative method and adaptations for mating*. Oxford University Press.

Roughgarden, J. and Pacala, S. (1989). Taxon cycle among *Anolis* lizard populations: review of evidence. In *Speciation and its consequences*, (ed. D. Otte and J. A. Endler), pp. 403–32. Sinauer, Sunderland, MA.

Schoener, T. W. and Schoener, A. (1983). The time to extinction of a colonizing propagule of lizards increases with island area. *Nature*, **302**, 332–4

Williams, E. E. (1972). The origin of faunas. Evolution of lizard congeners in a complex island fauna: a trial analysis. *Evol. Biol.*, **6**, 47–89.

Williams, E. E. (1983). Ecomorphs, faunas, island size, and diverse end points in island radiations of *Anolis*. In *Lizard ecology: studies of a model organism*, (ed. R. B. Huey, E. R. Pianka, and T. W. Schoener), pp. 326–70. Harvard University Press, Cambridge, MA.

Williams, E. E. (1989). A critique of Guyer and Savage (1986): cladistic relationships among anoles (Sauria: Iguanidae): are the data available to reclassify the anoles? In *Biogeography of the West Indies: past, present, and future*, (ed. C. A. Woods), pp. 433–77. Sandhill Crane, Gainesville, FL.

20

The evolution of body plans: HOM/*Hox* cluster evolution, model systems, and the importance of phylogeny

Axel Meyer

20.1 Introduction

Most evolutionary biologists wish to explain evolutionary patterns in the astonishing diversity among organisms. Developmental biologists ultimately hope to explain the developmental mechanisms that are at the basis for the wide array of Baupläne that characterize and differentiate phyla. In trying to understand biological diversification, albeit at different levels of inquiry, is where the interests of developmental biologists and evolutionary biologists meet. Moreover, developmental processes evolve just like other aspects of organisms. However, development is, in most evolutionary research, treated like a black box and most developmental biologists in turn typically do not recognize the potential contribution that evolutionary biology can make to the understanding of development. Historically, the connection between ontogeny and phylogeny was well appreciated and also reflected in the interchangeable use of the word 'evolution' (von Baer 1828, 1864; Haeckel 1866; Gould 1977). Research on the development–evolution connection lay dormant for over a century and was only recently re-established by evolutionary biologists considering 'developmental constraints' and the timing of developmental events as factors in shaping and constraining the evolution of adult phenotypes (for reviews see Baldwin 1902; Waddington 1957; Gould 1977; Alberch *et al.* 1979; Goodwin *et al.* 1983; Raff and Kaufman 1983; Arthur 1984; Northcutt 1990; Wray and Raff 1991, Hall 1992; Wray 1992, 1995; Wake 1995).

The recent establishment of powerful molecular genetics methods has allowed developmental biologists to identify developmental control genes and some of their interactions in early development (for example Nüsslein-Volhard and Wieschaus 1980; reviewed in Lawrence 1992). Because of the time-consuming and laborious need to establish baseline data on development, developmental investigations can typically focus on only a very small number of animal model systems. The major model organisms in developmental biology are mouse, frog, zebrafish, sea urchin, fly, and nematode (this is obviously an incomplete list and other widely studied species include, for example, the salamander and the leech). These models are widely spread across the evolutionary tree of animals (Fig. 20.1) and their phylogenetic relationships are relatively undisputed. Developmental patterns and processes that are established from these model systems are assumed to be typical for a much larger number of species, at least those within the clade to which the model belongs. For example, lessons from 'the vertebrate models' are thought to apply to all vertebrates including man.

Currently, many developmental biologists place much importance in the development–evolution connection and attempt to explain the evolution of diverse body plans by changes in the HOM/*Hox* cluster architecture (e.g. Akam *et al.* 1988, 1994; Holland *et al.* 1994; Ruddle *et al.* 1994; Patel 1994 and references therein).

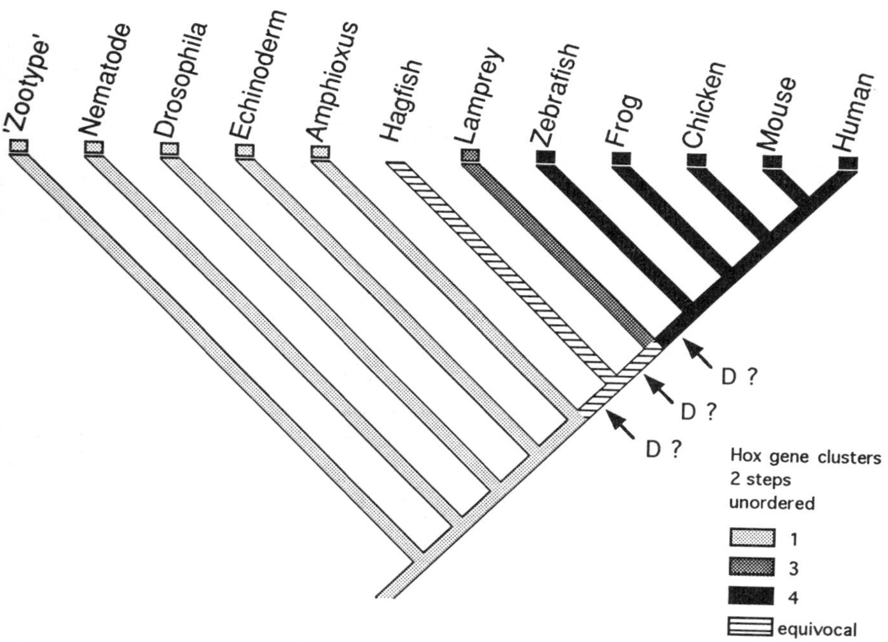

Fig. 20.1 Probable phylogenetic relationships among the major model systems used in developmental biology and some other crucial taxa for which the number of HOM/*Hox* clusters has been estimated or determined. At least two steps (evolutionary transitions) are required to go from the ancestral condition of a single HOM/*Hox* cluster to the presumed three HOM/*Hox* clusters in the lamprey and the presumed four-cluster condition in vertebrates. It is not known with certainty how many HOM/*Hox* clusters the hagfish and the lamprey (they are estimated to have three clusters) have, therefore the branch leading to the hagfish is drawn as 'equivocal'. This and several of the following figure are graphed and data analyzed with MacClade (Maddison and Maddison 1992). 'D' indicates where and when during metazoan evolution HOM/*Hox* clusters might have been duplicated.

In evolutionary biology, the comparative method has long been the favoured approach for addressing many different kinds of questions (for example on adaptation) (Ghiselin 1984; see references in Harvey and Pagel 1991; Brooks and McLennan 1991). The basic idea is to study the evolution of phenotypic characteristics of taxa (species or phyla) based on knowledge of the phylogenetic relationships among them. Therefore, at the core of the comparative method is a phylogeny, ideally firmly established and based on characters that are not going to be studied with the comparative approach. From this phylogeny the tracing of character evolution, developmental or otherwise, can be attempted. The modern use of the comparative method is based on the development of rigorous statistical and cladistic approaches to both the reconstruction of phylogenetic relationships and the study of character evolution (for example Swofford 1991; Maddison and Maddison 1992). The arrival of the polymerase chain reaction (PCR) as a powerful molecular method which greatly facilitates the gathering of molecular data for phylogenetic work coincided with the development of statistical methods in the comparative approach (Felsenstein 1985*b*; reviewed in Harvey and Pagel 1991; Brooks and McLennan 1991). Surprisingly, so far the comparative method in evolutionary biology has not been used to predict the evolution of developmental processes from evolutionary patterns of phylogenetic relationships (reviewed in Brooks and McLennan 1991; Harvey and Pagel 1991). Both major reviews of the comparative method consider the ontogeny–phylogeny connection only for the 'polarization' of characters, i.e. character states are treated as ancestral if they occur early, or derived if they occur later in development. The knowledge of the polarity of character state changes aids in the 'rooting' of phylogenetic trees.

Despite their recent interest in evolution, developmental biologists, typically have only made pairwise comparisons of developmental features; yet many of these comparisons have yielded highly interesting and often surprising results about evolutionary differences or similarities in development (for example *Drosophila* and *Tribolium* (Sommer and Tautz 1993) or nematodes (Sommer and Sternberg 1994); for similarity in early determination of polarity in *Drosophila* and *Caenorhabditis* see references in Kimble (1994)). However, pairwise comparisons have inherent limitations and do not provide nearly as much information as comparisons between more than two taxa, in an explicitly phylogenetic context (Garland and Adolph 1994). Figure 20.2 outlines one way in which a comparison involving more than two taxa in a phylogenetic context can be much more powerful than a pairwise comparison. A pairwise comparison between any two taxa (i.e. species or phyla for example) (from A to E) would not be able to establish whether two traits (for example the relative timing and domain of expression of two homeobox genes) evolved at the same time or in two consecutive steps, and, if the latter, in which order (Fig. 20.2a, b). However the comparison of the taxa under consideration (A–C) with more distantly related ones (D + E) might allow one to determine the sequence of evolutionary events. In this example (Fig. 20.2c) the comparisons of character state distributions between species of the two clades (A–C, D + E) might allow one to do so. This example suggests that in a first event the 'black box' evolved (in the common ancestor of both clades) followed in a second step by the evolution of the 'white box' in the common ancestor of the clade A–C.

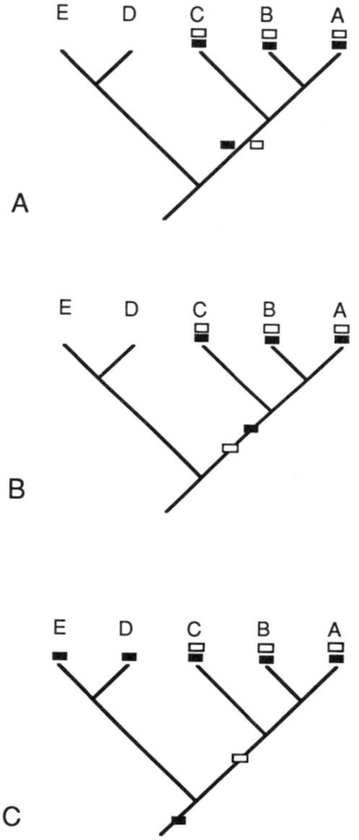

Fig. 20.2 Hypothetical phylogeny of five taxa (A–E). By comparing the character state distributions between more than two taxa some hypotheses about the origin of two traits are less likely than others. The phylogeny and distribution of character states suggests that the 'black' trait evolved before the 'white' trait which evolved along the branch shared by A–C and their common ancestor. C: This information would rule out hypothesis A (the evolution of both traits at the same time in the common ancestor of A–C) and hypothesis B (the consecutive evolution of the 'black' and 'white' trait (in no particular order) along the branch leading to taxa A–C).

There are few studies that included phylogenetic and ontogenetic information for more than two species (DeSalle and Grimaldi 1993; Luk *et al.* 1994; Patel 1994; Wray and Bely 1994). A small number of excellent phylogeny-based developmental studies exist comparing developmental patterns and the evolution of body plans in arthropods (Patel 1994; Akam *et al.* 1994 and references therein) and echinoderms (Raff 1992; Wray and Bely 1994, and references therein); these studies tend to be concerned with distantly related species. However, it is known that developmental mechanisms can differ dramatically between even closely related species, such as in direct and indirect developing sea urchins and amphibians (Elinson 1990; Wray and Raff 1991; Raff 1992; Jeffrey and Swalla 1992; Wray 1992, 1995; Henry & Martindale 1994). There are known instances where fundamentally different embryological trajectories result in phenotypically similar adults (for example in congeneric sea urchins (Raff 1992; Wray and Raff 1991)) and conversely, similar developments can result in strikingly different adult phenotypes, as is illustrated by the many cases of large morphological differences among closely related species. Therefore, the importance of comparative developmental work among closely related species (such as different species of zebrafish) in an explicitly phylogenetic context was stressed recently (Meyer *et al.* 1995).

Developmental biologists have made much progress by incorporating evolutionary thinking into their research agenda. However, the acceptance of evolution's relevance for the understanding of development has been somewhat incomplete. There are several ways in which knowledge about the evolution of, and specifically the phylogenetic relationships among, model systems can aid in the understanding of developmental processes (Kellog and Shaffer 1993, Meyer *et al.* 1993, 1995). Here I wish to point out that the comparative method can predict the likely condition in common ancestors, might permit the reconstruction of intermediate stages, might be able to determine the historical sequence of evolutionary events, and has the capacity to falsify or support hypotheses about the evolution of developmental processes.

The combination of (1) the dramatic advances in developmental biology, (2) the elaboration of statistical tests in the comparative method, and (3) the power of molecular datasets for phylogeny reconstruction, might now significantly facilitate progress on the understanding of the development–evolution connection. Unfortunately, many phylogenetic hypotheses about the relationships of animal phyla are still hotly debated. Strongly supported phylogenetic estimates must underlie all further work on the ontogeny–phylogeny connection if we hope to establish a causal relationship between ontogenetic changes, i.e. in evolution of HOM/*Hox* clusters, and the evolution of body plans.

19.2 Bauplan evolution and phylogeny of chordates: is there a correlation with the evolution of the HOM/*Hox* cluster architecture?

Homeobox (HOM/*Hox*) genes are found in all metazoans, for example in platyhelminth (Kenyon and Wang 1991; Bartels *et al.* 1993) and annelid worms (Dick and Buss 1994), cnidarians (Schierwater *et al.* 1991), and even plants (for review see De Robertis 1994; Gehring 1994). They code for a class of transcription factors defined by a helix-turn-helix motif with a 183-nucleotide core sequence that are involved in the regulation of developmental genes in animals. There are (at least) 38 *Hox* genes in mouse and human which are organized into four clusters (termed A–D or 1–4) of up to 13 members per cluster (reviewed in Scott 1992; Holland 1992; Garcia-Fernandez and Holland 1994; De Robertis 1994; Gehring 1994) (Fig. 20.3). Additionally, numerous individual homeobox genes that are not part of the HOM/*Hox* cluster are also present in most animals' genomes.

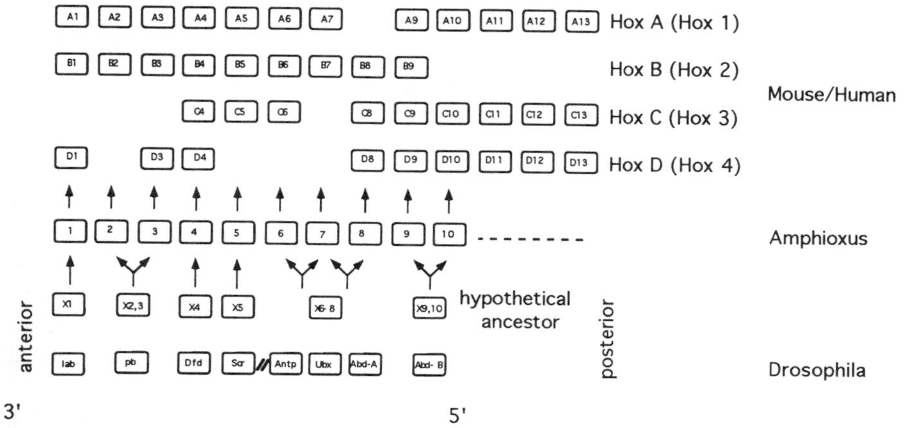

Fig. 20.3 Architecture of the HOM/*Hox* clusters of *Drosophila*, amphioxus, the mouse, and the presumed ancestral condition of the cluster (Redrawn after Garcia-Fernandez and Holland 1994.)

The major model systems in developmental biology differ in their number of HOM/Hox clusters (Figs. 20.1, 20.3), but how and when during evolution the number of clusters increased from the single ancestral cluster to the derived tetrapod condition of four is not known (see Fig. 20.1). The relative position of individual *Hox* genes in the HOM/*Hox* gene cluster defines, among other features, the sequence of transcription during development of the anterior–posterior axis, sensitivity to retinoic acid, and expression and relative positions of developing structures in the embryo (Scott and Carroll 1987; McGinnis and Krumlauf 1992; Slack *et al.* 1993; Marshall *et al.* 1994; Warren *et al.* 1994 and references therein). The increasing complexity in the homeobox cluster number and architecture has been hypothesized to be related to the increasing complexity in body plans among phyla of animals.

The assumed omnipresence of homeobox genes arranged into a HOM/*Hox* cluster of common architecture in animals recently led to the suggestion that a common, defining (in cladistic terminology, synapomorphic) character of all animals and their hypothetical ancestor (the 'zootype') is the presence of this *Hox* gene cluster (Slack *et al.* 1993) (Fig. 20.1). The 'zootype' concept, and that of the existence of phylotypic stages, are gaining in acceptance also outside the developmental biology community (for example Minelli and Schram 1994).

During metazoan evolution there were probably only three homeobox genes in the ancestral metazoan homeobox cluster (Akam *et al.* 1988; Kappen *et al.* 1989; Murtha *et al.* 1991; Holland 1992; Schubert *et al.* 1993). It is not known exactly how many and which class of homeobox genes made up the single ancestral chordate HOM/*Hox* cluster (Garcia-Fernandez and Holland 1994). It probably included five (Schubert *et al.* 1993) or six homeobox genes (Garcia-Fernandez and Holland 1994) (Fig. 20.3). Based on sequence comparisons it has been established that during deuterostome evolution several tandem duplications of the most 5′ group of homeobox genes (from a single gene that is homologous to the *Drosophila Abd-B*) occurred. Echinoderms might have only one *Abdominal B* (*Abd-B*) related gene (cognate group 9) and amphioxus (a cephalochordate) has at least two (cognate groups 9 and 10) (Garcia-Fernandez and Holland 1994) in their single cluster. All vertebrates that have been investigated have up to five *Abd-B* related genes, increasing the total number of cognate groups to 13 and the cluster number to four (see for example Pavel and Stellwag 1994) (Fig. 20.3). Not all four clusters contain the same number (due to independent deletions and duplications of genes) or members of the same *Hox* 'cognate groups', and only group 4 and group 9 homeobox genes are found in all four homeobox gene clusters (Fig. 20.3). Therefore, the exact homologies between homeobox genes of the derived vertebrate condition and the ancestral chordate one, or even more distantly related insect and worm clusters, are not entirely clear (see for example Garcia-Fernandez and Holland (1994) and references therein).

The substantial similarity, in terms of relative timing and position of expression, of mouse, human, and *Drosophila* homeobox genes suggests that HOM/*Hox* cluster and function is strikingly conserved over huge evolutionary time spans (Graham *et al.* 1989). Evidence for the astonishingly conserved positional information during development of homeobox genes also comes from transgenic experiments. When mouse genes (*Hoxb-6*) were expressed in *Drosophila*, ectopic mutations (antennal legs) were induced (Malicki *et al.* 1990) showing that anterior–posterior axis determination in flies and vertebrates are similarly controlled and conserved. Elegant 'promoter swap' experiments in mice provide another similar approach (Lufkin *et al.* 1991). Here, the control of the expression of *Hoxd-4* gene was placed under the control of the promoter of the *Hoxa-1* gene in transgenic mice. The effect was that the expression of *Hox-4.2* (and probably also those of several downstream effectors), which typically occurs in the cervical vertebrae of mice, was moved anteriorly to a region including the occipital bones which were homeotically transformed into structures that resembled cervical vertebrae. Other characteristic homeotic transformations were discovered in earlier classic experiments on transgenic mice (involving *Hox-1.1*) and retinoic acid treated mice (Kessel *et al.* 1990 and references therein).

It seems paradoxical that, at the DNA level and at the level of gene order, the homeobox genes and HOM/*Hox* cluster architecture are evolutionarily so conserved, yet they are often said to be the cause of, or at least correlated with, the diversity of body plans that differentiate phyla of animals (Akam *et al.* 1988; Holland 1992; see references in Akam *et al.* 1994). The cause for Bauplan evolution therefore cannot simply be the gene order of homeobox genes within a cluster, which, as far as is known, is conserved across species in different phyla with profoundly different body plans. It is more likely that the increase in HOM/*Hox* cluster number (and variation in the number of homeobox genes per cluster), changes in function of individual homeobox genes, co-option into new roles, changes in the regulation of expression of homeobox genes, and increased complexity in the nexus of communication between these individual transcription factors are all partly responsible for increased complexity of body plans during evolution (Akam *et al.* 1988; Holland 1992; references in Akam *et al.* 1994).

The increasing complexity at the genetic level in terms of both numbers of genes per cluster and number of clusters could be responsible for the increasing complexity of development and adult morphology throughout evolution in several ways. Gene duplications, which free up the old or the new copy of a gene, or group of genes, to take on a new function are possibly one of the major forces of molecular evolution that can lead to the evolution of new function and novelty (Ohno 1970; Zuckerkandl 1994; Walsh 1995). Duplications of homeobox genes would free up these transcription factors to take on new functions. Alternatively, the regulatory control of the expression of genes is also likely to be involved, and may be an even more important force in evolution than gene duplications (see review in Wilson *et al.* 1977).

Within the phylum Chordata, it was suggested that serial homology of fins in fishes and the origin of the tetrapod limb might be due to the ectopic expression and duplication of some *Abd-B* related genes (Ahlberg 1993; Tabin and Laufer 1993). Moreover, it was suggested that the evolutionary origin and transition from paired fins to the tetrapod pentadactyl limb is related to the above-mentioned cluster duplications and tandem duplications from a single to a final number of five *Abd-B* related genes per cluster (cognate groups 9–13) (Tabin 1992; but see Coates 1994; Favier *et al.* 1995). These hypotheses could be addressed in a phylogenetic framework since the duplications of these genes should not post-date the evolutionary origin of fins and pentadactyl limbs—if they are really causally related to the origin and increasing complexity of paired appendages. However, for this set of hypotheses to be tested much more information on the phylogeny of chordates, homeobox cluster architecture, and the mode and timing of cluster duplications during vertebrate evolution remains to be collected (e.g. Fig. 20.1). Since the homeobox cluster has been mapped in only a single cephalochordate it is unclear in which of several possible ways the postulated duplications of *Abd-B* related genes and the *Hox* gene clusters occurred.

During the evolution of chordates the ancestral chordate cluster was duplicated (in at least two duplication events) to the vertebrate condition of four clusters. The duplications from the one ancestral chordate cluster in amphioxus to the four clusters in all (?) vertebrates must have occurred after the evolution of cephalochordates (Garcia-Fernandez and Holland 1994). PCR-based approaches applied to lampreys suggest that they have at least three clusters (reviewed in Ruddle *et al.* 1994) (Fig. 20.1). It is not clear which of the four vertebrate clusters (A–D) is the most ancient and most closely related to the ancestral chordate cluster (Garcia-Fernandez and Holland 1994) (Fig. 20.2). There are 15 possible relationships (bifurcating trees) among the four clusters that relate the four gene clusters to an ancestral one (Felsenstein 1978). Therefore, there would be 15 possible bifurcating relationships (discounting the possibility that more clusters first evolved and then were later deleted) if the four clusters arose by individual duplication events (similar to speciation or bifuraction events). However, only two evolutionary events would be required if whole-genome duplication events (increases in ploidy) caused the increase in HOM/*Hox* cluster number from the single-cluster ancestral condition to the presumed typical four-cluster vertebrate condition, with two clusters being intermediate. There are other duplicated genes up- and downstream of these clusters (for example keratin-coding genes) that are duplicated (R. Krumlauf, personal communication). This would seem to support the hypotheses that two whole-genome duplication events during the evolution of the chordates led to the presumed typical four-cluster condition in tetrapods. However, if HOM/*Hox* cluster evolution proceeded by whole-genome duplications then, if taxa with three HOM/*Hox* clusters were to be found, it would seem to imply that the fourth cluster was subsequently lost. For example, if agnathan fish are monophyletic, and if the estimate of three clusters in lampreys is correct then probably one cluster was lost independently in lampreys. In this case, two duplication events must be postulated after the splitting off of amphioxus from the stem leading to higher chordates, and hagfish possibly already possessed four HOM/*Hox* clusters (Fig. 20.1). Alternative scenarios could be constructed based on the model of cluster evolution, and the phylogeny of chordates. However, since it is not firmly established (i) how many HOM/*Hox* clusters hagfish and lampreys have, (ii) how the cluster duplications and possible deletions occurred, and (iii) what the phylogeny of chordates is, these alternative scenarios remain just speculations.

A preliminary phylogenetic analysis using *Drosophila*, amphioxus, and the four mouse clusters based on cognate group 9 amino acid sequences of the homeobox domain and flanking regions from Garcia-Fernandez and Holland's study (1994), analysed with PROTPARS in PAUP (Swofford 1991), found only weak support for one of the 15 possible relationship (Fig. 20.4). This very preliminary analysis suggests a gene tree relationship between the four clusters in which the B cluster is the most ancestral, the C + D clusters are the most derived and the A cluster is the next most closely related one to the C + D group (Fig. 20.4). The amino acids of group 4 *Hox* genes, which is the only other group of *Hox* genes that is present in all four vertebrate HOM/*Hox* clusters, did not allow any resolution since two equally parsimonious solutions were found, the consensus of which was a completely unresolved tree. Unfortunately, the *Hox* genes are highly conserved and do not contain much phylogenetic information at the DNA or amino acid level. Future phylogenetic work on this question will need to consider the inclusion of more extensive and more variable flanking regions.

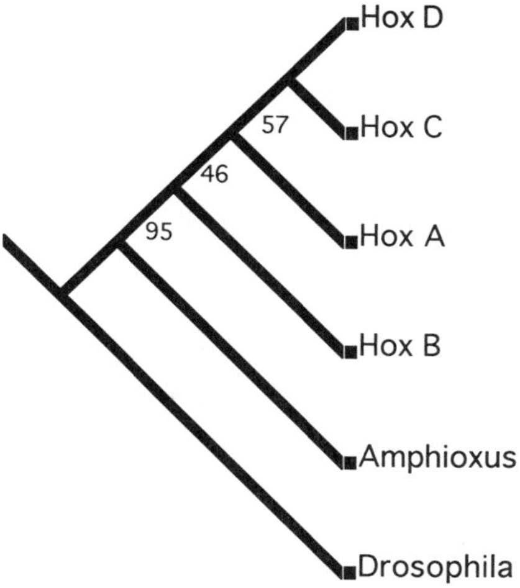

Fig. 20.4 The preliminary gene tree supported most strongly by a parsimony analysis. Numbers indicate bootstrap values for chordates (Felsenstein 1985) in 100 replications with PAUP (Swofford 1991).

In addition to the importance of phylogenetic knowledge (also see below) alternative models of character evolution need to be considered (an incomplete list is provided in Fig. 20.5) when one wants to attempts the reconstruction of historical events using the comparative method. Several alternative models of character evolution exist, the most unrestrictive is 'unordered' parsimony which permits the change from one character state to any other in a single evolutionary event (Fig. 20.5). Other more restrictive models of character change than 'unordered' parsimony, such as 'Dollo' parsimony, will always require more explanations, i.e. more steps in the reconstruction of evolutionary history (Figs. 20.5 and 20.6). For example, in 'ordered parsimony' the evolution from one character state to another can occur only in an ordered sequence in which evolution has to proceed through intermediate stages. In 'Dollo parsimony' it is assumed that a particular derived character state can only be gained once throughout the history of a group, but that it can be lost independently several times. As such 'Dollo parsimony' is less restrictive than 'irreversible parsimony' where no reversals at all are tolerated (Fig. 20.5). The 'Dollo' model might apply to the evolution of HOM/*Hox* clusters such that duplications of whole clusters or the whole genome are unlikely evolutionary events that may have occurred only once during evolutionary history, but that the loss or inactivation of clusters or individual elements of the cluster are more likely to occur repeatedly. If such a model is invoked for the evolution of the HOM/*Hox* clusters then a different (more complex) explanation might be necessary than if the gain and loss of a particular character states were equally likely (see the simple example in Fig. 20.6).

These alternative scenarios of HOM/*Hox* cluster evolution can be tested by further sequencing of homeobox clusters in crucial taxa such as echinoderms, tunicates, hagfish, and lampreys. Once the phylogeny of these species is well established (see below) a likely model of evolutionary change in *Hox* clusters

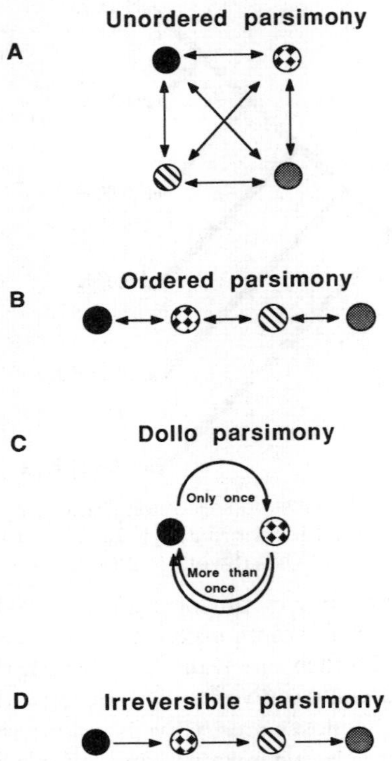

Fig. 20.5 Alternative models of evolutionary change. A: 'unordered parsimony' (also called Fitch parsimony) here changes between all observed character states can be made in a single step. B: 'ordered parsimony' (also called Wagner parsimony) requires that an order is maintained between transitions between different character states. Reversals are allowed. C: Dollo parsimony stipulates that derived character states only evolve once during the evolutionary history of a group, but that they can be lost more than once. D: irreversible parsimony (also called Camin–Sokal parsimony) requires that once a character state change occurred it cannot be reversed to a more ancestral state, but that it can change into a even more derived condition.

can be deduced. Depending on the outcome in these projects, other primitive fishes such as chondrostean and chondrichthyans may also need to be investigated for their HOM/*Hox* cluster architecture to further understand how HOM/*Hox* cluster evolution is linked to the increasing complexity of body plans in chordates. Current knowledge of homeobox cluster evolution, both in terms of within- and between-cluster evolution, is still too sketchy and too concentrated on a very small number of animal model systems to allow us to predict the mode and timing of deletion and duplication events during chordate evolution (Fig. 20.1). Without more comparative data on maps and sequences of complete homeobox clusters from a wider range of organisms from the 'tree of life', hypotheses about the unanswered questions surrounding the relationship of homeobox cluster evolution and body plan evolution in chordates will remain highly speculative.

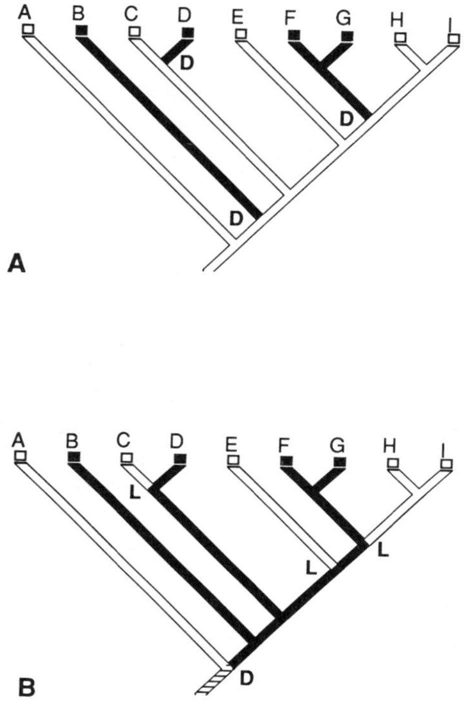

Fig. 20.6 A hypothetical phylogeny among taxa A–I. The reconstruction of the derived condition 'black' based on a model of A 'unordered parsimony' and B 'Dollo parsimony'. The A hypothesis requires only three steps, three independent duplications (D) of a trait. The B hypothesis assumes a more restrictive model of evolution 'Dollo parsimony' where a derived trait (the evolution from 'white' to 'black') can only occur once (D), but the loss of the trait (L) is allowed to occur more than once (here three times). A model of evolution that is more restrictive than the most simple one ('unordered parsimony') always requires more evolutionary explanations (steps).

19.3 The importance of phylogenetic knowledge and the evolution of HOM/*Hox* clusters

The phylogenetic relationships among most animal phyla are still largely unresolved, possibly due to their presumed 'explosive' origin within a very short time period in the Cambrian (see Conway-Morris 1993, 1994*a,b*; Bergström 1994; Phillipe *et al.* 1994). Unfortunately, even the phylogeny of deuterostomes is still uncertain and new phylogenetic hypotheses are constantly being suggested. Most recently, the phylogenetic position of lophophorates (a presumed deuterostome phylum) was called into question (Halanych *et al.* 1995) and they were placed with some groups of protostomes, rendering the deuterostomes an unnatural (paraphyletic) group. In general, knowledge of phylogeny is crucial if we ever hope to understand the relationship between ontogeny and phylogeny.

As already mentioned, a firmly established phylogeny based on which the tracing 'up and down the tree' of character evolution, developmental or otherwise, can be conducted is necessary. A brief example might serve to outline how uncertainty in phylogenetic relationships can affect the power of hypothesis testing in the study of the development–evolution relationship. The evolutionary relationships of the two living groups of jawless (agnathan) fishes, lampreys and hagfish, to each other and to other chordates (Figs. 20.1 and 20.7) are still debated. It should be noted that there is support for both the paraphyly (Fig. 20.1) and the monophyly (Fig. 20.7) hypotheses of agnathan relationships (reviewed by Forey and Janvier 1993). Monophyly has been suggested by the recent phylogenetic analysis of 18S rRNA data (Stock and Whitt 1991), but traditionally paraphyly is rather strongly supported by several kinds of phenotypic datasets (reviewed in Forey and Janvier 1993). These comparisons underscore that we need to know the answer to this phylogenetic question in addition to knowing the actual homeobox cluster architecture of these two crucial species. Without this knowledge we could not decide if the lamprey condition is intermediate (in terms of homeobox cluster evolution as well as morphological evolution between the hagfish and the jawed-vertebrate condition (in the case of paraphyly) or possibly independently derived for some more advanced vertebrate features (in the case of agnathan monophyly). Depending on the (phylogenetic as well as *Hox* cluster architecture) results for these two primitive groups of vertebrates, groups of primitive cartilaginous and bony fish will possibly have to be investigated as well.

The lamprey (*Petromyzon marinus*) appears to already have three to four HOM/*Hox* clusters, based on a homeobox PCR-based survey (Pendleton *et al.* 1993) (Figs. 20.1 and 20.7) whereas the number of HOM/*Hox* clusters in the hagfish is unknown. PCR-based approaches that use degenerate PCR primers recognizing homeobox motifs are a powerful labour-saving shortcut; however, it appears that the estimates of the number of homeobox clusters and homology assignments to cognate groups of PCR clones within homeobox clusters may not always be accurate with this technique. For example, the cluster number in amphioxus had been estimated to be two based on this technique (Pendelton *et al.* 1993), whereas it was later shown to be just one (Garcia-Fernandez and Holland 1994). Ideally, the lamprey estimate should be confirmed by genomic DNA mapping of the kind that was conducted to get the information for amphioxus (Garcia-Fernandez and Holland 1994).

19.4 The future of the investigation of the development–evolution relationship

Wolpert (1994) argues that in the next 20 years we will come to understand how development constrains and directs the form of organisms and that some of this understanding will come from the study of homology. By that he means similarities and generalities about development will be gathered by comparisons from a wide variety of organisms. This seems an optimistic but attainable goal, judging by the surprising similarities that emerge in developmental processes and control systems in astonishingly different model organisms. The sensational discovery of *eyeless* (Halder *et al.* 1995) which was interpreted to be a master control gene for the development of eyes in flies seemed to suggest that this gene is also instrumental in the development of eyes of all other animals from which a homologue of *eyeless* has been discovered. This discovery might herald the advent of even more amazing discoveries of genes high up in the cascade or nexus of developmental control genes.

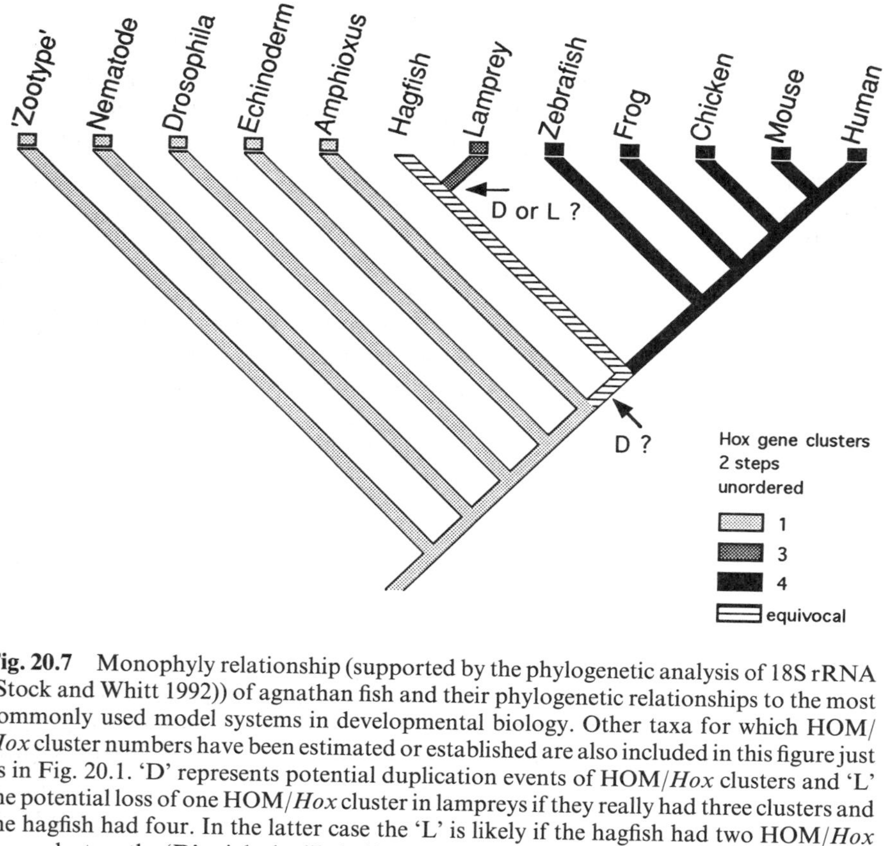

Fig. 20.7 Monophyly relationship (supported by the phylogenetic analysis of 18S rRNA (Stock and Whitt 1992)) of agnathan fish and their phylogenetic relationships to the most commonly used model systems in developmental biology. Other taxa for which HOM/*Hox* cluster numbers have been estimated or established are also included in this figure just as in Fig. 20.1. 'D' represents potential duplication events of HOM/*Hox* clusters and 'L' the potential loss of one HOM/*Hox* cluster in lampreys if they really had three clusters and the hagfish had four. In the latter case the 'L' is likely if the hagfish had two HOM/*Hox* clusters the 'D' might be likely if this phylogenetic hypothesis were correct.

In these studies, care has to be taken here, however, that homology and similarity of structure are not equated. Homology, is still an unsolved problem at the phenotypic as well as the genetic level (Patterson 1988). The issue of homology had been debated among evolutionary biologists for a long time and it proved notoriously intractable—no common definition has be agreed upon so far (Wake 1994; reviewed in Hall 1994). Homologous structures can be phenotypically very similar or quite different (for example hands in humans and wings in bats are homologous, but wings in birds are not considered to be homologous to wings in bats). Similarity in morphology may be due to common descent or due to convergence, the independent evolution of similar morphological structures. Convergence cannot be predicted and can only be determined if phylogenetic relationships are known. For most evolutionary biologists, similarity of developmental processes is not part of the definition of whether or not structures are homologous (Hall 1992, 1994). This is, for example, because the ontogenetic mechanisms of the formation and induction of the eye in salamanders and frogs are different yet these are indisputably homologous structures. However, developmental biologists seem to turn homology on its head by arguing that because, for example, *eyeless* may be pivotal in the formation of morphological structures that are distinctly different in shape and make-up, but that serve the same function, such as compound eyes in flies and camera eyes in vertebrates, that these structures are therefore homologous. To an evolutionary biologist a fly eye is not homologous to a vertebrate eye even if their ontogenies are controlled by an astonishingly ancient set of homologous genes.

Without a doubt, the homology debate will, through the exciting discovery of master control genes such as *eyeless*, receive new impetus from developmental biology. The combination of phylogenetics and developmental genetics will allow the determination of whether traits that are considered to be homologous arose by the same or a different developmental mechanism. Developmental and evolutionary biologists seem to ask different kinds of questions; however, their approaches can be reciprocally elucidating and it is to be regarded as a positive development that the evolutionary and developmental biological communities have initiated this important dialogue.

Acknowledgements

I thank Alexa Bely, Chris Lowe, David Wake, and Greg Wray for discussion and for sharing unpublished information. I am grateful to the National Science Foundation, USA, for financial support of my research (grants DEB-8918027, BSR-9107838 and BSR-9119867). I thank P. Harvey for the invitation to the Royal Society Discussion meeting on 'New Uses for New Phylogenies' which spawned my interest in this subject. I thank C. Nüsslein-Volhard for the invitation to the Ringberg Symposium on 'Development and Evolution' and all participants for stimulating discussion on some of the issues mentioned here.

References

Ahlberg, P. E. (1993). Coelacanth fins and evolution. *Nature*, **358**, 459.

Akam, M., Dawson, I., and Tear, G. (1988). Homeotic genes and the control of segment diversity. *Dev. Suppl.*, **104**, 123–33.

Akam, M., Averof, M., Castelli-Gair, J., Dawes, R., Falciani, F., and Ferrier, D. (1994). The evolving role of Hox genes in arthropods. *Dev. Suppl.*, **1994**, 209–15.

Alberch, P., Gould, S. J., Oster, G. F., and Wake, D. B. (1979). Size and shape in ontogeny and phylogeny. *Paleobiology*, **5**, 296–317.

Arthur, W. (1984). *Mechanisms of morphological evolution*. Wiley, Chichester.

Baer, K. E. von (1828). *Entwicklungsgeschichte der Thiere: Beobachtungen und Reflexion*. Bornträger, Königsberg.

Baer, K. E. von (1864). *Reden gehalten in wissenschaftlichen Versammlungen*. Karl Röttger St Petersburg.

Baldwin, J. M. (1902). *Development and evolution*. Macmillan, New York.

Bartels, J. L., Murtha, M. T., and Ruddle, F. H. (1993). Multiple *Hox*/HOM homeoboxes in platyhelminthes. *Mol. Phyl. Evol.*, **2**, 143–51.

Bergström, J. (1994). Ideas on early animal evolution. In *Early life on earth. Nobel Symposium no. 84*, pp. 460–6. Columbia University Press, New York.

Brooks, D. R. and McLennan, D. A. (1991). *Phylogeny, ecology, and behavior: a research program in comparative biology*. Chicago University Press.

Coates, M. I. (1994). The origin of vertebrate limbs. *Dev. Suppl.* **1994**, 169–80.

Conway-Morris, S. (1993). The fossil record and the early evolution of the Metazoa. *Nature*, **361**, 219–25.

Conway-Morris, S. (1994*a*). Early metazona evolution: first steps to an integration of molecular and morphological data. In *Early life on earth. Nobel Symposium no. 84*, pp. 450–9. Columbia University Press, New York.

Conway-Morris, S. (1994*b*). Why molecular biology needs palaeontology. *Devel. Suppl.* 1–13.

De Robertis, E. M. (1994*b*). The homeobox in cell differention and evolution. In *Guidebook to the homeobox genes*, (ed. D. Duboule), pp. 13–23. Oxford University Press.

DeSalle, R. and Grimaldi, D. (1993). Phylogenetic pattern and developmental process in *Drosophila*. *Syst. Biol.*, **42**, 458–75.

Dick, M. H. and Buss, L. W. (1994). A PCR-based survey of homeobox genes in *Ctenodrilus serratus* (Annelida: Polychaeta). *Mol. Phyl. Evol.*, **3**, 146–58.

Elinson, R. P. (1990). Direct development in frogs: wipping the recapitulationist slate clean. *Sem. Dev. Biol.*, **1**, 263–70.

Favier, B., Le Meur, M. Chambon, P., and Dollé, P. (1995). Axial skeleton homeosis and forelimb malfunctions in HOxd-11 mutant mice. *Proc. Natl. Acad. Sci. USA*, **92**, 310–14.

Felsenstein, J. (1978). The number of evolutionary trees. *Syst. Zool.*, **27**, 27–33.

Felsenstein, J. (1985*a*). Confidence limits on phylogenies: an approach using the bootstrap. *Evolution*, **39**, 783–91.

Felsenstein, J. (1985*b*). Phylogenies and the comparative method. *Am. Nat.*, **125**, 1–15.

Forey, P. and Janvier, P. (1993). Aganthans and the origin of jawed vertebrates. *Nature*, **361**, 129–34.

Garcia-Fernandez, J. and Holland, P. W. H. (1994). Archetypal organization of the amphioxus *Hox* gene cluster. *Nature*, **370**, 563–6.

Garland, T. jun and Adolph, S. C. (1994). Why not to do two-species comparative studies: limitations on inferring adaptation. *Physiol. Zool.*, **76**, 797–828.

Gehring, W. J. (1994). A history of the homeobox. In *Guidebook to the homeobox genes*, (ed. D. Duboule), pp. 3–10. Oxford University Press.

Ghiselin, M. T. (1984). *The triumph of the Darwinian method*. University of Chicago Press.

Goodwin, B. C., Holder, N., and Wylie, C. C. (ed.) (1983). *Development and evolution*. Cambridge University Press.

Gould, S. J. (1977). *Ontogeny and phylogeny*. Harvard University Press, Cambridge, MA.

Graham, A., Papalopulu, N., and Krumlauf, R. (1989). The murine and Drosophila homeobox gene complexes have common features of organization and expression. *Cell*, **57**, 367–78.

Haeckel, E. (1866). *Generelle Morphologie der Organismen: Allgemeine Grundzüge der organischen Formen-Wissenschaft, mechanisch begründet durch die von Charles Darwin reformierte Descendenz-Theorie*. George Riemer, Berlin.

Halanych, K. M., Bacheller, J. D., Aguinaldo, A. M. A., Liva, S. M., Hillis, D. and Lake, J. A. (1995). Evidence from 18S ribosomal DNA that the lophophorates are protostome animals. *Science*, **267**, 1641–3.

Halder, G., Callaerts, P., and Gehring, W. J. (1995). Induction of ectopic eyes by targeted expression of the *eyeless* gene in *Drosophila*. *Science*, **267**, 1788–92.

Hall, B. K. (1992). *Evolutionary developmental biology*. Chapman and Hall, London.

Hall, B. K. (1994). *Homology: the hierarchical basis of comparative biology*. Academic, San Diego.

Harvey, P. H. and Pagel, M. D. (1991). *The comparative method in evolutionary biology*. Oxford University Press.

Henry, J. Q. and Martindale, M. Q. (1994). Establishment of the dorsoventral axis in nemetean embryos: evolutionary considerations of spiralian development. *Dev. Genet.*, **15**, 64–78.

Holland, P. W. H. (1992). Homeobox genes in vertebrate evolution. *BioEssays*, **14**, 267–73.

Holland, P. W. H., Garcia-Fernández, J., Williams, N. A., and Sidow, A. (1994). Gene duplications and the origins of vertebrate development. *Dev. Suppl.* **1994**, 125–33.

Jeffrey, W. R. and Swalla, B. J. (1992). Evolution of alternate modes of development in ascidians. *BioEssays*, **14**, 219–26.

Kappen, C., Schughart, K., and Ruddle, F. H. (1989). Two steps in the evolution of Antennapedia-class vertebrate homeobox genes. *Proc. Natl. Acad. Sci. U.S.A.*, **86**, 5459–63.

Kellog, E. A. and Shaffer, H. B. (1993). Model organisms in evolutionary studies. *Syst. Biol.*, **42**, 409–14.

Kenyon, C. and Wang, B. (1991). A cluster of *Antennapedia*-class homeobox genes in a nonsegmented animal. *Science*, **253**, 516–17.

Kessel, M., Balling, R., and Gruss, P. (1990). Variations of cervical vertebrae after expression of a *Hox 1.1* transgene in mice. *Cell*, **61**, 301–8.

Kimble, J. (1994). An ancient molecular mechanism of establishing embryonic polarity? *Science*, **266**, 577–8.

Lawrence, P. A. (1992). *The making of a fly*. Blackwell, London.

Luk, S. K.-S., Kilpatrick, M., Kerr, K., and McDonald, P. M. (1994). Components acting in localization of bicoid mRNA are conserved among *Drosphila* species. *Genetics*, **137**, 521–30.

Lufkin, T., Dierich, A., LeMeur, M., Mark, M., and Chambon, P. (1991). Disruption of the Hox-1.6 homeobox gene results in defects in a region corresponding to its rostral domain of expression. *Cell*, **66**, 1105–19.

McGinnis, W. and Krumlauf, R. (1992). Homeobox genes and axial patterning. *Cell*, **68**, 283–302.

Maddison, W. P. and Maddison, D. R. (1992). *MacClade: analysis of phylogeny and character evolution. Version 3.01*. Sinauer, Sunderland, MA.

Malicki, J., Schugart, K., and McGinnis, W. (1990). Mouse *Hox-2.2* specifies thoracic segmental identity in *Drosophila* embryos and larvae. *Cell*, **63**, 961–7.

Marshall, H., Studer, M., Pöpperl, H., Aparicio, S., Kuroiwa, A., Brenner, S., and

Krumlauf, R. (1994). A conserved retinoic acid response element required for early expression of the homeobox gene *Hoxb-1*. *Nature*, **370**, 567–71.
Meyer, A., Biermann, C. H., and Orti, G. (1993). The phylogenetic position of the zebrafish (*Danio rerio*), a model system in developmental biology: an invitation to the comparative method. *Proc. R. Soc.*, **B252**, 231–36.
Meyer, A., Ritchie, P., and Witte, K-E. (1995). Predicting developmental processes from phylogenetic patterns: a molecular phylogeny of the zebrafish (*Danio rerio*) and its relatives. *Phil. Trans. R. Soc.*, **349**, 103–11.
Minelli, A. and Schram, F. R. (1994). Owen revisited: a reappraisal of morphology in evolutionary biology. *Bidragen tot de Dierkunde*, **64**, 65–74.
Murtha, M. T., Leckman, J. F., and Ruddle, F. H. (1991). Detection of homeobox genes in development and evolution. *Proc. Natl. Acad. Sci. USA*, **88**, 1071–15.
Northcutt, R. G. (1990). Ontogeny and phylogeny: a re-evaluation of conceptual relationships and some applications. *Brain Behav. Evol.*, **36**, 116–40.
Nüsslein-Volhard, C., and Wieschaus, E. (1980). Mutations affecting segment number and polarity in *Drosophila*. *Nature*, **287**, 795–803.
Ohno, S. (1970). *Evolution by gene duplication*. Springer, Berlin.
Patel, N. H. (1994). Developmental evolution: insights from studies of insect segmentation. *Science*, **266**, 581–9.
Patterson, C. (1988). Homology in classical and molecular biology. *Mol. Biol. Evol.*, **5**, 603–25.
Pavell, A. M. and Stellwag, E. J. (1994). Survey of Hox-like genes in the teleost *Morone saxatilis*: implications for evolution of the *Hox* gene family. *Mol. Mar. Biol. Biotech.*, **3**, 149–57.
Pendelton, J. W., Nagai, B. K., Murtha, M. T., and Ruddle, F. H. (1993). Expansion of the *Hox* gene family and the evolution of chordates. *Proc. Natl. Acad. Sci. USA*, **90**, 6300–4.
Phillipe, H., Chenuil, A., and Adoutte, A. (1994). Can the Cambrian explosion be inferred through molecular phylogeny? *Dev. Suppl.*, **1994**, 15–25.
Raff, R. A. (1992). Direct-developing sea urchins and the evolutionary reorganization of early development. *BioEssays*, **14**, 211–18.
Raff, R. A. and Kaufman, T. C. (1983). *Embryos, genes, and evolution. The developmental–genetic basis of evolutionary change*. Macmillan, New York.
Ruddle, F. H., Bentley, J. L., Murtha, M. T., and Risch, N. (1994). Gene loss and gain in the evolution of vertebrates. *Dev. Suppl.* **1994**, 155–61.
Schierwater, B., Murtha, M., Dick, M., Ruddle, F. H., and Buss, L. W. (1991). Homeoboxes in cniderians. *J. Exp. Zool.*, **260**, 413–16.
Schubert, F. R., Nieselt-Struwe, K., and Gruss, P. (1993). The antennapedia-type homeobox genes have evolved from three precursors separated early in metazoan evolution. *Proc. Natl. Acad. Sci. USA*, **4**, 143–7.
Scott, M. P. (1992). Vertebrate homeobox gene nomenclature. *Cell*, **171**, 551–3.
Scott, M. P. and Carroll, S. B. (1987). The segmentation and homeotic gene network in early *Drosophila* development. *Cell*, **51**, 689–98.
Slack, J. M. W., Holland, P. W. H., and Graham, C. F. (1993). The zootype and the phylotypic stage. *Nature*, **361**, 490–2.
Sommer, R. J. and Sternberg, P. W. (1994). Changes of induction and competence during the evolution of vulva development in nematodes. *Science*, **265**, 115–18.
Sommer, R. J. and Tautz, D. (1993). Involvement of an orthologue of the *Drosophila* pair-rule gene *hairy* in segment formation of the short germ-band embryo of *Tribolium* (Coleoptera). *Nature*, **361**, 448–50.
Stock, D. W. and Whitt, G. S. (1992). Evidence from 18S ribosomal RNA requences that lamprey and hagfishes form a natural group. *Science*, **257**, 787–9.

Swofford, D. L. (1991). *Phylogenetic analysis using parsimony (PAUP Version 3.1.1.)*. Illinois Natural History Survey, Champaign.

Tabin, C. J. (1992). Why we have (only) five fingers per hand: Hox genes and the evolution of paired limbs. *Development*, **116**, 289–96.

Tabin, C. J. and Laufer, E. (1993). Hox genes and serial homology. *Nature*, **361**, 692–3.

Waddington, C. H. (1957) *The strategy of genes*. Allen and Unwin, London.

Wake, D. B. (1994) Comparative terminology. *Science*, **265**, 268–9.

Wake, D. B. (1995). Evolutionary developmental biology—Prospects for an evolutionary synthesis at the developmental level. *Proc. Calif. Acad. Sci.* (In press.)

Walker, C. and Streisinger, G. (1983). Induction of mutations by γ-rays in pregonial germ cell of zebrafish embryos. *Genetics*, **103**, 125–36.

Walsh, J. B. (1995). How often do duplicated genes evolve a new function? *Genetics*, **139**, 421–8.

Warren, R. W., Nagy, L., Selegue, J., Gates, J. and Carroll, S. (1994). Evolution of homeotic gene regulation and function in flies and butterflies. *Nature*, **372**, 458–61.

Wilson, A. C., Carlson, S., and White, T. J. (1977). Biochemical evolution. *Ann. Rev. Biochem*, **46**, 573–639.

Wolpert, L. (1994). Do we understand development? *Science*, **266**, 571–2.

Wray, G. A. (1992). Rates of evolution in developmental processes. *Am. Zool.*, **32**, 123–34.

Wray, G. A. (1995). Puctuated evolution of embryos. *Science*, **267**, 1115–16.

Wray, G. A. and Bely, A. E. (1994). The evolution of echinoderm development is driven by several distinct factors. *Dev. Supp.*, **1994**, 97–106.

Wray, G. A. and Raff, R. A. (1991). The evolution of developmental strategy in marine invertebrates. *Trends Ecol. Evol.*, **6**, 45–50.

Zuckerkandl, E. (1994). Molecular pathways to parallel evolution: I. gene nexuses and their morphological of correlates. *J. Mol. Evol.*, **39**, 66–78.

Index